Adaptive Antennas and Phased Arrays for Radar and Communications

For a list of recent titles in the *Artech House Radar Series,*
turn to the back of this book.

Adaptive Antennas and Phased Arrays for Radar and Communications

Alan J. Fenn

Massachusetts Institute of Technology
Lincoln Laboratory

ARTECH
HOUSE

BOSTON | LONDON
artechhouse.com

Library of Congress Cataloging-in-Publication Data
A catalog record for this book is available from the U.S. Library of Congress.

British Library Cataloguing in Publication Data
A catalogue record for this book is available from the British Library.

ISBN 13: 978-1-59693-273-9

Cover design by Igor Valdman

© 2008 Massachusetts Institute of Technology, Lincoln Laboratory
All rights reserved.

This work was sponsored in part under Air Force Contract FA8721-05-C-0002. Opinions, interpretations, conclusions, and recommendations are those of the author and are not necessarily endorsed by the United States Air Force.
   Printed and bound in the United States of America. No part of this book may be reproduced or utilized in any form or by any means, electronic or mechanical, including photocopying, recording, or by any information storage and retrieval system, without permission in writing from the publisher. All terms mentioned in this book that are known to be trademarks or service marks have been appropriately capitalized. Artech House cannot attest to the accuracy of this information. Use of a term in this book should not be regarded as affecting the validity of any trademark or service mark.

10 9 8 7 6 5 4 3 2 1

To my family

Contents

	Preface	xv
1	**Adaptive Antennas and Degrees of Freedom**	**1**
1.1	Introduction	1
1.2	Adaptive Antenna Fundamentals	3
1.3	Consumption of Adaptive Antenna Degrees of Freedom	8
1.4	Derivation of the Applebaum-Howells Steady-State Adapted Weight Equation in Eigenspace	9
1.5	Conditions for Complete Consumption of N-Degrees of Freedom	18
1.6	Orthogonal Interference Sources	19
1.6.1	Derivation of an Interference Signal Matrix to Consume N Degrees of Freedom of an N-Channel Adaptive Nulling Array Antenna	19
1.7	Eigenvalues and INR for Two-Element Array with Two Interference Sources	22
1.8	Figure of Merit for Consumption of Degrees of Freedom	28
1.9	Gradient Search to Maximize Consumption of Adaptive Antenna Degrees of Freedom	32
1.10	Maximum Jammer Effectiveness, Numerical Results	35
1.10.1	Seven-Element Hexagonal Array	36
1.10.2	Seven-Element Ring Array	38

| 1.11 | Summary | 39 |
| 1.12 | Problem Set | 40 |

References   **41**

2 Array Mutual Coupling Effects on Adaptive Radar Clutter Suppression   43

2.1	Introduction	43
2.2	Displaced Phase Center Antenna Concept for Clutter Suppression	43
2.3	Theory	49
2.3.1	Array Analysis Including Mutual Coupling Effects	49
2.3.2	Calculation of Displaced Phase Center Antenna Clutter Cancellation	53
2.3.3	Derivation of Clutter Cancellation Factor	56
2.4	Results	59
2.4.1	Introduction	59
2.4.2	Subscale 96-Element Planar Phased Array	60
2.4.3	Subscale and Full-Scale Planar Phased Arrays	65
2.5	Summary	71
2.6	Problem Set	72

References   **73**

3 Focused Near-Field Technique for Evaluating Adaptive Phased Arrays   75

3.1	Introduction	75
3.2	Theory	79
3.2.1	Adaptive Nulling Formulation	79
3.3	Near-Field Formulation	82
3.3.1	Focused Near-Field Nulling Concept	82
3.3.2	Covariance Matrix for Near-Field Interference	83
3.3.3	Array Radiation Pattern in the Focused Near Field	84
3.3.4	Far-Field Formulation	87
3.3.5	Array Radiation Pattern	88
3.3.6	Near-Field Boundary	88

3.4	Results	89
3.4.1	Focused Linear Array Quiescent Conditions	89
3.4.2	Fully Adaptive Array Behavior	90
3.5	Summary	108
3.6	Problem Set	108
	References	**109**
4	**Moment Method Analysis of Focused Near-Field Adaptive Nulling**	**111**
4.1	Introduction	111
4.2	Theory	113
4.2.1	Focused Near-Field Nulling Concept	113
4.2.2	Adaptive Array Concepts	114
4.2.3	Moment Method Formulation	117
4.3	Results	120
4.3.1	Focused Array Quiescent Conditions	120
4.4	Sidelobe Canceller Adaptive Array Behavior	123
4.4.1	Nonstressing Interference: One Source	123
4.4.2	Stressing Interference: Two Sources	125
4.5	Summary	129
	References	**130**
5	**Focused Near-Field Testing of Multiphase-Center Adaptive Array Radar Systems**	**133**
5.1	Introduction	133
5.2	Near-Field/Far-Field Source Wavefront Dispersion	138
5.3	Focused Near-Field Testing Concept	139
5.4	Adaptive DPCA Radar Concept	141
5.5	Adaptive DPCA Array Formalism	143
5.6	Array Antenna/Source Modeling	147
5.7	DPCA Near-Field Source Distribution	149
5.8	Near-Field/Far-Field Simulations	150

5.9	Summary	157
5.10	Problem Set	158
	References	**159**

6 Experimental Testing of Focused Near-Field Adaptive Nulling — 161

6.1	Introduction	161
6.2	Adaptive Nulling System Description	162
6.2.1	Introduction	162
6.2.2	Adaptive Nulling Receiver	163
6.2.3	Receiver Channel Equalization	166
6.3	Prototype Phased Array and Simulation Model	169
6.4	Results	171
6.4.1	Equalization Performance	171
6.4.2	Focused Near-Field Nulling Performance	171
6.5	Summary	174
	References	**174**

7 Experimental Testing of High-Resolution Nulling with a Multiple Beam Antenna — 177

7.1	Introduction	177
7.2	MBA Design and Construction	178
7.3	Measured Results	184
7.4	Summary	188
	References	**189**

8 Phased Array Antennas: An Introduction — 191

8.1	Introduction	191
8.2	Theoretical Background	192
8.3	Example Measurements of Array Mutual Coupling and Array Element Gain Pattern	208
8.4	Summary	211

	References	212
9	**Monopole Phased Array Antenna Design, Analysis, and Measurements**	**215**
9.1	Introduction	215
9.2	Monopole Phased Arrays	217
9.3	Theory for Analysis of Finite Arrays of Monopoles	219
9.3.1	Introduction	219
9.3.2	Matrix Equations for the Array Element Currents	220
9.3.3	Array Input Impedance, Patterns, and Gain	224
9.4	Theory for Analysis of Infinite Arrays of Monopoles	229
9.4.1	Introduction	229
9.4.2	Derivation of the Near-Zone Radiated Electric Field	230
9.4.3	Derivation of the Induced Voltage at the Reference Element	238
9.4.4	Scan Input Impedance	241
9.4.5	The Element-Gain Pattern for an Infinite Array	241
9.5	Results	242
9.5.1	The Effects of Array Size for Ideal One-Quarter Wavelength Monopoles	242
9.5.2	121-Element Square Grid Monopole Array: Experiment and Theory	247
9.5.3	Hexagonal Lattice Infinite Array Results	252
9.6	Summary	255
9.7	Problem Set	256
	References	**257**
10	**Monopole Phased Array Field Characteristics in the Focused Near-Field Region**	**259**
10.1	Introduction	259
10.2	Monopole Phased Array Antenna Design	260
10.3	Theoretical Formulation for the Electric Field Components	261
10.4	Results	264
10.5	Summary	272
	References	**272**

11	**Displaced Phase Center Antenna Measurements Using Near-Field Scanning**	**275**
11.1	Introduction	275
11.2	Displaced Phase Center Antenna Clutter Cancellation	276
11.3	Displaced Phase Center Antenna Clutter Near-Field Measurement Technique	278
11.3.1	DPCA Test Array Description	279
11.3.2	Near-Field Scanner	281
11.4	Results	283
11.5	Summary	289
	References	**289**
12	**Low-Sidelobe Phased Array Antenna Measurements Using Near-Field Scanning**	**291**
12.1	Introduction	291
12.2	Theory	293
12.2.1	Planar Near-Field Scanning Formulation	293
12.2.2	Monopole Phased Array Near-Field Modeling Using the Method of Moments	298
12.3	Low-Sidelobe Phased Array Antenna Prototype	305
12.4	Near-Field Measurements System	307
12.5	Results	309
12.5.1	Selection of Theoretical Probe Model	309
12.6	Summary	317
12.7	Problem Set	319
	References	**319**
13	**Arrays of Horizontally Polarized Omnidirectional Elements**	**321**
13.1	Introduction	321
13.2	Theory	322
13.2.1	Array Gain and Element Gain	322
13.3	Loop-Fed Slotted Cylinder Antenna	322

13.4	Results	323
13.4.1	Single Element	323
13.4.2	Seven-Element Hexagonal Array	327
13.4.3	Linear Arrays	330
13.5	Summary	332
	References	**332**
14	**Finite Arrays of Crossed V-Dipole Elements**	**335**
14.1	Introduction	335
14.2	Theory	335
14.3	Dipole Element Prototypes Description	340
14.4	Measurements of Single Straight Dipole and V-Dipole Elements above a Ground Plane	341
14.5	V-Dipole Array Description	343
14.6	Measured Array Mutual Coupling Results	343
14.7	Center Element Gain Pattern Measurements and Simulations	346
14.8	Summary	348
	References	**349**
15	**Experimental Ultrawideband Dipole Antenna Array**	**351**
15.1	Introduction	351
15.2	Ultrawideband Dipole Array Design	351
15.3	Measured Results for Prototype Ultrawideband Array	356
15.4	Summary	357
	References	**358**
16	**Finite Rectangular Waveguide Phased Arrays**	**359**
16.1	Introduction	359
16.2	Array Formulation	361
16.3	Results	374
16.4	Summary	379

16.5	Problem Set	383
	References	**383**
	About the Author	**387**
	Index	**389**

Preface

Adaptive antennas and phased arrays, with rapidly scanned beams or multiple beams, are commonly suggested for radar and communications systems in ground-based, airborne, and spaceborne applications that must function in the presence of jamming and other sources of interference. This monograph is written primarily for practicing antenna engineers and graduate students in electrical engineering, and describes research on adaptive antennas and phased arrays that I have performed with my colleagues at MIT Lincoln Laboratory. A portion of the material in this book has been published previously only in Lincoln Laboratory reports, and some of the material has been published in the peer-reviewed literature. Most of the information in this book is not available in other books. No prior knowledge of adaptive antennas or phased arrays is needed to understand the content of this book, although some background in signal processing, electromagnetic theory, antennas, radar, and communications will be helpful to the reader.

The book begins with a discussion of the fundamentals of adaptive antennas pertaining to radar and communications systems, with an emphasis on consumption of adaptive array degrees of freedom from the jammer's viewpoint. Displaced phase center antenna array mutual coupling effects in the problem of adaptive suppression of radar clutter is discussed in Chapter 2. Next, in Chapters 3 through 5 a theoretical foundation for a focused near-field technique that can be used to quantify the far-field adaptive nulling performance of a large aperture adaptive phased array system is described. Simulations of focused near-field and focused far-field nulling performance for adaptive arrays are presented for arrays of isotropic elements in Chapter 3, for arrays including mutual coupling effects in Chapters 4 and 5, and for arrays with multiple phase centers in Chapter 5. Experimental testing of the focused near-field adaptive nulling technique for phased arrays is described in

Chapter 6. An experimental high-resolution multiple-beam adaptive-nulling antenna system is described in Chapter 7.

Chapters 8 through 16 then concentrate on phased array antenna development for a variety of array elements. Chapter 8 provides an introduction to phased array antenna theory. In Chapter 9, finite and infinite array analyses and measurements for periodic phased arrays of monopole elements are presented. Chapter 10 describes the focused near-field polarization characteristics of monopole phased arrays as related to adaptive array testing in the near field. Next, in Chapter 11 a test bed phased array that implements the displaced phase center antenna technique, as related to the analysis presented in Chapter 2, is described along with the planar near field testing technique that is used to assess adaptive clutter cancellation performance. The planar near field scanning method for measuring low-sidelobe radiation patterns of phased arrays is described in Chapter 12. Experimental arrays of horizontally polarized loop-fed slotted cylinder antennas (Chapter 13), dual-polarized dipole arrays (Chapter 14), and ultrawideband dipole arrays (Chapter 15) are described. In Chapter 16, rectangular waveguide arrays are analyzed by the method of moments.

Some of the data contained in this monograph are due to the efforts of a number of colleagues at MIT Lincoln Laboratory, most notably H.M. Aumann and F.G. Willwerth. Technical discussions with A.J. Simmons, L.J. Ricardi, E.J. Kelly, G.N. Tsandoulas, R.W. Miller, D.H. Temme, S.C. Pohlig, and J.R. Johnson are sincerely appreciated. The software support of L. Niro, D.S. Besse, S.E. French, and D.L. Washington is also greatly appreciated. For the material presented in Chapter 7, B.M. Potts and W.C. Cummings developed the 127-beam distributed MBA concept, A.R. Dion developed the MBA lenses, J.C. Lee developed the feed horns, D.C. Weikle developed the RF nulling weight network, and R.J. Burns organized the multiple beam antenna measurements Technical discussions with Professor B.A. Munk of The Ohio State University ElectroScience Laboratory were very helpful to the infinite array analysis presented in Chapter 9. Assistance in preparation of the manuscript is due to Ms. Loretta Wesley. Madeline Riley coordinated the preparation of the graphics.

1
Adaptive Antennas and Degrees of Freedom

1.1 INTRODUCTION

The subject of adaptive antenna systems has been explored extensively since the 1950s [1-13]. The primary function of any adaptive nulling receive antenna system for radar [14, 15] or communications [3-5] applications is to minimize (null) the received power from one or more interference sources located in the antenna field of view as depicted in Figure 1.1. In the case of a radar application, the adaptive antenna system may need to suppress radar clutter [15] as well as to cancel interference from intentional jamming or unintentional interference. Additionally, the adaptive nulling system must have constraints that allow the system to detect radar targets or to receive signals from desired communications users with an adequate signal-to-noise ratio (SNR). The ability of the adaptive nulling antenna system to perform these functions is governed by a number of factors, some of which include the number of adaptive array elements (or nulling channels), antenna aperture diameter, nulling bandwidth, and number, strength, and locations of interfering sources.

An adaptive antenna can be, for example, a microwave or millimeter-wave phased array antenna [16-20] consisting of multiple antenna elements in which the main beam is electronically steered over the field of view. Alternately, the adaptive antenna can be a multiple beam antenna (MBA) [21, 22] consisting of multiple horn antennas feeding a parabolic reflector antenna system or lens antenna system [23].

A typical adaptive nulling antenna system block diagram is shown in Figure 1.2. The adaptive antenna generally has degrees of freedom in the

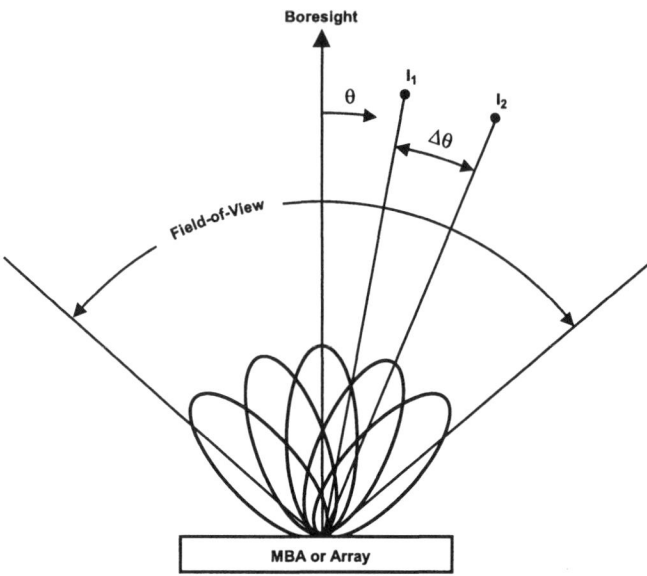

Figure 1.1 Geometry for interference sources in the field of view of either a multiple beam antenna or a phased array antenna.

form of amplitude and phase or time delay weighting of multiple channels to adjust its radiation pattern response in the direction of the interference. In this book, a degree of freedom for adaptive weighting refers to a complex weight (amplitude/phase pair). Interference can be due to in-band signals from other communications or radar systems or from intentional jamming. Depending on the desired architecture, adaptive antenna systems commonly will use low-noise amplifiers close to the receiving antenna elements, followed by analog filtering, analog weighting at the initial radio frequency (RF) stage or at an intermediate frequency (IF) stage, and downconversion to baseband and analog-to-digital conversion and subsequent digital signal processing. An adaptive antenna system can be referred to as either partially adaptive or fully adaptive depending on whether some or all of the available antenna elements or channels are used in the formation of pattern nulls, respectively. This book will describe examples of both partially adaptive and fully adaptive antennas, with a heavy emphasis on arrays.

This chapter deals primarily with investigating the degrees of freedom for adaptive arrays operating in the presence of far-field interference, in which the receive antenna elements are assumed to be simple isotropic receiving antennas [24-26]. Practical adaptive antenna system effects such as array mutual coupling, polarization, multipath, channel mismatch (frequency

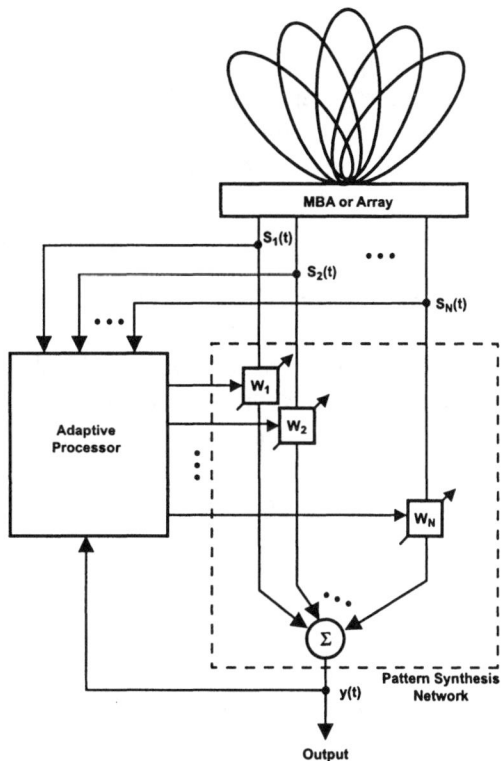

Figure 1.2 Block diagram for an N-channel adaptive antenna system. An adaptive weighting network is used to synthesize a desired radiation pattern.

dispersion), and near-field effects are ignored in this chapter but are considered in Chapters 2 through 6. Chapter 7 provides an example of a multiple-beam antenna adaptive nulling system. In the subsequent portion of the book, Chapters 8 to 16 are concerned with phased array antenna design, analysis, and testing.

1.2 ADAPTIVE ANTENNA FUNDAMENTALS

A fully adaptive antenna has N elements and N adjustable complex weights denoted $w(t) = [w_1(t), w_2(t), \ldots, w_N(t)]^T$ (where T means transpose) that can be set to synthesize a desired radiation pattern at N points. In other words, the adaptive antenna has N degrees of freedom in shaping the antenna radiation pattern. For example, $N-1$ pattern nulls (or minima) can be synthesized while simultaneously maintaining some antenna gain in a desired direction. The array elements or multiple beam antenna feeds receive a complex signal vector (a voltage matrix) denoted $\boldsymbol{S}(t) = [S_1, S_2, \ldots, S_N]^T$ that is used in the adaptive signal processor to adjust the complex weights as a function of time to suppress interference. The output $y(t)$ of the weighted signals, at the summing junction indicated in Figure 1.2, can be used as a feedback signal in the signal processor to determine the adaptive weights as a function of time. The output signal (voltage) can be expressed as the matrix product of the weight vector transpose times the signal vector as [6]

$$y(t) = \boldsymbol{w}^T(t)\boldsymbol{S}(t) = \boldsymbol{S}^T(t)\boldsymbol{w}(t) \tag{1.1}$$

In the adaptive nulling literature, the output signal voltage is sometimes expressed in an alternate mathematical form as the matrix product of the weight vector conjugate transpose times the signal vector as

$$y(t) = \boldsymbol{w}^\dagger(t)\boldsymbol{S}(t) = \boldsymbol{S}^\dagger(t)\boldsymbol{w}(t) \tag{1.2}$$

where $\dagger$ means complex conjugate transpose.

The expected value (denoted $\boldsymbol{E}$) or mean value of the adaptive antenna output power at the summing junction can be expressed as

$$\boldsymbol{E}|y(t)|^2 = \boldsymbol{E}(y^*(t)y(t)) \tag{1.3}$$

where $*$ means complex conjugate. First, substituting (1.1) in (1.3) yields

$$\boldsymbol{E}|y(t)|^2 = \boldsymbol{E}[(\boldsymbol{w}^T\boldsymbol{S})^*(\boldsymbol{S}^T\boldsymbol{w})] = \boldsymbol{w}^\dagger \boldsymbol{E}[\boldsymbol{S}^*\boldsymbol{S}^T]\boldsymbol{w} = \boldsymbol{w}^\dagger \boldsymbol{E}[\boldsymbol{S}\boldsymbol{S}^\dagger]\boldsymbol{w} = \boldsymbol{w}^\dagger \boldsymbol{R}\boldsymbol{w} \tag{1.4}$$

where $\boldsymbol{R} = \boldsymbol{E}[\boldsymbol{S}\boldsymbol{S}^\dagger]$ is the channel covariance matrix. The channel covariance matrix can consist, for example, of contributions from radar target signal (and clutter), or communications user signal, jamming and interference, and receiver noise. Using the alternate form of the output signal voltage, that is, (1.2),

$$\boldsymbol{E}|y(t)|^2 = \boldsymbol{E}(y(t)y^*(t)) = \boldsymbol{E}[(\boldsymbol{w}^\dagger(t)\boldsymbol{S}(t))(\boldsymbol{w}^T(t)\boldsymbol{S}^*(t))] \tag{1.5}$$

but,

$$\boldsymbol{w}^T(t)\boldsymbol{S}^*(t) = \boldsymbol{S}^\dagger \boldsymbol{w} \tag{1.6}$$

Thus, it again follows that

$$\boldsymbol{E}|y(t)|^2 = \boldsymbol{w}^\dagger \boldsymbol{E}[\boldsymbol{S}\boldsymbol{S}^\dagger]\boldsymbol{w} = \boldsymbol{w}^\dagger \boldsymbol{R}\boldsymbol{w} \tag{1.7}$$

Therefore, the same covariance matrix is obtained for either definition of the output signal voltage ((1.1) or (1.2)).

This book primarily focuses on a covariance-matrix-based approach for adaptive nulling, and primarily analyzes adaptive nulling under steady-state conditions. The reader is referred to references such as Monzingo and Miller [6], Compton [7], and others [8-13] for detailed discussions of adaptive array algorithms.

Figure 1.3 shows the general case of M interference sources in the field of view of an N-element adaptive array (or the feeds of an MBA). If there are $N - 1$ interference sources distributed within the antenna field of view, there are $N - 1$ distinct nulls available to adaptively suppress them. After adaption, the SNR achieved depends on the null depths achieved and the amount of reduction in gain (if any) to a desired direction (such as the radar target angle or user angle). Although $N - 1$ interference sources might be present, depending on their angular separations, $N - 1$ or fewer degrees of freedom will be utilized in adaptively forming the pattern nulls. As this chapter will show, sources grouped closely in terms of the nulling antenna's half-power beamwidth will tend to use fewer degrees of freedom compared to sources widely separated. The degree to which the interference sources are cancelled can be quantified by computing the interference-to-noise ratio (INR) before and after adaption for the adaptive antenna. From the jammer or countermeasures designer viewpoint, it is expected that jamming sources grouped in certain, preferred, geometries can reduce the amount of cancellation compared to when the same number of sources are distributed at random. To gain a better understanding of the ability to suppress interference in adaptive antennas, it is useful to investigate how interference sources consume degrees of freedom.

This chapter describes and quantifies the consumption of degrees of freedom for adaptive antennas. In investigating the relative capabilities of adaptive nulling array antenna designs for radar or communications systems, it is useful to define a formidable set of jammers (sometimes referred to as electronic countermeasures (ECM) to quantify how the electronic counter-countermeasures (ECCM) capability of the system resists a serious attempt to jam. One way to quantify this capability is to define the specific adaptive nulling characteristics of the ECCM system and then introduce a number of jammers of selected effective isotropic radiated power (EIRP), and randomly or systematically distribute these jammers in the field of view of the adaptive antenna. The results of numerical simulations or prototype system measurements can produce an estimate of the maximum jammer power that can be tolerated for the various assumed numbers of jammers. These results are highly scenario dependent because they depend strongly on whether the location of a desired communications user or radar target is close to or far from a jammer. A random distribution of interference sources or closely

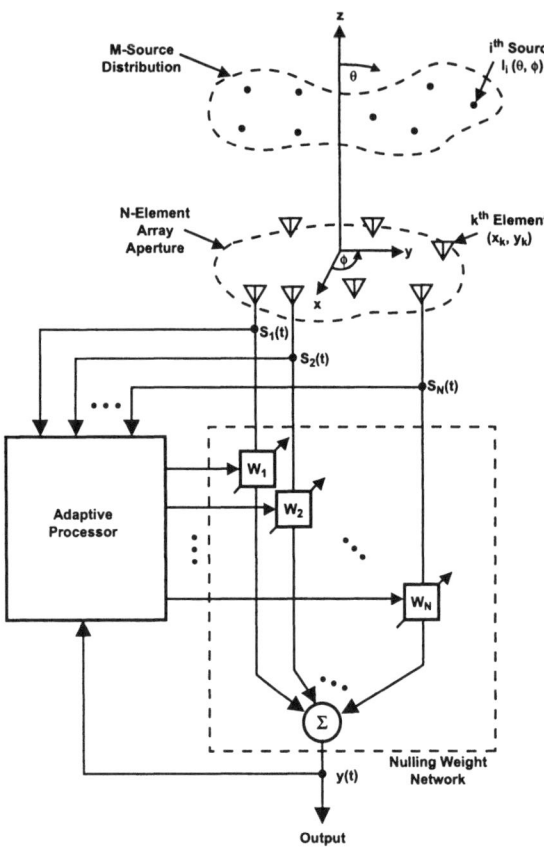

Figure 1.3 Distribution of M interference sources in the field of view of an adaptive nulling array antenna. © 1985 IEEE [24].

spaced group of interference sources usually results in the array using only some of its degrees of freedom to form adaptive nulls, because not all of the degrees of freedom are being fully consumed. An alternate mathematical approach that is taken in this chapter is to maximize the effectiveness of a given number of fixed power level jammers by appropriately positioning them such that the consumption of degrees of freedom is maximized. A motivation for this analysis is the particular case where the array elements are in either a thinned or an irregularly spaced configuration [27]. For periodic or regularly spaced arrays, it is straightforward to compute analytically an effective jammer configuration that overcomes the nulling capability of the adaptive array. However, analytic solutions for irregular or thinned arrays are difficult if not impossible to obtain.

The following definition of complete consumption of degrees of freedom is adopted in this book. *For an N-element adaptive antenna, complete consumption of all N degrees of freedom occurs when the antenna does not adapt to the incident fields from a geometric arrangement of interference sources. In other words, no adaptive pattern shaping can take place when all degrees of freedom are consumed.* Clearly, for the jammers to be effective they must be sufficiently large to reduce the SNR below a minimum desired value. If this happens, in the case of a communications system, the link will not be available to any user within the antenna field of view.

A discussion concerning consumption of degrees of freedom for adaptive arrays has been given by Gabriel [3]. He investigated an eight-element linear array for a variety of interference source distributions. In particular, he related consumption of degrees of freedom to the number of interference covariance matrix eigenvalues greater than unity. (Here, an eigenvalue equal to unity is taken to be equal to the receiver noise level.) For each eigenvalue greater than unity (due to jamming) it was shown that the associated eigenvector contributed to the formation of pattern nulls. The terms *capturing* or *consuming* degrees of freedom were used in describing the adaptive array response to the assumed source scenarios. Results were given for various numbers, power levels, and positions of interference sources. The effect of nulling bandwidth was also investigated. The eigenvalues were shown for each of the source scenarios. Also, the degradation in SNR after adaptation was given. The examples considered caused very little degradation in the steady state adapted SNR (less than about 5 dB), which means that the degrees of freedom were not severely taxed. Mayhan et al. [5] used orthogonal interference signals to maximize $N - 1$ degrees of freedom of N-element arrays. Analytic expressions for the source coordinates can be derived using this procedure for triangular or rectangular lattice arrays, but for thinned or random arrays an analytic solution may not be possible. A general approach is desirable to obtain solutions for irregular arrays, and this is accomplished by the numerical methods presented in this chapter.

This chapter theoretically investigates the problem of how to distribute interference sources such that consumption of the adaptive array degrees of freedom is maximized. From the adaptive array viewpoint, this type of interference source arrangement represents a maximally stressed environment for the given number of jammers. This approach may be useful both in evaluating the performance of adaptive arrays and in determining the minimum number of nulling channels required in a particular adaptive nulling system design. The method described here is general in the sense that the number of interference sources and their power levels and the number of array elements and their layouts are arbitrary. To initially facilitate describing the concept of consuming degrees of freedom, a two-element array with two interference sources is investigated in detail – pertinent properties of the covariance matrix and the degrees of freedom are discussed. A figure of merit and a gradient search are then derived, which are used to find the particular configuration of M interference sources that maximizes consumption of the degrees of freedom for an arbitrary N-channel adaptive antenna. Examples of N-element arrays with their associated N-source maximally stressed interference configurations are given.

1.3 CONSUMPTION OF ADAPTIVE ANTENNA DEGREES OF FREEDOM

The goal of this section is to provide a basic understanding of how the degrees of freedom of an adaptive nulling antenna are consumed by undesired interference sources. It is tempting to associate the number of interference sources that are nulled with the number of degrees of freedom that are consumed; however, such an approach is generally not satisfactory, because the number of degrees of freedom consumed depends on the interference source spacing and the adaptive antenna geometry. As will be shown, a better method is to compute the eigenvalues of the interference covariance matrix and examine their amplitude spread [5] as a function of the source spacing – the spread of the eigenvalues is demonstrated here for an array antenna. The Applebaum-Howells analog servo-control-loop processor (in the steady state) is used as the adaptive nulling algorithm in this chapter; however, the results are expected to be fundamental to the performance of any adaptive antenna system.

A measure of the susceptibility to interference for an adaptive antenna is the number and spread of the eigenvalues of the covariance matrix that is formed by the cross-correlation of interference signals received as input at each antenna port. The spatial location and strength of the interference source (or sources) affects the eigenvalues and the interference-to-noise ratio prior to adaption. The quiescent (before adaption) radiation pattern of the adaptive antenna factors heavily in the initial value of the interference-to-noise ratio. The eigenvalues depend in part on the radiation pattern and location of each element of either an adaptive phased array or each feed in a multiple-beam antenna. The covariance matrix eigenvalues are independent of scan angle when the channel covariance matrix is formed prior to array weighting or beam forming. In cases where the channel covariance matrix is formed after array element weighting (for example, in a sidelobe cancellation system as discussed in Chapter 3), the eigenvalues will depend on the array quiescent radiation pattern.

A basic adaptive nulling system block diagram for N-channels is shown in Figure 1.2. The signals $S_1(t), S_2(t), \ldots, S_N(t)$ (signal vector $\boldsymbol{S}(t)$) as a function of time t are received by either the feeds of an MBA or the elements of a phased array. These signals are fed into the adaptive signal processor that performs the cross-correlation operation (to determine $\boldsymbol{R}$, for example, as in (1.7)) either by analog circuitry or by digital signal processing. The adaptive signal processor controls the weights $w_1(t), w_2(t), \ldots, w_N(t)$ as a function of time such that the output signal $y(t)$ from the summing junction does not contain (ideally) any incident interference power. It is assumed here that interference signals only are sensed by the processor. This assumption means that desired signals have significantly less power over the nulling bandwidth than the interference signals and avoids the problem of nulling desired user signals, as desired in a communications system.

1.4 DERIVATION OF THE APPLEBAUM-HOWELLS STEADY-STATE ADAPTED WEIGHT EQUATION IN EIGENSPACE

The Applebaum-Howells analog servo-control-loop processor [1, 2] is commonly used in determining a set of complex weights $\boldsymbol{w}$ for nulling interference while maintaining a desired signal-to-noise ratio for radar targets or communications users. For this nulling processor, the time-dependent adaptive weight matrix equation is given by [6]

$$\tau_o \frac{d\boldsymbol{w}_a}{dt} + [\boldsymbol{I} + \mu \boldsymbol{R}]\boldsymbol{w}_a = \boldsymbol{w}_o \qquad (1.8)$$

where τ_o is a filter time constant, $\boldsymbol{w}_a$ is the adaptive weight vector, $\boldsymbol{I}$ is the identity matrix, $\boldsymbol{R}$ is the channel covariance matrix, μ is the effective loop gain that provides the threshold for sensing signals, and $\boldsymbol{w}_o$ is a weight vector (steering vector) that gives a desired quiescent radiation pattern in the absence of interference sources.

Setting the time derivative $\frac{d\boldsymbol{w}_a}{dt}$ equal to 0 in (1.8), it follows that, under steady-state conditions,

$$[\boldsymbol{I} + \mu \boldsymbol{R}]\boldsymbol{w}_a = \boldsymbol{w}_o \qquad (1.9)$$

Solving (1.9), the steady-state adaptive weights are expressed as the following column vector

$$\boldsymbol{w}_a = [\boldsymbol{I} + \mu \boldsymbol{R}]^{-1}\boldsymbol{w}_o \qquad (1.10)$$

where $^{-1}$ denotes matrix inverse.

The adaptive weight vector given by (1.10) yields the maximum signal-to-noise ratio (SNR), assuming that the desired signal is absent during the adaptation process and/or does not contribute to the covariance matrix R. For an N-channel adaptive nulling processor $[I + \mu R]$ is an $N \times N$ matrix.

For convenience, it can be assumed that the adaptive antenna channel weights w are normalized such that

$$w_1^2 + w_2^2 + \cdots + w_N^2 = 1 \tag{1.11}$$

In decibels, the weight vector can be expressed as

$$w_{\text{dB}} = 20\log_{10}|w| \tag{1.12}$$

Enforcing (1.11) prevents the situation where the adaptive antenna turns off to reduce the interference, that is, the null solution $w = 0$. Also, it is assumed that the nulling system is designed so that desired signals are not sensed by the loops. For example, in a radar system, the timing between the transmit pulses and the receive processing is adjusted so that the transmit signal is not present during reception. Thus, desired signals, from radar targets or communications users, ideally do not influence the adaptive weight settings.

The interference-to-noise ratio (INR) is computed from the interference covariance matrix and weight vector w as

$$\text{INR} = \frac{w^\dagger \cdot R \cdot w}{w^\dagger \cdot w} \tag{1.13}$$

The cancellation (C) achieved by an adaptive nulling antenna is defined as the ratio of the interference-to-noise ratio before and after adaption, that is,

$$C = \frac{\text{INR (before adaption)}}{\text{INR (after adaption)}} \tag{1.14}$$

It is useful to relate w to the eigenvalues λ_k and eigenvectors e_k of R such that [28]

$$R \cdot e_k = \lambda_k e_k, \quad k = 1, 2, \ldots, N \tag{1.15}$$

where it is assumed that the eigenvectors are normalized to unity amplitude.

The covariance matrix elements are defined to be

$$R_{pq} = \frac{1}{f_2 - f_1} \int_{f_1}^{f_2} S_p(f) S_q^*(f) df \tag{1.16}$$

where $f_2 - f_1$ is the nulling bandwidth (denoted B) and $S_p(f)$ and $S_q(f)$ are the received voltages in the pth and qth channels, respectively, over the nulling bandwidth. The fractional nulling bandwidth (FBW) is defined as

$$\text{FBW} = (f_2 - f_1)/f_o = \frac{B}{c}\lambda_o \tag{1.17}$$

where f_o is the center frequency, and λ_o is the wavelength at the center frequency, and c is the speed of light.

Noting that $R_{p,q} = R_{q,p}^*$ in (1.16), R is recognized as a Hermitian matrix, that is, $R^\dagger = R$. According to the spectral theorem [28], a Hermitian matrix such as R, can be decomposed in eigenspace as

$$R = \sum_{k=1}^{N} \lambda_k e_k e_k^\dagger \qquad (1.18)$$

or as

$$R = \sum_{k=1}^{N} \lambda_k P_k \qquad (1.19)$$

where

$$P_k = e_k e_k^\dagger \qquad (1.20)$$

is referred to as the kth projection matrix as described by Strang [28]. That is, P_k is the projection onto the eigenspace for λ_k and, since the eigenvectors have unit length, it follows that

$$\sum_{k=1}^{N} P_k = I \qquad (1.21)$$

The $N \times N$ covariance matrix R has a number of useful properties. Since R is Hermitian, all of the eigenvalues are real. The sum of the diagonal entries of R (known as the trace of R) is equal to the sum of the N eigenvalues. The product of the N eigenvalues of R is equal to the determinant of R [28].

It is useful to compare (1.16) and (1.19) and observe that the eigenvalues are proportional to power. In decibels, the eigenvalues can be expressed as

$$\lambda_{k\mathrm{dB}} = 10 \log_{10} \lambda_k \qquad (1.22)$$

An equivalent expression for the channel covariance matrix is [28]

$$R = Q \Lambda Q^\dagger \qquad (1.23)$$

where Q is a transformation matrix whose columns are the eigenvectors of R, and $\Lambda = \mathrm{diag}[\lambda_1, \lambda_2, \ldots, \lambda_n]$. Setting (1.18) equal to (1.23) yields

$$Q \Lambda Q^\dagger = \sum_{k=1}^{N} \lambda_k e_k e_k^\dagger \qquad (1.24)$$

Using (1.18), (1.9) can now be written in terms of the weight vector w as

$$[I + \mu \sum_{k=1}^{N} \lambda_k e_k e_k^\dagger]w = w_o \qquad (1.25)$$

or

$$w + \mu \sum_{k=1}^{N} \lambda_k e_k <e_k^\dagger, w> = w_o \qquad (1.26)$$

where $<e_k^\dagger, w> = e_k^\dagger w$ is a complex scalar. Note that in the summation in (1.26), the adapted weight vector w has been projected onto the kth eigenvector e_k. Next, take the product of $e_l^\dagger$, $l = 1, 2, \ldots, N$ with (1.26), which yields

$$<e_l^\dagger, w> + \mu \sum_{k=1}^{N} \lambda_k <e_l^\dagger, e_k><e_k^\dagger, w> = <e_l^\dagger, w_o>, \quad l = 1, 2, \ldots, N. \qquad (1.27)$$

The eigenvectors are orthonormal which means that

$$<e_l^\dagger, e_k> = 1, \quad k = l \qquad (1.28)$$

and

$$<e_l^\dagger, e_k> = 0, \quad k \neq l \qquad (1.29)$$

from which it follows that (1.27) reduces to

$$<e_l^\dagger, w> + \mu \lambda_l <e_l^\dagger, w> = <e_l^\dagger, w_o> \quad l = 1, 2, \ldots, N \qquad (1.30)$$

Factoring $<e_l^\dagger, w>$ on the left side of (1.30) and substituting k for l yields

$$<e_k^\dagger, w> = \frac{<e_k^\dagger, w_o>}{1 + \mu \lambda_k} \qquad (1.31)$$

Substituting (1.31) into (1.26) gives the desired result for the adapted weight vector in eigenspace

$$w = w_a = w_o - \sum_{k=1}^{N} \frac{\mu \lambda_k}{1 + \mu \lambda_k} <e_k^\dagger, w_o> e_k \qquad (1.32)$$

Equation (1.32) shows how the quiescent weight vector is modified in the presence of interference sources. The scalar quantity $<e_k^\dagger, w_o>$ is the projection of the eigenvector e_k on the quiescent weight vector w_o. For values of $\mu \lambda_k < 1$, corresponding to weak interference sources, the adapted weight vector would be relatively unchanged from w_o. However, when the interference sources have a large effective isotropic radiated power (EIRP), the eigenvalues are large (that is, $\mu\lambda_k > 1$), and the adapted weight vector tends to be more significantly varied from w_o. This variation in weight vector is due to the influence of the $(\mu\lambda_k/(1+\mu\lambda_k)) < e_k^\dagger, w_o >$ factor. The negative summation essentially removes any projections of interference source eigenvectors from the quiescent weight vector, which causes the complex adaptive antenna pattern to have a null (or nulls) in the direction of the interference as discussed in the next paragraph.

Each of the vectors in (1.32) are complex weights that can be applied to the adaptive antenna. From (1.32) and using the principle of superposition, the adaptive far-field (or near-field) pattern can be expressed as

$$P(\theta, \phi; w) = P_o(\theta, \phi; w_o) - \sum_{k=1}^{N} \frac{\mu\lambda_k}{1+\mu\lambda_k} < e_k^\dagger, w_o > P_k(\theta, \phi; e_k) \quad (1.33)$$

where $P(\theta, \phi; w)$ is the adaptive radiation pattern due to the weight vector w, $P_o(\theta, \phi; w_o)$ is the quiescent radiation pattern due to the weight vector w_o, and $P_k(\theta, \phi; e_k)$ is the kth eigenvector radiation pattern due to the weight vector e_k. In decibels, the radiation pattern can be expressed as

$$P(\theta, \phi)_{\text{dB}} = 20 \log_{10} |P(\theta, \phi)| \quad (1.34)$$

Equation (1.33) shows how the quiescent radiation pattern is modified in the presence of interference sources. The scalar quantity $< e_k^\dagger, w_o >$ is the projection of the kth eigenvector onto the quiescent antenna weight vector. If, for example, a single interference source, that gives rise to a single eigenvalue λ_1 and eigenvector e_1, is located on a null of the quiescent radiation pattern, then the quantity $< e_1^\dagger, w_o >= 0$. Since the interference source is already in a null, no further adaption is necessary and this means that the adaptive weight is equal to the quiescent weight, that is $w = w_o$. However, if an interference source is located on a sidelobe of the quiescent radiation pattern (for example, see Figure 1.4), the inner product of e_1 with w_o will be nonzero. This nonzero inner product or projection is then weighted by the quantity $\mu\lambda_1/(1+\mu\lambda_1)$ and is subtracted from w_o. For a large value of λ_1 corresponding to a strong interference source, the product $\mu\lambda_1$ is much greater than unity and this means that $\mu\lambda_1/(1+\mu\lambda_1) \approx 1$. Similarly, a weak interference source that has $\mu\lambda_1$ much less than unity results in $\mu\lambda_1/(1+\mu\lambda_1) \approx 0$. If another interference source sufficiently separated (by at least one half-power beamwidth of the adaptive antenna) from the first is included in the interference scenario, a second large eigenvalue, λ_2 can occur. Two terms ($k = 1, 2$) would then be significant in (1.33).

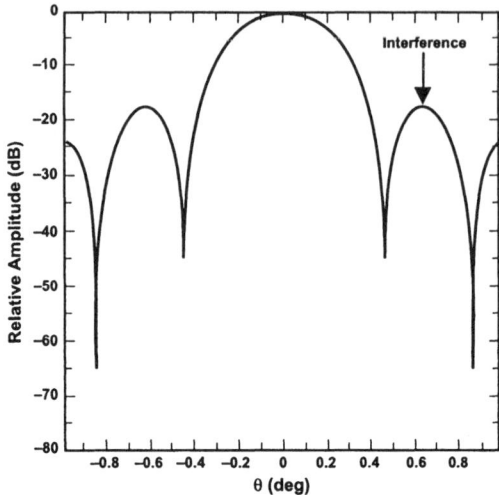

Figure 1.4 Radiation pattern of an adaptive nulling antenna before adaption.

From these examples, it is clear that strong interference sources will generally cause a larger change in the quiescent weight vector than weak sources. Equation (1.33) is dependent on both interference source location and on interference power level. Fundamentally, the quiescent radiation pattern is modified by removing any projections of interference source eigenvectors on w_o, and this is the basic mechanism by which the adaptive antenna forms a null (or nulls) in the direction of the interference.

It is instructive to examine, in terms of the adaptive antenna radiation patterns, how the interference covariance matrix eigenvalues and eigenvectors are related to adaptive nulling. Typically, for a single strong interference source, the covariance matrix has one large eigenvalue (compared to the quiescent receiver noise level) and a corresponding eigenvector. Suppose that a narrowband interference signal is incident in the direction of the adaptive antenna's first sidelobe. For example, the radiation pattern shown in Figure 1.4 is prior to adaptive nulling and is the pattern obtained for a uniformly illuminated 150-wavelength diameter circular aperture. For such a uniformly illuminated circular aperture, the far-field radiation pattern is of the form $J_1(u)/u$, where $u = (\pi D/\lambda_o)\sin\theta$ and the one-half power beamwidth in degrees is $58\lambda_o/D$, where D is the aperture diameter and λ_o is the wavelength [29]. The main beam of this adaptive antenna (which could be implemented, for example, as a multiple beam antenna) is pointed at broadside with beamwidth $0.39°$ and first sidelobe peak occurring at $\theta = 0.625°$. Without discussing details of the adaptive antenna design at this point, conceptually the eigenvector corresponding to the single large eigenvalue of the interference covariance matrix can be considered a weight vector. For an interference source arriving from the angle $\theta = 0.625°$, the eigenvector can be computed, and if it is applied as a weight vector, the radiation pattern shown in Figure 1.5 results. The eigenvector weight creates a beam pointed at the interference source direction. The eigenvector beam peak amplitude is adjusted to be equal to the unadapted or quiescent radiation pattern sidelobe level at the interference source location. When the eigenvector radiation pattern is subtracted from the quiescent radiation pattern, a null results at the interference source angle and the main beam remains pointed at broadside as shown in the adapted pattern in Figure 1.6.

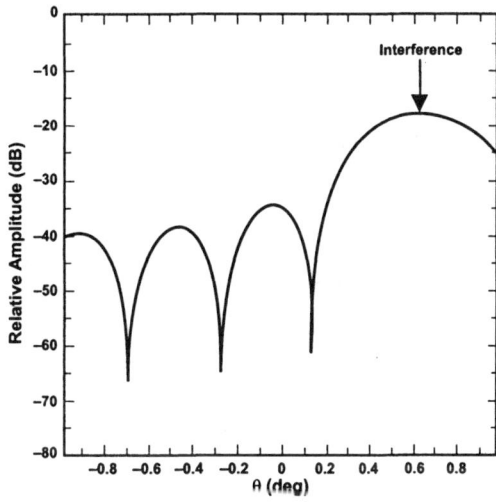

Figure 1.5 Radiation pattern using the dominant eigenvector as an antenna weight.

When there are multiple interference sources present, it is tempting to associate the number of interference sources nulled to the number of degrees of freedom consumed in the adaptive antenna nulling process. However, counting the number of sources nulled and equating that number to the degrees of freedom consumed is not always accurate, as demonstrated in the following conceptual examples given in Figures 1.7 and 1.8.

In the first example shown in Figure 1.7, consider the case where the adaptive antenna forms a directional radiation pattern to a single communications user at a known location in the antenna field of view. If there are two interference sources (denoted I_1, I_2) in the field of view, the adaptive antenna can create a null region that intersects both interference sources. Since discrete (point) nulls are not created in this case, it is not possible in this case,

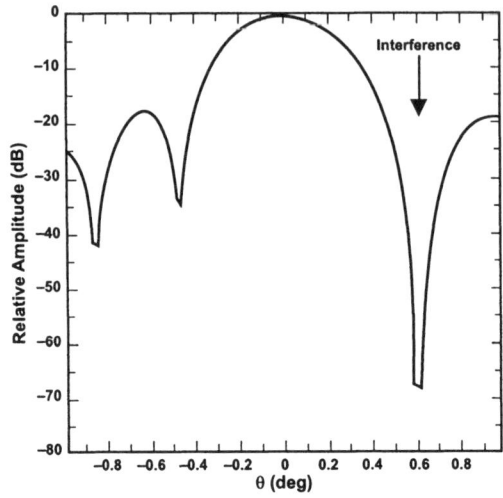

Figure 1.6 Radiation pattern of an adaptive nulling antenna after adaption.

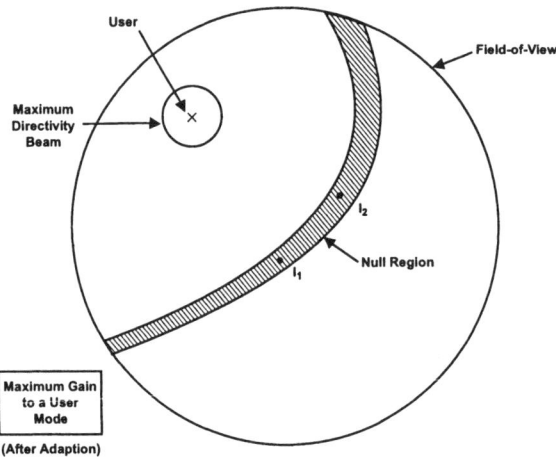

Figure 1.7 Typical adaptive null-region shape for a communications link where the user is in a known location.

by inspection of the adapted radiation pattern, to count the number of degrees of freedom consumed.

A second example, shown in Figure 1.8, is for the case where uniform pattern coverage is desired over a communications field of view [21]. This case corresponds to a communications link where the system users have unknown locations. Suppose that there are two interference sources in the

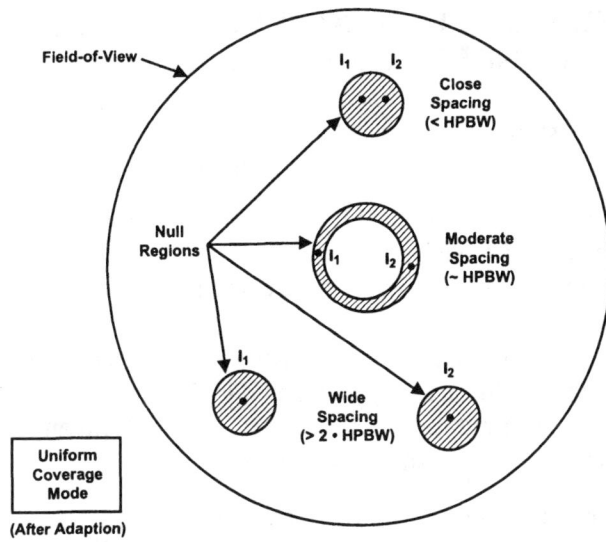

Figure 1.8 Typical adaptive null-region shapes for various interference source spacings for a communications link where the users are in unknown locations.

antenna field of view. Some typical examples of the shape of the adaptive null region(s) formed for various sources spacings are depicted in Figure 1.8. For close spacing less than one half-power beamwidth, both interference sources can be contained by a single circular-shaped null region, which suggests that one degree of freedom is consumed. For a source spacing close to one half-power beamwidth, the null region can be ring shaped and it is not clear how many degrees of freedom are consumed. When the sources are spaced greater than two half-power beamwidths apart, two separate circular-shaped null regions can be formed, and it could be surmised that two degrees of freedom have been consumed.

From this discussion, counting the number of interference sources that are nulled and setting this number equal to the number of degrees of freedom consumed by the adaptive process can be misleading. A more general approach is to quantify the eigenvalues of the interference covariance matrix. As will be shown in this chapter, a count of the number of large eigenvalues present in the interference covariance matrix is an accurate way to quantify the number of degrees of freedom consumed in the adaptive process. This analysis is investigated here for an adaptive array antenna.

1.5 CONDITIONS FOR COMPLETE CONSUMPTION OF N-DEGREES OF FREEDOM

N-degrees of freedom of an N-channel adaptive nulling antenna (array or MBA) can be completely consumed when the following conditions are met:

1. There are N uncorrelated equal-power interference sources in the antenna field of view.
2. The interference power is large enough to be sensed by the nulling system.
3. The interference covariance matrix has N equal (or comparable) eigenvalues that are large compared to the receiver noise level.
4. The adaptive nulling antenna system fails to form any adaptive nulls in the interference source directions.

Let it be assumed that a set of N uncorrelated equal power interference sources can be adjusted in position so as to obtain N equal eigenvalues, that is, $\lambda_1 = \lambda_2 = \cdots = \lambda_N = \lambda$. Then the covariance matrix given by (1.18) reduces to

$$R = \lambda \sum_{k=1}^{N} e_k e_k^\dagger \qquad (1.35)$$

but since the eigenvectors are orthogonal it can be shown that [28] (see (1.21))

$$\sum_{k=1}^{N} e_k e_k^\dagger = I \qquad (1.36)$$

Thus, the covariance matrix in this case reduces to

$$R = \lambda I \qquad (1.37)$$

Substituting (1.37) into (1.10) yields

$$w_a = \frac{1}{1 + \mu\lambda} w_o \qquad (1.38)$$

Equation (1.38) means that the adapted weight vector is a constant times the quiescent weight vector. For large values of $\mu\lambda$ the adaptive antenna would tend to turn off. However, the normalization given by (1.11) actually makes $w_a = w_o$. In other words, the radiation pattern does not adaptively change its shape from the quiescent pattern, and no adaptive cancellation is possible. In practice, for a very stressing interference environment, the restriction that all eigenvalues be equal can be relaxed somewhat, as long as they are all large compared to noise.

1.6 ORTHOGONAL INTERFERENCE SOURCES

Interference source configurations are considered in [5], where each source is geometrically arranged to consume one complete degree of freedom per source. These sources are termed *orthogonal* because the dot product of the received voltages, in the N-channels of the array, due to one source with the received voltages due to a second source is zero. That is, let S_i be the received signal vector with amplitude proportional to the square root of the incident power (denoted P_i), that is, $\sqrt{P_i}$ and phase Ψ_{ki} at the kth element from source i, such that $S_{ki} = \sqrt{P_i}e^{j\Psi_{ki}}$. Similarly, let S_j be the received signal vector with amplitude $\sqrt{P_j}$ and phase Ψ_{kj} at the kth element for source j, such that $S_{kj} = \sqrt{P_j}e^{j\Psi_{kj}}$. Then for an N-element array and orthogonal sources

$$S_i^\dagger \cdot S_j = \begin{cases} NP_i, & i = j \\ 0, & i \neq j \end{cases} \quad (1.39)$$

The case of subjecting N-element linear and planar arrays to $N-1$ orthogonal sources was investigated in [5]. Such source distributions require that $N-1$ degrees of freedom be allocated for spatial nulling. This was found to be useful in evaluating the wide-bandwidth performance of adaptive arrays. Several planar arrays and their associated orthogonal source distributions were shown; however, the case of N sources was not addressed. $N-1$ orthogonal-source distributions were shown for a three-element linear array, a seven-element triangle array, and a seven-element rotated double triangle array, and the cancellation as a function of frequency and nulling bandwidth was computed for various numbers of tapped delay lines and tap spacings. In Section 1.7, (1.39) will be applied to the case of a two-element array to illustrate the effectiveness of orthogonal sources in completely consuming the available degrees of freedom.

1.6.1 DERIVATION OF AN INTERFERENCE SIGNAL MATRIX TO CONSUME N DEGREES OF FREEDOM OF AN N-CHANNEL ADAPTIVE NULLING ARRAY ANTENNA

In this section, a mathematical proof is given that shows that the optimum configuration of N equal power uncorrelated interference sources that consumes the N degrees of freedom of an N-channel adaptive nulling array antenna requires a unitary signal matrix.

In the complex case, a unitary matrix has orthonormal columns and has the following properties

$$U^\dagger U = UU^\dagger = I \quad (1.40)$$

or
$$U^\dagger = U^{-1} \tag{1.41}$$

If there are N uncorrelated interference sources (in the presence of an N channel adaptive antenna), each generating a signal vector $S_1^i, S_2^i, \ldots, S_N^i$, where $i = 1, 2, \ldots, N$, the covariance matrix can be written as a summation of the individual covariance matrices, that is,

$$R = R_1 + R_2 + \cdots + R_N \tag{1.42}$$

where

$$R_1 = S_1 S_1^\dagger, \quad R_2 = S_2 S_2^\dagger, \quad \cdots, R_N = S_N S_N^\dagger \tag{1.43}$$

Another, more convenient, way to express the signal vector in the case of N sources is as an extended $N \times N$ matrix of signal column vectors

$$S = [\, S_1 \quad S_2 \quad \cdots \quad S_N \,] \tag{1.44}$$

In the case where the interference signals form a unitary signal matrix, the resulting $N \times N$ covariance matrix that is formed from the received interference signals is a diagonal matrix, as will now be shown.

As mentioned earlier, the concept of orthogonal interference sources has been discussed by Mayhan et al. [5]. By definition, each of these sources consumes one complete degree of freedom of the adaptive antenna. A necessary condition is that the sources must be separated angularly by at least one half-power beamwidth of the adaptive antenna. The received signal matrix for orthogonal sources has orthogonal columns, and this produces a covariance matrix with eigenvalues that are proportional to the incident power, as discussed in the following paragraphs.

With little loss of generality, an adaptive array can be assumed to be composed of an array of isotropic point receive antenna elements. (In the general case, the array elements will have a radiation pattern that must be taken into account, as well as polarization and mutual coupling effects between the array elements.) These isotropic elements are assumed to be located in the xy plane such that the kth array element has arbitrary rectangular coordinates (x_k, y_k). Further, it is assumed that the ith interference source with incident power P_i is located at a large distance in the far-field of the array, at the spherical coordinate system angles (θ_i, ϕ_i). Then the $k i$th element of the received signal matrix (S) is given by

$$S_{ki} = \sqrt{P_i} e^{j(2\pi/\lambda) \sin \theta_i (x_k \cos \phi_i + y_k \sin \phi_i)} \tag{1.45}$$

where $k = 1, 2, \ldots, N$ and $i = 1, 2, \ldots, N$.

Strictly speaking, the incident power P_i is really a power density in watts per meter squared, but it is assumed for convenience that the effective receive aperture of isotropic array elements normalizes P_i so that it has units of power (watts).

The covariance matrix for narrowband interference can be expressed as

$$R = SS^\dagger + I \tag{1.46}$$

Repeating (1.18), that is,

$$R = \sum_{k=1}^{N} \lambda_k e_k e_k^\dagger \tag{1.47}$$

consider the following illustrative examples of the covariance matrix expressed by (1.46) and (1.47) under different source conditions. Assume first that interference sources are not present, that is, $P_i = 0$, $i = 1, 2, \ldots, N$, so that $S = 0$ is the null matrix and $R = I$. For this null-signal case the eigenvalues are all equal to unity (the receiver noise level), that is,

$$\lambda_1 = \lambda_2 = \cdots, \lambda_N = 1 \tag{1.48}$$

Assume now that N large equal-power interference sources are present, thus

$$P_1 = P_2 = \cdots = P_N = P \tag{1.49}$$

where P is the power received at each array element due to an interference source. These sources produce covariance matrix eigenvalues that are large compared to receiver noise, that is, $\lambda \gg 1$. To completely consume N degrees of freedom requires that the covariance matrix must have N identical large eigenvalues, such that

$$\lambda_1 = \lambda_2 = \cdots = \lambda_N = \lambda = (PN + 1) \tag{1.50}$$

That is, the eigenvalues are proportional to the incident interference power. Then, from (1.37), the following relation is obtained

$$R = SS^\dagger = \lambda I \tag{1.51}$$

Next let

$$S_n = \frac{S}{\sqrt{\lambda}} = \frac{S}{\sqrt{PN+1}} \tag{1.52}$$

denote the normalized optimum signal matrix. Substituting (1.52) in (1.51) yields

$$S_n S_n^\dagger = I \tag{1.53}$$

Comparing (1.53) with (1.40) demonstrates that S_n must be a unitary matrix with orthonormal columns, thus the signal matrix S is an orthogonal matrix, meaning that the sources are orthogonal.

1.7 EIGENVALUES AND INR FOR TWO-ELEMENT ARRAY WITH TWO INTERFERENCE SOURCES

Consider a two-element array of isotropic receive antenna elements with element spacing d and two interference sources (I_1, I_2) positioned at angles (θ_1, θ_2) with incident power (P_1, P_2) relative to thermal noise. It will be shown in this section how these two sources can be configured to completely consume two degrees of freedom of the adaptive nulling process.

Choosing element number 1 as a zero-degree phase reference, the received signal at the kth element due to the ith source can be expressed as

$$S_{ki} = \sqrt{P_i} e^{j2\pi(d/\lambda)(k-1)\sin\theta_i} \tag{1.54}$$

where $k = 1, 2$ and $i = 1, 2$. Assuming a narrow nulling bandwidth and uncorrelated sources, the covariance matrix relative to thermal noise is computed from

$$\boldsymbol{R}_n = \boldsymbol{S}\boldsymbol{S}^\dagger + \boldsymbol{I} \tag{1.55}$$

where the identity matrix ($\boldsymbol{I}$) represents the thermal noise level and $\boldsymbol{S}$ is of the form of (1.44). Substituting (1.54) in (1.55) yields

$$\boldsymbol{R}_n = \begin{bmatrix} P_1 + P_2 + 1 & P_1 e^{-j\Psi_1} + P_2 e^{-j\Psi_2} \\ P_1 e^{j\Psi_1} + P_2 e^{j\Psi_2} & P_1 + P_2 + 1 \end{bmatrix} \tag{1.56}$$

where $\Psi_1 = 2\pi(d/\lambda)\sin\theta_1$ and $\Psi_2 = 2\pi(d/\lambda)\sin\theta_2$.

The signal vectors in (1.54) are orthogonal according to (1.39) when $\boldsymbol{S}_1^\dagger \cdot \boldsymbol{S}_2 = 0$ which yields

$$\sqrt{P_1 P_2} + \sqrt{P_1 P_2} e^{j2\pi(d/\lambda)(\sin\theta_2 - \sin\theta_1)} = 0 \tag{1.57}$$

Equation (1.57) is satisfied when

$$e^{j2\pi(d/\lambda)(\sin\theta_2 - \sin\theta_1)} = -1 \tag{1.58}$$

Setting $2\pi(d/\lambda)(\sin\theta_2 - \sin\theta_1) = \pm\pi, \pm 3\pi, \pm 5\pi, \cdots$, one can solve for the orthogonal angles (θ_1, θ_2). The case where the spacing d is less than or equal to one-half wavelength yields

$$\theta_2 = \sin^{-1}(\sin\theta_1 \pm \frac{\lambda}{2d}), \quad \frac{d}{\lambda} \leq \frac{1}{2} \tag{1.59}$$

Equation (1.59) shows that the two orthogonal source locations are independent of interference power. Equation (1.59) is plotted in Figure 1.9 for a few element spacings, $d = \lambda/4, \lambda/3, \lambda/2$. For one-quarter wavelength spacing, there are only two orthogonal source pairs: ($\theta_1 = 90°, \theta_2 = -90°$) and ($\theta_1 = -90°, \theta_2 = 90°$). For $d = \lambda/3$ there are many solutions; however, both θ_1 and θ_2 are restricted to $|\theta| > 30°$. For the case where $d = \lambda/2$, when $\theta_1 = 0$ there are two solutions, $\theta_2 = \pm 90°$ – for all other values of θ_1 only one unique value of θ_2 represents an orthogonal source angle.

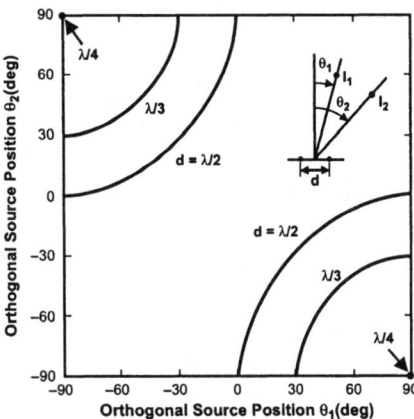

Figure 1.9 Orthogonal source pair locations for two interference sources in the field of view of a two-element array with various element spacings. © 1985 IEEE [24].

For discussion purposes, the case where $d = \lambda/2$ will be investigated in detail. Of interest is to determine for two orthogonal sources what relative power levels minimize the amount of adaptive cancellation achieved. Minimizing the amount of cancellation by (1.14) means that $C = 1$ is desired. Equivalently, in this situation, the INR (given by (1.13)) after adaption will not change from its value before adaption.

Let it be assumed, for example, that the array quiescent weight vector gives omnidirectional pattern coverage. Then, initially only one array element is on, so

$$w_o = [1, 0]^T \qquad (1.60)$$

Substituting (1.56) and (1.60) in (1.13), the quiescent INR for the two-element array/two source case is computed to be

$$\text{INR(before adaption)} = P_1 + P_2 + 1 \qquad (1.61)$$

The INR after adaption is computed by using (1.10) and (1.56) in (1.13). Of interest is the change in INR or cancellation for two orthogonal interference sources as a function of the source power levels.

The narrowband eigenvalues of the covariance matrix are found by solving $\det(\boldsymbol{R} - \lambda \boldsymbol{I}) = 0$ using Equation (1.56) for $\boldsymbol{R}$. The two narrowband eigenvalues are computed to be

$$\lambda_1 = 1 + P_1 + P_2 + \sqrt{P_1^2 + P_2^2 + 2P_1 P_2 \cos(\Psi_1 - \Psi_2)} \qquad (1.62)$$

$$\lambda_2 = 1 + P_1 + P_2 - \sqrt{P_1^2 + P_2^2 + 2P_1 P_2 \cos(\Psi_1 - \Psi_2)} \qquad (1.63)$$

If desired, the eigenvectors can be found by substituting the eigenvalues given by (1.62) and (1.63), and the covariance matrix given by (1.56) in (1.15) and solving the two simultaneous equations.

It is useful at this point to relate nulling bandwidth to the narrow-band assumption in (1.55). As was shown in [4, 5], the eigenvalues of $\boldsymbol{R}$ are a function of the nulling bandwidth. To show bandwidth effects, consider a single interference source with incident power $P_1 = 40$ dB relative to thermal noise. First, using (1.62) and (1.63) with $P_1 = 10,000$ and $P_2 = 0$, the narrow-band eigenvalues are computed to be $\lambda_1 = 43.0$ dB, $\lambda_2 = 0$ dB, which are independent of the angle of incidence (θ_i). Next, using (1.16) to compute the covariance matrix for fractional nulling bandwidths ranging from 0 to 10 percent and incident angles $\theta_i = 0°, 30°, 60°, 90°$, Table 1.1 lists the resulting eigenvalues. For these cases, the first eigenvalue is always 43 dB. The second eigenvalue increases with increasing angle of incidence from broadside and also with increasing bandwidth. In terms of degrees of freedom, the largest eigenvalue can be interpreted as one degree of freedom being completely consumed. The smaller eigenvalue represents partial consumption of the remaining degree of freedom due to the nonzero nulling bandwidth. For bandwidths not exceeding 3 or 4 percent, there is only a relatively small contribution to λ_2. This choice is somewhat arbitrary; however, for the small diameter arrays considered in this chapter a bandwidth of less than approximately 4 percent is termed narrow as far as the eigenvalues and hence the degrees of freedom are concerned. In the numerical examples presented during the remainder of this chapter, a narrow bandwidth of 3.3 percent is used. The term narrow band is a function of aperture size (length L) according to $BL\sin\theta/c < 1$, where c is the speed of light. For an array many wavelengths long, bandwidth has a more significant effect.

The value of μ in (1.10) is typically chosen according to [2]

$$\frac{DR}{\lambda_{max}} \qquad (1.64)$$

where DR is the dynamic range of the nulling processor and λ_{max} is the maximum eigenvalue of $\boldsymbol{R}$. For example, assuming a dynamic range of, say

Table 1.1
Eigenvalues Computed for Two-Element Array with $\lambda/2$ Element Spacing and a Single Interferer Versus Nulling Bandwidth

Nulling Bandwidth (%)	Ordered Eigenvalues (dB)							
	$\theta_1 = 0°$		$\theta_1 = 30°$		$\theta_1 = 60°$		$\theta_1 = 90°$	
	λ_1	λ_2	λ_1	λ_2	λ_1	λ_2	λ_1	λ_2
0	43	0	43	0	43	0	43	0
1	43	0	43	0.4	43	1.2	43	1.5
2	43	0	43	1.5	43	3.5	43	4.2
3	43	0	43	2.9	43	5.8	43	6.7
4	43	0	43	4.2	43	7.7	43	8.8
5	43	0	43	5.5	43	9.4	43	10.5
6	43	0	43	6.7	43	10.8	43	12.0
7	43	0	43	7.8	43	12.1	43	13.3
8	43	0	43	8.8	43	13.2	43	14.4
9	43	0	43	9.7	43	14.2	43	15.4
10	43	0	43	10.5	43	15.0	43	16.2

30 dB, and a maximum eigenvalue of 43 dB, then the loop gain constant would be $\mu = 1000/20,000 = 0.05$.

Consider now the two element array with spacing $d = \lambda/2$ and arbitrarily a fairly narrow nulling bandwidth of 3.3 percent. From Figure 1.9 (or from (1.59)), two orthogonal source positions are chosen to be $\theta_1 = -30°$, $\theta_2 = +30°$. The incident power level of source number 1 is chosen to be fixed at $P_1 = 40$ dB and P_2 is varied over the range 0 to 60 dB. The computed INR before and after adaption is shown in Figure 1.10(a), and it is seen that 0 dB cancellation (C = 1) occurs when $P_2 = 40$ dB. For this example, when $P_1 = P_2$ the antenna no longer adapts, and for $P_2 < P_1$ or $P_2 > P_1$ the antenna does adapt. The interference sources maximally stress the array's degrees of freedom in accordance with the definition given in Section 1.2, when the source power levels are equal. (Note of course that increasing the power of source number 2 without bound has an equivalent effect of producing a large INR after adaption.)

It is now appropriate to compute the eigenvalues (λ_1, λ_2) of the interference covariance matrix. With sources I_1 and I_2 fixed at $\theta = \pm 30°$ and with $P_1 = 40$ dB and P_2 variable as before, the two eigenvalues are shown in Figure 1.10(b). As the two source power levels become more similar, the difference between the eigenvalues decreases. For $P_1 = P_2 = 40$ dB the two eigenvalues are equal. In a comparison of Figures 1.10(a) and 1.10(b), it is important to note that when $\lambda_1 = \lambda_2 = 43$ dB, the cancellation is 0 dB. It can be inferred from (1.13) that because the INR is the same before and after

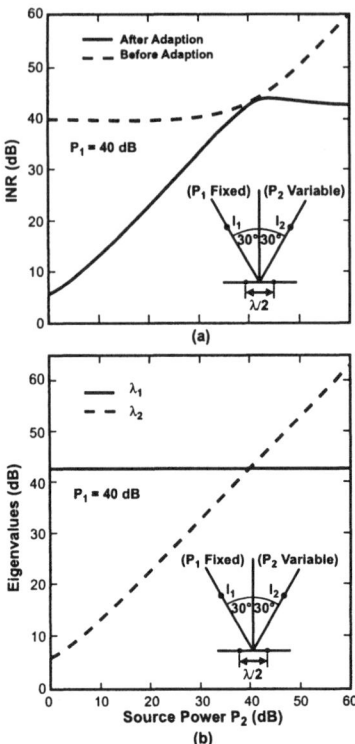

Figure 1.10 Two-element adaptive array behavior for two orthogonal interference sources as a function of source power P_2, with $P_1 = 40$ dB and 3.3 percent nulling bandwidth. Array element spacing is $\lambda/2$. (a) INR, and (b) covariance matrix eigenvalues. © 1985 IEEE [24].

adaption, the adapted weight vector is equal to the quiescent weight vector. This is indeed the case as could be shown by substituting (1.56) in (1.10) with $P_1 = P_2$, $\theta_1 = -30°$, $\theta_2 = 30°$.

For two equal-power sources it is useful to show how the adapted INR or cancellation varies with source spacing. This example is depicted in Figure 1.11(a) with $P_1 = P_2 = 40$ dB and FBW = 0.033 (that is, 3.3 percent nulling bandwidth), where source number 1 is fixed at $\theta = -30°$ and source number 2 varies in angular position over the range $-90° < \theta_2 < 90°$. The only occurrence of 0 dB cancellation (two degrees of freedom completely consumed) is at the orthogonal position $\theta_2 = 30°$, as expected from (1.59). The maximum or best cancellation occurs when the two sources share the same position ($\theta_1 = \theta_2 = -30°$). The eigenvalue behavior can be used in describing these results. The two eigenvalues (λ_1, λ_2) are plotted in Figure 1.11(b) as source number 2 varies in angular position. When the two sources are colocated at $\theta = -30°$ there is only one dominant eigenvalue ($\lambda_1 = 46$ dB, $\lambda_2 = 5.3$ dB), meaning that only one degree of freedom is consumed. At the orthogonal positions, $\theta = \pm 30°$, the eigenvalues are equal ($\lambda_1 = \lambda_2 = 43$ dB) so that two degrees of freedom are completely consumed. For other source spacings some cancellation is achieved and it is seen that the eigenvalues are unequal, suggesting that two degrees of freedom are engaged but not fully consumed.

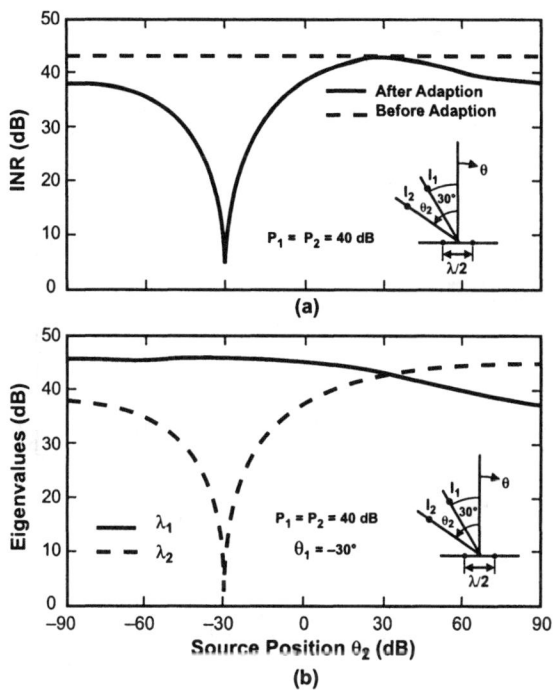

Figure 1.11 Two-element adaptive array behavior for two equal power sources ($P_1 = P_2 = 40$ dB) with source number 1 fixed at $\theta_1 = -30°$ and source number 2 position variable. Nulling bandwidth is 3.3 percent. (a) Interference-to-noise ratio. (b) Covariance matrix eigenvalues. © 1985 IEEE [24].

In contrast, if the two source powers are unequal, a 0 dB cancellation ratio is not achieved for all source spacings. The INR cancellation for this case is shown in Figure 1.12(a), where source number 1 with power $P_1 = 40$ dB is fixed at $\theta = -30°$ and source number 2 with power $P_2 = 30$ dB varies in position over $-90° < \theta_2 < 90°$. A minimum cancellation of 7 dB occurs at the orthogonal position $\theta_2 = 30°$ and a maximum cancellation of 37 dB occurs at $\theta_2 = \theta_1 = -30°$. The corresponding eigenvalues are shown in Figure 1.12(b), and the minimum eigenvalue spread ($\lambda_1 - \lambda_2$) is

10 dB (corresponding to the source power spread) occurring at the orthogonal position $\theta_2 = 30°$. Clearly, the cancellation has been limited depending on the position of source number 2. Some cancellation is always possible for this case because the two eigenvalues are unequal.

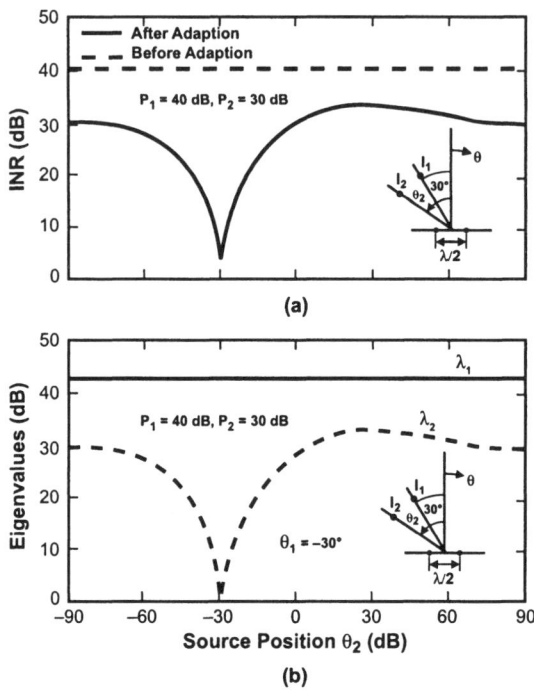

Figure 1.12 Two-element adaptive array behavior for two unequal power sources ($P_1 = 30$ dB, $P_2 = 40$ dB) with source number 1 fixed at $\theta_1 = -30°$ and source number 2 position variable. Nulling bandwidth is 3.3 percent. (a) INR, and (b) covariance matrix eigenvalues. © 1985 IEEE [24].

1.8 FIGURE OF MERIT FOR CONSUMPTION OF DEGREES OF FREEDOM

It was shown in the previous section that N degrees of freedom of an N element adaptive antenna are fully consumed (and no adaptive cancellation is possible) when the interference covariance matrix has N equal eigenvalues. This fully consumed condition occurs when the normalized signal matrix is unitary. For an arbitrary distribution of interference sources, the signal matrix is readily computed. To determine how close this matrix is to a unitary matrix, a figure of merit (a quantitative measure) is necessary.

An assumption is made here that to maximize consumption of N degrees of freedom, it is desired to find the source distribution that minimizes the difference between the signal matrix S and an associated unitary matrix denoted by A (even if it is not possible in all cases to reduce this difference to zero). A figure of merit is now derived that is utilized by a gradient search (given in the next section) to make S the closest approximation to A.

To proceed, the signal matrix S can be written as the sum of the unitary matrix A plus a residual matrix r, that is,

$$S = A + r \qquad (1.65)$$

It is desired to minimize the residual r, so that S is the best approximation to the unitary matrix A. This can be done in the least squares sense by considering the following minimization problem [30]

$$\min \|r\|^2 = \min \|S - A\|^2 \qquad (1.66)$$

where $\|\cdot\|$ denotes norm such that

$$\|r\|^2 = \sum_i \sum_j |r_{ij}|^2 \qquad (1.67)$$

By the singular value decomposition theorem [30] any matrix, such as S, can be decomposed as

$$S = V \Sigma T^\dagger \qquad (1.68)$$

where

$$\Sigma = \mathrm{diag}(\sigma_i), i = 1, 2, \ldots, N \qquad (1.69)$$

where $|\sigma_i| = \sqrt{\lambda_i}$ are the positive square roots of the eigenvalues for $SS^\dagger$, V is the unitary eigenvector matrix for $SS^\dagger$, and T is the unitary eigenvector matrix for $S^\dagger S$.

Substituting (1.68) in (1.66) yields

$$\|V \Sigma T^\dagger - A\|^2 = \mathrm{minimum} \qquad (1.70)$$

Next, use the unitary matrix properties that [30]

$$\|M\|^2 = \|U_1 M\|^2 = \|M U_2\|^2 = \|U_1 M U_2\|^2 \qquad (1.71)$$

where U_1 and U_2 are unitary matrices; and M is an $N \times N$ matrix.

Premultiplying by $V^\dagger$ and postmultiplying by T, (1.70) becomes

$$\|V \Sigma T^\dagger - A\|^2 = \|V^\dagger V \Sigma T^\dagger T - V^\dagger A T\|^2 \qquad (1.72)$$

Since V and T are unitary matrices, that is,

$$V^\dagger V = I = T^\dagger T \tag{1.73}$$

it follows that

$$V^\dagger V \Sigma T^\dagger T = I\Sigma I = \Sigma \tag{1.74}$$

Substituting (1.74) in (1.72) yields

$$||V\Sigma T^\dagger - A||^2 = ||\Sigma - V^\dagger AT||^2 \tag{1.75}$$

Now let,

$$P = V^\dagger AT \tag{1.76}$$

so that (1.75) can now be expressed as

$$||V\Sigma T^\dagger - A||^2 = ||\Sigma - P||^2 = \text{minimum} \tag{1.77}$$

Now P is unitary since

$$P^\dagger P = T^\dagger A^\dagger V V^\dagger AT = T^\dagger A^\dagger AT = T^\dagger T = I \tag{1.78}$$

Hence,

$$\sum_{j=1}^{N} |P_{ij}|^2 = 1 \tag{1.79}$$

that is, the sum of the absolute value squared of the matrix elements along any column of P is unity. Also, because P is unitary, the diagonal elements of P are constrained by

$$|P_{ii}| \leq 1 \tag{1.80}$$

Another useful property of the unitary matrix P is that

$$\sum_{i=1}^{N}\sum_{j=1}^{N} |P_{ij}|^2 = N \tag{1.81}$$

Since the matrix Σ is diagonal it is logical that, to minimize $||\Sigma - P||^2$, the optimum P must also be diagonal (this is the only way that the off-diagonal elements of $(\Sigma - P)$ can be equal to zero). Therefore, by (1.79) and (1.80), P must be the identity matrix, that is,

$$P = I \tag{1.82}$$

It must now be shown that of all unitary matrices, $P = I$ is the closest unitary matrix to Σ. To show this relation, it is required to prove that

$$||\Sigma - I||^2 \leq ||\Sigma - U||^2 \tag{1.83}$$

for all unitary matrices U. Performing the norm on the left- and right-hand sides of the earlier inequality yields

$$||\Sigma - I||^2 = \sum_{i=1}^{N}(\sigma_i - 1)^2 = \sum_{i=1}^{N}\sigma_i^2 - 2\sum_{i=1}^{N}\sigma_i + N \qquad (1.84)$$

and

$$||\Sigma - U||^2 = \sum_{i=1}^{N}\sum_{j=1}^{N}|U_{ij}|^2 + \sum_{i=1}^{N}\sigma_i^2 - \sum_{i=1}^{N}\sigma_i(U_{ii} + U_{ii}^*) \qquad (1.85)$$

Using (1.81), the double summation for the norm operation of the unitary matrix U, denoted $|U_{ij}|^2$, is equal to N, so that (1.85) reduces to

$$||\Sigma - U||^2 = N + \sum_{i=1}^{N}\sigma_i^2 - \sum_{i=1}^{N}\sigma_i(U_{ii} + U_{ii}^*) \qquad (1.86)$$

Substituting (1.84) and (1.86) in (1.83) yields

$$\sum_{i=1}^{N}\sigma_i^2 - 2\sum_{i=1}^{N}\sigma_i + N \leq N + \sum_{i=1}^{N}\sigma_i^2 - \sum_{i=1}^{N}\sigma_i(U_{ii} + U_{ii}^*) \qquad (1.87)$$

Clearly, if the unitary matrix U is other than the identity matrix, then by (1.80) and (1.81)

$$(U_{ii} + U_{ii}^*) < 2 \qquad (1.88)$$

and the inequalities of (1.87) and (1.83) are satisfied. Therefore, $P = I$ is the optimum matrix that minimizes (1.77).

Using the left-hand side of (1.87), (1.77) can now be written as

$$||\Sigma - P||^2 = \sum_{i=1}^{N}\sigma_i^2 - 2\sum_{i=1}^{N}\sigma_i + N = \text{minimum} \qquad (1.89)$$

In (1.89) the summation of σ_i^2, N, and $||\Sigma - P||^2$ are all positive, hence $||\Sigma - P||^2$ is minimized when

$$F = \sum_{i=1}^{N}\sigma_i = \sum_{i=1}^{N}\sqrt{\lambda_i} = \text{maximum} \qquad (1.90)$$

Equation (1.90) is the desired figure of merit (denoted by F), that is, when the interference sources are located so that the sum of the square roots of the eigenvalues computed from the interference covariance matrix $SS^\dagger$ is maximized, the signal matrix S is the best approximation to the unitary matrix A.

Consider an example, now, that demonstrates that the figure of merit, given by (1.90), can be used to identify the most stressing configuration of two interference sources. Assume an arbitrary N-element array of isotropic point sources, with equal power (P) incident from two interference sources with two different sets of spacings. In the first set, the two sources are angularly separated by approximately one half-power beamwidth such that there are two equal eigenvalues $\lambda_1 = \lambda_2 = \lambda$ associated with the covariance matrix. Thus, two degrees of freedom are consumed. It is assumed here that λ is much larger than the quiescent noise level ($\lambda_q = 1$). The ($N - 2$) remaining eigenvalues are equal to unity. From (1.90) the figure of merit for case number 1 is

$$F_1 = \sum_{i=1}^{N} \sqrt{\lambda_i} = \sqrt{\lambda_1} + \sqrt{\lambda_2} + (N - 2) \quad (1.91)$$

or

$$F_1 = 2\sqrt{\lambda} + (N - 2) \quad (1.92)$$

Next, assume that the two interference sources are located at the same angular position. In this case, there is only one eigenvalue and it is 3 dB larger than the eigenvalue for a single interference source [1]. That is,

$$\lambda_1 = 2\lambda, \lambda_2 = 1, \lambda_3 = 1, \cdots, \lambda_N = 1 \quad (1.93)$$

and one degree of freedom is consumed. For this case (case #2), the figure of merit is computed to be

$$F_2 = \sum_{i=1}^{N} \sqrt{\lambda_i} = \sqrt{2}\sqrt{\lambda} + (N - 1) \quad (1.94)$$

and by (1.90), since F_1 is greater than F_2, and configuration number 1 has a more stressing effect as interference than configuration number 2. In other words, this comparison demonstrates the ability of the figure of merit, given by (1.90), to identify a more stressing set of interference sources.

The next section discusses a gradient-search technique that uses the previous figure of merit given by (1.90) to find maximally stressing interference source configurations for arbitrary adaptive antenna arrays.

1.9 GRADIENT SEARCH TO MAXIMIZE CONSUMPTION OF ADAPTIVE ANTENNA DEGREES OF FREEDOM

Since a unitary interference signal matrix implies complete consumption of N degrees of freedom, it is reasonable to suppose that the interference signal matrix must be the closest approximation to a unitary matrix in order to maximize consumption of the degrees of freedom. The closest approximation to a unitary interference signal matrix for an arbitrary array can be realized by employing a gradient search.

Let the received signal at the kth array element be given by

$$S_{kij} = \sqrt{P_i} e^{j\pi(D/\lambda)\sin\theta_{ij}(x'_k \cos\phi_{ij} + y'_k \sin\phi_{ij})} \tag{1.95}$$

where (θ_{ij}, ϕ_{ij}) are the coordinates for the ith interference source with power P_i and the subscript j being the jth configuration of interference sources. (Note: the imaginary exponent $j = \sqrt{-1}$ should not be confused with the integer subscript $j = 1, 2, \ldots, J$.) In (1.95) D is the array diameter and (x'_k, y'_k) are the array element coordinates relative to $D/2$.

The figure of merit for the jth interference source configuration is given from (1.90) as

$$F_j = \sum_{i=1}^{N} \sqrt{\lambda_{ij}} \tag{1.96}$$

The optimum interference source configuration, from a given set of J source configurations, occurs when F_j is maximized, that is,

$$F_{\text{opt}} = \max(F_j), \quad j = 1, 2, \ldots, J \tag{1.97}$$

Assume now that M interference sources are distributed within the antenna field of view as depicted in Figure 1.13. The ith interference source from the jth source configuration, has position coordinates (u_{ij}, v_{ij}) where

$$u_{ij} = \pi \frac{D}{\lambda} \sin\theta_{ij} \cos\phi_{ij} \tag{1.98}$$

$$v_{ij} = \pi \frac{D}{\lambda} \sin\theta_{ij} \sin\phi_{ij} \tag{1.99}$$

Assuming an initial configuration of interference sources, the sources are to be moved until the optimum figure of merit is achieved. From (1.16), (1.18), (1.95), (1.98), and (1.99), observe that

$$\lambda_{ij} = \lambda_{ij}((u_{1j}, v_{1j}), (u_{2j}, v_{2j}), \ldots, (u_{Mj}, v_{Mj})) \tag{1.100}$$

that is, each eigenvalue is a function of the positions of all M sources. It is desired to find the new positions for the interference sources such that the figure of merit increases most rapidly. That is, select directions such that the directional derivative is maximized at (u_j, v_j) [31].

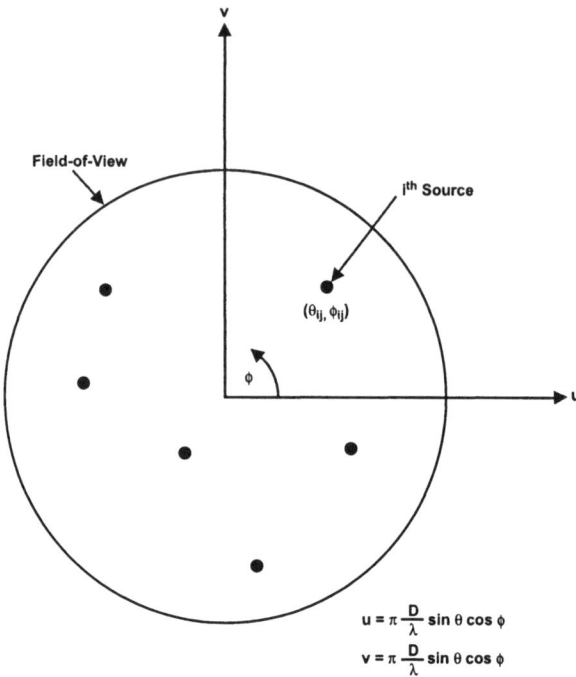

Figure 1.13 Distribution of interference sources in the adaptive antenna field of view.

The directional derivative (denoted by $D(\cdot)$) of the figure of merit is given by

$$D(F_j) = \sum_{i=1}^{M} \left(\frac{\partial F_j}{\partial u_{ij}} r_{uij} + \frac{\partial F_j}{\partial v_{ij}} r_{vij} \right) \quad (1.101)$$

where ∂ means partial derivative and r_{uij}, r_{vij} are the (u, v) directions for which F_j is increasing most rapidly. The directions, r_{uij}, r_{vij}, are constrained by

$$\sum_{i=1}^{N} (r_{uij}^2 + r_{vij}^2) = 1 \quad (1.102)$$

It is desired to maximize $D(F_j)$ (given by (1.101) subject to (1.102)).

The details of such a maximization are given in [31] and so only the result is stated here. The partial derivatives in (1.101) are replaced by finite differences, and the new positions of the $(j+1)$th source configuration are computed from the jth configuration according to

$$u_{i(j+1)} = u_{ij} + \Delta u \, r_{uij} \quad (1.103)$$

$$v_{i(j+1)} = v_{ij} + \Delta v \, r_{vij} \tag{1.104}$$

where

$$r_{uij} = \frac{\Delta F_{uij}}{\sqrt{\sum_{i=1}^{M}(\Delta F_{uij})^2 + (\Delta F_{vij})^2}} \tag{1.105}$$

$$r_{vij} = \frac{\Delta F_{vij}}{\sqrt{\sum_{i=1}^{M}(\Delta F_{uij})^2 + (\Delta F_{vij})^2}} \tag{1.106}$$

where

$$\Delta F_{uij} = F_j(u_{ij} + \Delta u; v_{ij}) - F_j(u_{ij} - \Delta u; v_{ij}) \tag{1.107}$$

$$\Delta F_{vij} = F_j(v_{ij}; v_{ij} + \Delta v) - F_j(v_{ij}; u_{ij} - \Delta u) \tag{1.108}$$

and Δu and Δv are the maximum step sizes.

1.10 MAXIMUM JAMMER EFFECTIVENESS, NUMERICAL RESULTS

The gradient search and figure of merit, described in the previous sections, are utilized in this section to maximize jammer effectiveness for planar arrays. Two example planar arrays, hexagonal and circular ring, are shown in Figure 1.14. For convenience, seven isotropic elements are assumed in each array and each array has a one-wavelength diameter at the center frequency. The nulling bandwidth is arbitrarily chosen to be 3.3 percent, so the bandwidth is narrow as far as consuming degrees of freedom is concerned. Both equal-power and unequal-power source distributions are investigated.

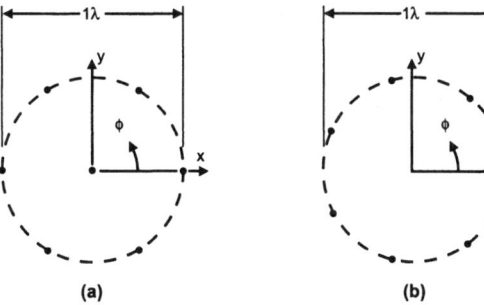

Figure 1.14 Seven-element planar array configurations. (a) Hexagonal, and (b) uniform ring. © 1985 IEEE [24].

1.10.1 SEVEN-ELEMENT HEXAGONAL ARRAY

Consider now, the seven-element hexagonal array shown in Figure 1.14(a) first with seven interference sources with equal incident power ($P_1 = P_2 = \cdots = P_7 = 40$ dB). Assuming, for example, a processor dynamic range of 30 dB the loop gain constant is computed from (1.64) as $\mu = DR/\lambda_{max} = 0.014$, where the maximum eigenvalue used is computed from $1 + NP$ with $N = 7$ and $P = 10,000$. Also, it is assumed, for example, that the quiescent weight vector is chosen as uniform coverage with only one element on in the absence of interference, that is, $w_o = (1, 0, 0, 0, 0, 0, 0)^T$. For the gradient search to be unbiased, an initial grouping of closely spaced sources is chosen so that they consume only one degree of freedom. The initial source arrangement was a hexagonal cluster with diameter $0.4°$. The eigenvalues computed for this initial configuration are given in Table 1.2. The nonzero eigenvalues are $\lambda_1 = 56.9$ dB, $\lambda_2 = 10.7$ dB, and $\lambda_3 = 10.7$ dB. The existence of one dominant eigenvalue λ_1 implies that only one degree of freedom is completely consumed. Two other degrees of freedom (λ_2, λ_3) have been slightly engaged due to the small but finite source cluster diameter and the nonzero bandwidth.

Next, the gradient search was performed using small increments $\Delta u = \Delta v = 0.15$ in (1.103) and (1.104) and a converged source distribution is obtained after approximately 125 iterations. The figure of merit given by (1.96) was used to determine source movement. After the gradient search, the source positions and the eigenvalues are listed in Table 1.2. The sources are arranged hexagonally and rotated about $10.8°$ with respect to the array. All seven degrees of freedom are completely consumed because the seven eigenvalues are nearly equal. As shown in Table 1.2, the INR before and after adaption is unchanged resulting in a 0 dB cancellation ratio. This means that the adapted radiation pattern is the same as the quiescent pattern and no adaptive nulls are formed. Thus, the seven orthogonal sources represent a maximally stressed environment for this array.

To show the effect of reducing the number of completely consumed degrees of freedom by one, source number 7 in Table 1.2 is now turned off. This means that there are only six orthogonal sources in the field of view of the hexagonal array; the resulting ordered eigenvalues are shown in Table 1.3. Clearly, there are six nearly equal eigenvalues, and one eigenvalue (λ_7) is partially engaged. In other words, only six degrees of freedom are completely consumed so that six degrees of freedom can be used for nulling, with one degree of freedom available for providing gain to a desired direction. The INR before adaption is computed to be 47.8 dB and after adaption the INR is 12.0 dB. Thus, a cancellation of 35.8 dB is achieved, and so the removal of just one orthogonal source has allowed the array to significantly improve the INR. Next, to investigate the effect of having different source

Table 1.2
Eigenvalues and INR Before and After Gradient Search for Seven Equal-Power Sources in the Field of View of the Seven-Element Hexagonal Array

	Before Gradient Search					After Gradient Search				
	Source Locations (deg)		Source Powers	Ordered Eigenvalues			Source Locations (deg)		Source Powers	Ordered Eigenvalues
i	θ_i	ϕ_i	P_i (dB)	j	λ_j (dB)	i	θ_i	ϕ_i	P_i (dB)	j	λ_j (dB)
1	0	0	40	1	56.9	1	0	0	40	1	48.6
2	0.2	0	40	2	10.7	2	61.5	10.8	40	2	48.6
3	0.2	60	40	3	10.7	3	61.5	70.8	40	3	48.4
4	0.2	120	40	4	0	4	61.5	130.0	40	4	48.4
5	0.2	180	40	5	0	5	61.5	−169.2	40	5	48.4
6	0.2	−120	40	6	0	6	61.5	−109.2	40	6	48.3
7	0.2	−60	40	7	0	7	61.5	−49.2	40	7	48.3
	INR = 48.5 dB Before Adaption						INR = 48.4 dB Before Adaption			
	INR = 0.0 dB After Adaption						INR = 48.4 dB After Adaption			

powers, the gradient search is applied for seven unequal power interference sources. The seven sources are assumed to range in power from 40 dB to 10 dB in 5-dB increments. The initial source coordinates are the same as were given previously, and the initial ordered eigenvalues are listed in Table 1.4. Since the sources are grouped in a small cluster, there is only one dominant eigenvalue. A cancellation of 41.6 dB is achieved for this arrangement of unequal power sources, which consumes only one degree of freedom. After the gradient search is invoked the converged source configuration obtained is listed in Table 1.4. The final source coordinates are essentially the same as the equal-power source case, as was given in Table 1.2. The eigenvalues are spread over approximately a 30-dB range which coincides with the spread in source power. The eigenvalues suggest that one degree of freedom is completely consumed and that the other six are only partially consumed. The cancellation is only 20.1 dB for this case. Thus, for this particular gradient search, seven equal-power sources were more effective in terms of consuming degrees of freedom than were seven unequal-power sources. The two examples considered so far have been for equal numbers of array elements and interference sources. Although not shown, the gradient search has been used successfully to maximize consumption of degrees of freedom for cases involving fewer and greater numbers of interference sources than the number of array elements.

Table 1.3
Eigenvalues and INR for Six Orthogonal Sources in the Field of View of the Seven-Element Hexagonal Array

	Source Locations		Source Powers	Ordered Eigenvalues	
i	θ_i	ϕ_i	P_i (dB)	j	λ_j (dB)
1	0°	0°	40	1	48.6
2	61.5°	10.8°	40	2	48.5
3	61.5°	70.8°	40	3	48.4
4	61.5°	130.8°	40	4	48.4
5	61.5°	−169.2°	40	5	48.4
6	61.5°	−109.2°	40	6	48.3
				7	11.9

INR = 47.8 dB Before Adaption
INR = 12.0 dB After Adaption

Table 1.4
Eigenvalues and INR Before and After Gradient Search for Seven Unequal-Power Sources in the Field of View of the Seven-Element Hexagonal Array

Before Gradient Search						After Gradient Search					
Source Locations (deg)		Source Powers	Ordered Eigenvalues			Source Locations (deg)		Source Powers	Ordered Eigenvalues		
i	θ_i	ϕ_i	P_i (dB)	j	λ_j (dB)	i	θ_i	ϕ_i	P_i (dB)	j	λ_j (dB)
1	0	0	40	1	50.1	1	1.0	33.2	40	1	48.5
2	0.2	0	35	2	3.1	2	61.3	9.8	35	2	43.5
3	0.2	60	30	3	1.3	3	61.1	70.7	30	3	38.5
4	0.2	120	25	4	0	4	60.2	130.3	25	4	33.5
5	0.2	180	20	5	0	5	57.8	−169.8	20	5	28.5
6	0.2	−120	15	6	0	6	59.0	−108.3	15	6	23.5
7	0.2	−60	10	7	0	7	61.6	−48.6	10	7	19.1

INR = 41.6 dB Before Adaption
INR = 0.0 dB After Adaption

INR = 41.6 dB Before Adaption
INR = 21.5 dB After Adaption

1.10.2 SEVEN-ELEMENT RING ARRAY

As an example of an irregularly spaced two-dimensional array, consider the seven-element uniform circular ring array (with 1λ diameter) shown in Figure 1.14(b) and assume it is desired to arrange seven equal-power interference sources such that their effectiveness is maximized. Since the elements do not lie on a regular grid there was no intuitively obvious source configuration that would accomplish maximum source effectiveness. Also,

Table 1.5
Eigenvalues and INR Before and After Gradient Search for Seven Equal-Power Sources in the Field of View of the Seven-Element Uniform Circular Array

Before Gradient Search					After Gradient Search				
Source Locations (deg)		Source Powers	Ordered Eigenvalues		Source Locations (deg)		Source Powers	Ordered Eigenvalues	
i	θ_i	ϕ_i	P_i (dB)	j	λ_j (dB)	i	θ_i	ϕ_i	P_i (dB)	j	λ_j (dB)

i	θ_i	ϕ_i	P_i (dB)	j	λ_j (dB)	i	θ_i	ϕ_i	P_i (dB)	j	λ_j (dB)
1	0	0	40	1	56.9	1	4.6	23.0	40	1	50.5
2	0.2	0	40	2	11.4	2	56.9	−4.9	40	2	50.5
3	0.2	60	40	3	11.4	3	56.5	55.8	40	3	49.3
4	0.2	120	40	4	0.0	4	57.6	114.6	40	4	48.9
5	0.2	180	40	5	0.0	5	90.0	158.2	40	5	46.6
6	0.2	−120	40	6	0.0	6	90.0	−106.5	40	6	46.4
7	0.2	−60	40	7	0.0	7	58.4	−63.2	40	7	41.8
INR = 48.5 dB Before Adaption						INR = 48.4 dB Before Adaption					
INR = 6.4 After Adaption						INR = 46.0 dB After Adaption					

investigation of the orthogonality conditions in (1.39) failed to yield a closed form solution; thus, a gradient search was in order. The initial arrangement of sources used is the same as before. For this initial grouping of sources there is only one dominant eigenvalue, as shown in Table 1.5. The adapted cancellation for this source distribution is computed to be 42.1 dB. After the gradient search, the sources have the configuration and the ordered eigenvalues as listed in Table 1.5. There is an 8.7-dB spread in the eigenvalues, meaning that not all seven degrees of freedom are completely consumed. This implies (for this gradient search) that there is no configuration of seven equal-power sources that completely consumes seven degrees of freedom of the circular array in Figure 1.14(b). Even so, the cancellation is only 2.4 dB because the degrees of freedom have been severely taxed.

1.11 SUMMARY

In this chapter the interference covariance matrix eigenvalues have been used to represent the N degrees of freedom of an adaptive antenna. When all N eigenvalues are equal, it was shown that no adaptive cancellation can occur, and this condition corresponds to a unitary signal matrix. Sources arranged such that N equal eigenvalues are produced represent a maximally stressed interference environment for the N-element adaptive nulling array's degrees of freedom.

The concept of orthogonal interference sources was used to describe sources that consume one complete degree of freedom per source. A two-element array with two interference sources was investigated in detail. The eigenvalues and adaptive cancellation were shown as a function of source spacing and source power. One result is that only when the two sources are orthogonal with equal power does the array fail to provide any adaptive cancellation.

For an arbitrary adaptive antenna with a given interference signal matrix, a figure of merit was derived to determine how to minimize the difference between the signal matrix and a desired unitary matrix. The figure of merit is used in a gradient search to automatically rearrange a given source distribution to a distribution that minimizes this difference and thus, it is assumed, maximizes consumption of degrees of freedom. The gradient search was applied to two seven-element planar arrays, hexagonal and circular ring. For the hexagonal array it was shown numerically that seven interference sources could be arranged to completely consume all seven degrees of freedom, provided the sources have equal incident power. The array could not adaptively cancel the interference power for this case. Seven unequal-power sources were also considered, and while the gradient search maximized consumption of degrees of freedom it could not find a source distribution that completely consumed all seven degrees of freedom. For a seven-element uniform ring array, the gradient search was able to distribute seven equal-power sources such that a small spread in the eigenvalues occurred and little cancellation was possible.

The gradient search has been shown to be an effective method for maximizing consumption of degrees of freedom by a given number, power level, and distribution of sources. This method is useful for evaluating the performance of adaptive arrays (and multiple beam antennas as well) in general and determining the minimum number of adaptive nulling channels necessary for a particular adaptive nulling system. The numerical examples considered in this chapter utilized a narrow nulling bandwidth. The results apply to wider bandwidths because increasing the bandwidth tends to consume even more degrees of freedom.

1.12 PROBLEM SET

1.1 Derive (1.56) by using (1.54) and (1.55).

1.2 From Section 1.7, using the orthogonality condition of the signal vectors, determine the possible orthogonal source pair angles (θ_1, θ_2) when the spacing between two isotropic point receive antenna elements is $d = 1.0\lambda$.

1.3 Assume a two-element array of isotropic point receive antenna elements with $\lambda/2$ element spacing. If one interference source is located at $\theta_1 = 60°$, at what angle (θ_2) would an orthogonal source be located?

1.4 Using (1.56), verify (derive) the eigenvalue relations as given by (1.62) and (1.63) for a two-element array with two interference sources.

1.5 Using (1.62) and (1.63), compute and plot the narrow-band eigenvalues λ_1 and λ_2 for a two-element array with $\lambda/2$ element spacing with two interferers (I_1, I_2) when $P_1 = 40$ dB, $P_2 = 40$ dB, $\theta_1 = 60°$, and θ_2 varies from $0°$ to $180°$.

1.6 Using (1.56) in (1.10), with $P_1 = P_2$, $\theta_1 = -30°$, and $\theta_2 = 30°$, show that $w_a = w_o$; in other words, show the adaptive array weight is equal to the quiescent weight.

References

[1] Howells, P.W., "Explorations in Fixed and Adaptive Resolution at GE and SURC," *IEEE Trans. Antennas Propagat.*, Vol. 24, No. 5, 1976, pp. 575-584.

[2] Applebaum, S.P., "Adaptive Arrays," *IEEE Trans. Antennas Propagat.*, Vol. 24, No. 5, 1976, pp. 585-598.

[3] Gabriel, W.F., "Adaptive Arrays – An Introduction," *Proc. IEEE*, Vol. 64, 1976, pp. 239-271.

[4] Mayhan, J.T., "Some Techniques for Evaluating the Bandwidth Characteristics of Adaptive Nulling Systems," *IEEE Trans. Antennas Propagat.*, Vol. 27, No. 3, 1979, pp. 363-373.

[5] Mayhan, J.T., A.J. Simmons, and W.C. Cummings, "Wide-Band Adaptive Antenna Nulling Using Tapped Delay Lines," *IEEE Trans. Antennas Propagat.*, Vol. 29, No. 6, 1981, pp. 923-936.

[6] Monzingo, R.A., and T.W. Miller, *Introduction to Adaptive Arrays*, New York: Wiley, 1980.

[7] Compton, Jr., R.T., *Adaptive Antennas, Concepts and Performance*, Upper Saddle River, NJ: Prentice-Hall, 1988.

[8] Weiner, M.M., *Adaptive Antennas and Receivers*, Boca Raton, FL: CRC Press, 2006.

[9] Chandran, S., (ed.), *Adaptive Antenna Arrays: Trends and Applications*, Berlin: Springer-Verlag, 2004.

[10] Manolakis, D.G., V.K. Ingle, and S.M. Kogon, *Statistical and Adaptive Signal Processing: Spectral Estimation, Signal Modeling, Adaptive Filtering, and Array Processing*, Norwood, MA: Artech House, 2005.

[11] Nitzberg, R., *Adaptive Signal Processing for Radar*, Norwood, MA: Artech House, 1992.

[12] Farina, A., *Antenna-Based Signal Processing Techniques for Radar Systems*, Norwood, MA: Artech House, 1992.

[13] Hudson, J.E., *Adaptive Array Principles*, New York: Peter Peregrinus, 1981.

[14] Skolnik, M.I., *Introduction to Radar Systems*, 3rd ed., New York: McGraw-Hill, 2001.

[15] Skolnik, M.I., (ed.), *Radar Handbook*, 2nd ed., New York: McGraw-Hill, 1990, pp. 12.1-13.27.

[16] Mailloux, R.J., *Phased Array Antenna Handbook*, Norwood, MA: Artech House, 1994.

[17] Hansen, R.C., *Phased Array Antennas*, New York: Wiley, 1998.

[18] Hansen, R.C., (ed.), *Significant Phased Array Antenna Papers*, Dedham, MA: Artech House, 1973.

[19] Oliner, A.A., and G.H. Knittel, (eds.), *Phased Array Antennas*, Dedham, MA: Artech House, 1972.

[20] Amitay, N., V. Galindo, and C.P. Wu, *Theory and Analysis of Phased Array Antennas*, New York: Wiley, 1972.

[21] Mayhan, J.T., "Nulling Limitations for a Multiple-Beam Antenna," *IEEE Trans. Antennas Propagat.*, Vol. AP-24, No. 6, 1976, pp. 769-779.

[22] Mayhan, J.T., "Adaptive Nulling with Multiple-Beam Antennas," *IEEE Trans. Antennas Propagat.*, Vol. 26, No. 2, 1978, pp. 267-273.

[23] Johnson, R.C., and H. Jasik, *Antenna Engineering Handbook*, 2nd ed., New York: McGraw-Hill, 1984, Ch. 22, pp. 22-1-22-20.

[24] Fenn, A.J., "Maximizing Jammer Effectiveness for Evaluating the Performance of Adaptive Nulling Array Antennas," *IEEE Trans. Antennas Propagat.*, Vol. 33, No. 10, 1985, pp. 1131-1142.

[25] Fenn, A.J., "Interference Sources and Degrees of Freedom in Adaptive Nulling Antennas," Technical Report 604, Lincoln Laboratory, Massachusetts Institute of Technology, May 12, 1982.

[26] Fenn, A.J., "Consumption of Degrees of Freedom in Adaptive Nulling Array Antennas," Technical Report 609, Lincoln Laboratory, Massachusetts Institute of Technology, October 12, 1982.

[27] Mayhan, J.T., "Thinned Array Configurations for Use with Satellite-Based Adaptive Antennas," *IEEE Trans. Antennas Propagat.*, Vol. AP-28, 1980, p. 846.

[28] Strang, G., *Linear Algebra and Its Applications*, New York: Academic, 1976, pp. 211-227.

[29] Kraus, J.D., *Antennas*, 2nd ed., New York: McGraw-Hill, 1988, p. 570.

[30] Noble, B., and J.W. Daniel, *Applied Linear Algebra*, Upper Saddle River, NJ: Prentice-Hall, 1977, pp. 323-330.

[31] Zahradnik, R.L., *Theory and Techniques of Optimization for Practicing Engineers*, New York: Barnes and Noble, 1971, pp. 118-124.

2

Array Mutual Coupling Effects on Adaptive Radar Clutter Suppression

2.1 INTRODUCTION

In airborne or spaceborne radar systems applications, moving target indicator (MTI) techniques are used to distinguish scatterers containing a velocity component above some minimum detectable velocity (MDV) from those scatterers moving at speeds less than the MDV. One MTI technique is known as displaced phase center antenna (DPCA). With adaptive DPCA, which falls in the category of space-time adaptive processing (STAP), ground clutter is canceled by utilizing appropriate signal generation/processing and two (or more) independent well-matched radiation patterns generated by an antenna with displaced phase centers [1-10]. This chapter describes displaced phase center antennas and the effect of array mutual coupling on clutter cancellation.

2.2 DISPLACED PHASE CENTER ANTENNA CONCEPT FOR CLUTTER SUPPRESSION

The displaced phase center antenna concept has particular application in the design of a space-based radar (SBR) system [3-7]. Clutter cancellation of 40-50 dB is often discussed in the SBR context. A two-phase center DPCA system is depicted in Figure 2.1, where a moving target and a moving spaceborne DPCA radar platform are shown. Here, the full antenna aperture is used for two successive pulse transmissions and, on receive, two displaced portions of the aperture are used. The phase center displacement between the

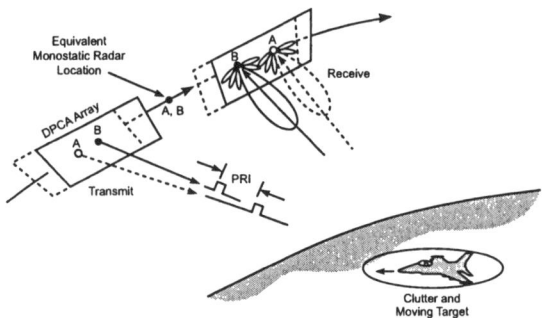

Figure 2.1 Two-phase center DPCA radar platform showing transmit and receive phase centers for consecutive pulses.

receive apertures is adjusted to compensate for the platform velocity. Thus, for two pulses separated in time by one pulse repetition interval (PRI) the first reception occurs at the forward phase center. During a PRI, because of the phase center displacement, the clutter is effectively observed as stationary; however, during this interval the target moves. Due to this movement, the target has a relative phase shift. There is no such phase shift from the clutter during this time. In simple terms, when the signals received by the two phase centers are subtracted, the clutter is significantly canceled and the corresponding target scattering return depends on the amount of target phase shift that occurs in one PRI. The amount of clutter cancellation achieved is limited by how well the two phase center radiation patterns are matched, in amplitude and phase, primarily over the main beam. The main beam pattern match is affected, in part, by array geometry and scan conditions (due to array element mutual coupling), and hardware tolerances (such as the errors in the transmit/receiver (T/R) modules and the beamformer). In this chapter, the effects of array mutual coupling on clutter cancellation are considered. Other sources of error (antenna deformation, receiver channel mismatch, T/R module amplitude and phase errors, platform scattering effects, and so on) must be analyzed to make an overall assessment of clutter cancellation for a particular DPCA system.

The displacement (separation) between the phase centers of the two receive antennas is denoted by Δ. To achieve the desired clutter cancellation, this separation distance must be related to the PRI of the radar waveform by the "DPCA phase center condition"

$$\Delta = 2V_p T, \tag{2.1}$$

where T is the PRI and V_p is the platform speed [8]. To eliminate the blind speeds associated with a simple pulse-train waveform, a practical system will base target detection on a noncoherent combination of many coherent processing intervals, each with a slightly different PRI, chosen according to a suitable algorithm. Thus, a range of phase center separations, typically varying by a factor of two between largest and smallest, must be provided by the antenna system to ensure target detection.

Movement of the phase centers in a displaced phase center antenna is accomplished by using amplitude illumination control. Consider Figure 2.2, which depicts two examples for displaced receive apertures – overlapped and split. As the phase center displacement increases, the size of each phase center subaperture decreases. Hence, the width of the receive antenna main beam increases as the phase center displacement increases. Consider Figure 2.3, which shows a corporate-fed phased array for application to DPCA. Phase center displacement in a phased array antenna is conveniently effected by means of attenuators in the transmit/receive modules. In the receive portion of the module, low-noise amplifiers with an appropriate gain and noise figure

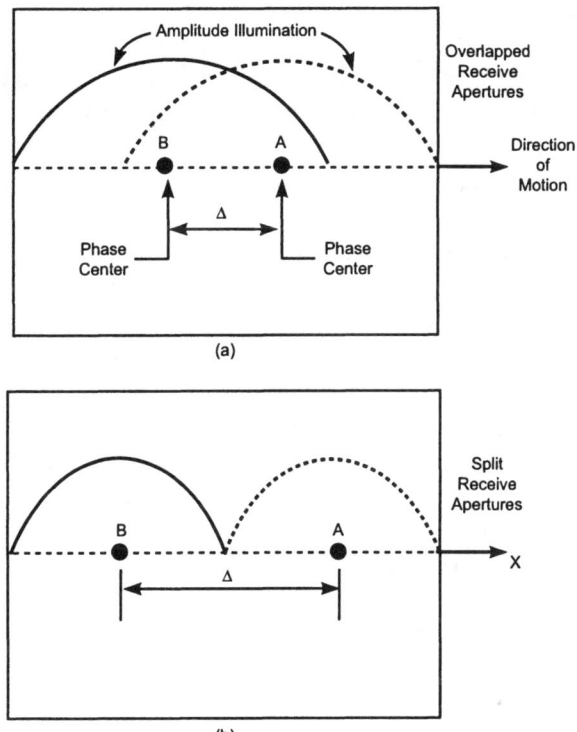

Figure 2.2 Amplitude illumination for two-phase center DPCA: (a) overlapped receive apertures and (b) split receive apertures.

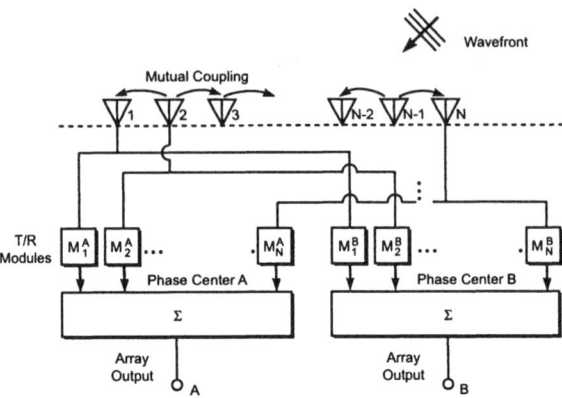

Figure 2.3 Corporate feed for two receive phase centers.

are used to establish a desired system noise figure and sensitivity, prior to the attenuators. Unused elements in each phase center are "turned off" to a large value of attenuation (say, 50 dB or more), typically in a matched load impedance. After performing amplitude and phase weighting in each module, the received signals in each phase center are summed in a coherent power combiner (corporate feed) as shown in Figure 2.3. Notice that array mutual coupling is represented by the curved arrows emerging from the various array elements.

It has already been mentioned that array module errors will not be treated in this chapter, but they will be briefly addressed in Chapter 5. If there was no array mutual coupling as well, then the radiation patterns generated by two ideal (undeformed) subapertures would be identical. However, array mutual coupling is always present in a practical antenna, and this coupling leads to mismatch in the resulting displaced phase center antenna aperture illumination functions and, thus, mismatch of the receive patterns as well. The reason for this mismatch can be easily seen when referring to Figure 2.4, where the two subarrays are fully split apart. In this situation, there is a mirror symmetry in the subarray geometries with respect to the boundary line between the forward and trailing phase centers. For example, the forward phase center has as many terminated elements to its left as the trailing subarray has to its right. Thus, the subarrays are not symmetrically surrounded by terminated elements. In an infinite array, even with mutual coupling, finite-sized forward and trailing subarrays would produce radiation patterns that are perfectly matched. Since, in practice, the subarrays are embedded within a finite array, array edge effects will cause the forward and trailing subarray radiation patterns to become mismatched. Guard bands of passively terminated elements will tend

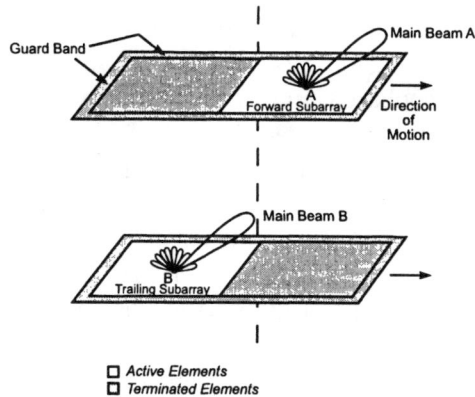

Figure 2.4 Illustration of asymmetry in array mutual coupling environment for forward and trailing phase centers.

to restore the symmetry of the two (or more) phase centers. Guard bands are commonly used to provide impedance matching to edge elements of a phased array antenna as well.

An experimental subscale SBR DPCA phased array was designed and fabricated for ground-based prototype testing and demonstration of DPCA clutter cancellation. The active portion of the antenna consists of 96 monopole elements arranged in a hexagonal lattice having eight rows and twelve columns. The design of this array was determined using the analysis techniques presented in this chapter and in Chapter 9. Two guard bands of passively terminated monopole elements surround the 96-element active array to reduce both array edge effects and finite ground plane effects. Using a near-field measurement technique described in Chapter 11, DPCA testing of this antenna has been performed.

A number of questions can be raised regarding the possible influence of phased array antenna design on DPCA clutter cancellation. Fundamental design parameters for a DPCA phased array are:

- Size of array;
- Type of array antenna elements;
- Array lattice;
- Number of passively terminated element rows (guard bands);
- Phase center displacement;
- Scan sector.

To quantify the effects of varying these design parameters experimentally would require a great deal of time and effort and is impractical for large arrays. In contrast, theoretical calculations involving changes in the array design parameters can readily be accomplished. In order to gain a better understanding of how phased array antenna design affects DPCA clutter cancellation performance, a detailed theoretical study was performed. Both subscale and full-scale SBR antenna designs are examined in this chapter. Only corporate-fed phased arrays will be examined here. Space-fed lens DPCA phased arrays are possible, but are not mentioned further here. Antenna array elements typically considered for a low-altitude spaceborne radar application are monopoles (broadside null) and dipoles (broadside maximum). These antenna elements are readily analyzed using the method of moments (refer to Chapters 9 and 14) [7, 11, 12]. Figure 2.5 shows monopole and dipole antenna elements and their simulated center element patterns in an array. In computing these radiation patterns, an 11-row by 11-column array with a square lattice with $\lambda/2$ element spacing has been assumed (refer to Chapter 9 for further discussions of the monopole array). Notice that the monopole has a null at broadside compared to the dipole which has a peak at broadside. The useable scan sector for an array of monopoles is from about 30° to 60° from broadside. In contrast, the dipole array scan sector covers about 0° to 60° from broadside. Monopoles are sometimes considered for a low-altitude space-based radar system where the broadside null is actually useful

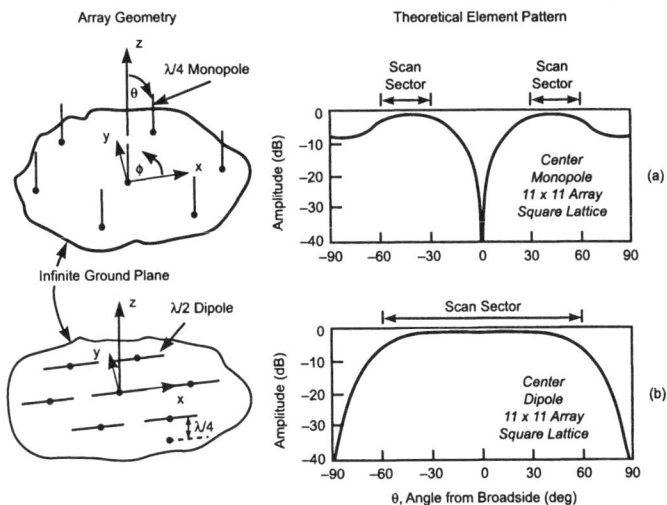

Figure 2.5 Center element radiation patterns for two types of array antenna elements. The array size in each case is 11×11 with the center element driven and the surrounding elements terminated in matched loads. (a) Monopoles and (b) dipoles.

in reducing the nadir (high grazing angle) clutter. The method of moments is known to be accurate in predicting the performance of monopole and dipole arrays. For example, the input impedance and radiation patterns of monopole phased arrays have been computed, using the method of moments, and are in good agreement with measurements as described in Chapter 9.

This chapter is organized as follows: In Section 2.3, a description of array antenna analysis and adaptive DPCA array theory are given. In Section 2.4, results are presented for subscale and full-scale spaceborne radar DPCA arrays. The effects of scan angle, phase center displacement, type of radiating element, and number of passively terminated guard bands are examined in detail. A subscale SBR 96-element planar array is described and analyzed in detail. Subscale and full-scale SBR arrays, up to 16 meters in maximum length, are investigated. The results presented in this chapter make it possible to place an *upper bound* (due to array mutual coupling effects) on the DPCA cancellation that is possible with a given size array antenna. The situation of more than two phase centers is readily handled under the theoretical framework presented here; however, only two-phase center DPCA is investigated in this chapter.

2.3 THEORY

2.3.1 ARRAY ANALYSIS INCLUDING MUTUAL COUPLING EFFECTS

As mentioned earlier, in this chapter both dipole and monopole arrays over an infinite ground plane will be analyzed. Bandwidth effects will not be addressed in this chapter, and so optimization of the element design is not under consideration. The two basic array elements, $\lambda/2$ dipole and $\lambda/4$ monopole (where λ is the wavelength), are depicted in Figure 2.6. These elements are assumed to be thin cylindrical wire antennas close to resonance, which allow a convenient formulation by the method of moments. Chapter 9 presents a detailed mutual coupling (moment method) formulation for finite arrays of monopole elements. A similar formulation can be given for dipole arrays. The mathematical details of the moment method formulation will be covered only briefly in this chapter.

In the method of moments [13], boundary conditions are used to find the antenna response (current) to a given excitation (voltage). The excitation here is the amplitude and phase incident at each element of the phased array. Due to mutual coupling or the mutual impedance between array elements, the actual illumination achieved will be different from the theoretically desired

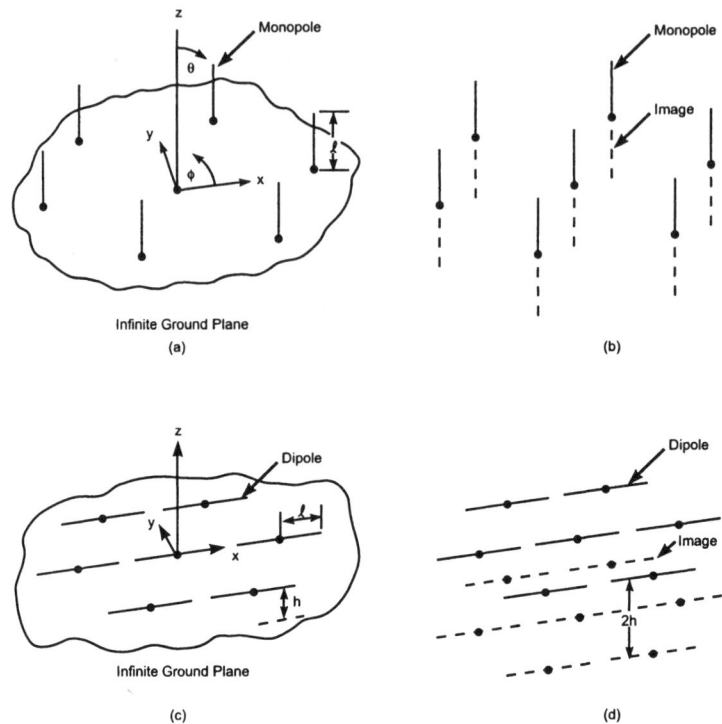

Figure 2.6 Geometry for finite array antennas. (a) Monopole array over ground plane, (b) equivalent dipole array with ground plane removed and monopole images included, (c) horizontal dipole array over ground plane, and (d) equivalent dipole array with ground plane removed and dipole images included.

illumination. It is this deviation from the ideal illumination that DPCA clutter-cancellation performance will depend on.

For convenience, it is assumed that there is one unknown complex current per element of the array, and the current distribution is assumed to be piecewise sinusoidal. This current distribution is used as moment method basis and testing functions. When the basis functions and testing functions are the same, this is known as a Galerkin's formulation. For a piecewise-sinusoidal Galerkin's moment method formulation, the mutual impedance between array elements is readily computed [13, 14].

The geometry for a finite array of dipoles or monopoles over ground plane is shown in Figure 2.6. Standard spherical coordinate angles (θ, ϕ) are used to describe the observation position for far-field pattern computation. The ground plane is located in the $z = 0$ plane. The dipoles are $\hat{x}$ polarized, and the monopoles are $\hat{z}$ polarized. Using image theory, the ground plane can be removed from the analysis. For a monopole array, an equivalent dipole array results (Figure 2.6(b)). For a dipole array, removal of the ground plane is accomplished by adding an image array of dipoles (Figure 2.6(d)) with the currents 180° out of phase with respect to the primary (actual) dipoles. In theory, the monopole radiates (or receives) only the E_θ electric field component. In the $\phi = 0°$ plane, the principal polarization for the dipole is the E_θ component. In the $\phi = 90°$ plane, the dipole principal polarization is the E_ϕ component. However, for a radar application, the slant-linear polarization (total electric field) is of interest. In free space, the radiation pattern of a vertical dipole is well known and is expressed (in primed coordinates) as

$$E_{\theta'} = j60 i_n P_{\theta'}(\theta') \tag{2.2}$$

where i_n is the complex terminal current for the nth element and $P_{\theta'}$ is the isolated-element normalized pattern, which is given by

$$P_{\theta'} = \frac{\cos(\beta l \cos \theta') - \cos(\beta l)}{\sin(\beta l) \sin \theta'} \tag{2.3}$$

where l is the dipole half-length, and $\beta = 2\pi/\lambda$ is the phase constant. It has been assumed that the vertical dipole is $\hat{z}'$ oriented with θ' being the angle measured from the axis of the dipole, as shown in Figure 2.7. Equation (2.2) is the required expression used to compute the isolated element pattern for either a monopole array or dipole array. This expression is the desired slant-linear polarization component of the electric field. To compute the array pattern including mutual coupling effects it is necessary to compute the array terminal currents i_n defined in (2.2).

Consider Figure 2.8 which shows a diagram for two elements (m,n) of an array operating in the transmit mode. The driving source is a voltage generator in series with a generator impedance. The generator and impedance load are connected to the terminals of a radiating antenna element that has a self-impedance and a mutual impedance with respect to every element of the array.

Let Z represent the mutual impedance matrix for the array. Referring to Figure 2.8, Z is expressed as

$$Z = Z^{o.c.} + Z_L I \tag{2.4}$$

where $Z^{o.c.}$ is the open-circuit mutual impedance matrix for the array, I is the identity matrix, and Z_L is the load impedance. Define v as the voltage excitation matrix of the array. Then the array element terminal currents, denoted i, are found by solving the system of equations written in matrix form as

$$v = Z \cdot i \tag{2.5}$$

Upon calculating the array terminal currents, by $i = Z^{-1}v$, the array radiation pattern including mutual coupling is expressed as

$$E_{\theta'}^{array} = j60 P_{\theta'} \cdot \mathrm{AF} \tag{2.6}$$

where $P_{\theta'}$ is given by (2.1) and AF is the array factor, which is given by

$$\mathrm{AF} = \sum_{n=1}^{N} i_n e^{j\beta(\sin\theta(x_n \cos\phi + y_n \sin\phi) + z_n \cos\theta)} \tag{2.7}$$

where (x_n, y_n, z_n) are the coordinates for the terminals of the nth element and i_n is the moment method terminal current for the nth element, which includes array mutual coupling effects.

Finally, the relation between the observation angles (θ, ϕ) and θ' is determined as follows: The dot product of the unit vectors in the dipole direction and the radial direction is, simply,

$$\hat{z}' \cdot \hat{r} = \cos\theta' \tag{2.8}$$

Next, the unit vector in the radial direction is expressed in terms of the rectangular coordinate system unit vectors as

$$\hat{r} = \sin\theta \cos\phi \, \hat{x} + \sin\theta \sin\phi \, \hat{y} + \cos\phi \, \hat{z} \tag{2.9}$$

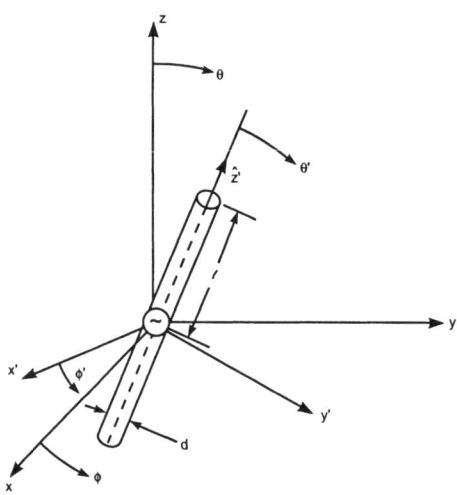

Figure 2.7 Geometry for arbitrarily oriented dipole antenna in free space.

For $\hat{z}$-directed dipoles (monopole plus image) $\hat{z}' = \hat{z}$ so that (2.7) and (2.8) produce

$$\theta'_{\text{monopole}} = \theta \tag{2.10}$$

Similarly, for $\hat{x}$-directed dipoles it follows that

$$\theta'_{\text{dipole}} = \cos^{-1}(\sin\theta\,\cos\phi) \tag{2.11}$$

2.3.2 CALCULATION OF DISPLACED PHASE CENTER ANTENNA CLUTTER CANCELLATION

To compute the clutter cancellation capability of two DPCA radiation patterns, it is necessary to form the pattern correlation matrix

$$M = \begin{bmatrix} M_{11} & M_{12} \\ M_{21} & M_{22} \end{bmatrix} \tag{2.12}$$

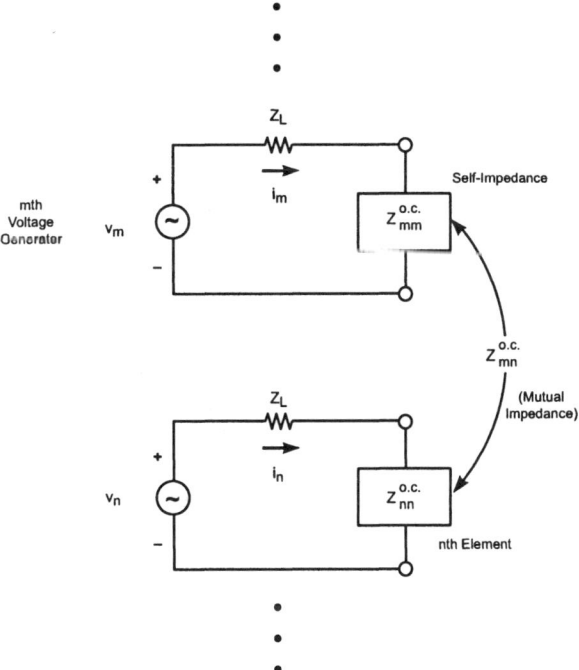

Figure 2.8 Diagram depicting self and mutual impedances involved in determining the array currents i_m and i_n.

where the correlation between two channels is expressed as

$$M_{ij} = \int \int |E_o(\theta, \phi)|^2 E_i(\theta, \phi) E_j^*(\theta, \phi) e^{j(\psi_x + \psi_y)} A(\theta, \phi) d\theta d\phi \quad (2.13)$$

where (θ, ϕ) are standard spherical coordinates; $i = 1, 2$; $j = 1, 2$; $E_o(\theta, \phi)$ is the electric field pattern of the transmitting antenna; $E_1(\theta, \phi)$ and $E_2(\theta, \phi)$ are the electric field patterns of the two receiving antennas (* denotes conjugate), $A(\theta, \phi)$ is a weighting function that depends on the radar waveform, the clutter model, and the geometry of the problem. In this chapter, it is assumed that $A(\theta, \phi) = 1$ so the clutter cancellation is dependent on the antenna pattern match only. The electric field patterns $E_1(\theta, \phi)$ and $E_2(\theta, \phi)$ are measured or computed (as in this chapter) with respect to their assumed phase center positions, and are taken to be the geometric centers of the excited portions of the respective apertures. The phase functions ψ_x and ψ_y represent the effect of differences (offset errors), in the longitudinal and transverse directions, between the separation of these assumed phase centers and the actual values required to meet the DPCA condition mentioned earlier (see (2.1)). The phase functions are expressed in terms of these spatial differences by the following equations

$$\psi_x = \frac{2\pi}{\lambda} \Delta_x \sin\theta \cos\phi \quad (2.14)$$

$$\psi_y = \frac{2\pi}{\lambda} \Delta_y \sin\theta \sin\phi \quad (2.15)$$

where Δ_x and Δ_y are the departures of the phase center separations, in the x (longitudinal) and y (transverse) directions, from the values required by the DPCA condition.

This chapter assumes that the phase centers are aligned with the direction of platform motion. Since the antenna excitations are normally symmetric in the transverse direction, Δ_y is usually zero. The other component, Δ_x, can be nonzero for a variety of reasons. For example, the radar PRI may be incorrectly matched to the phase center separation. An example more relevant to this study occurs if the geometrical centers are not representative of the actual phase centers of the receive patterns. This phase center error will actually be the case when no guard bands are used, since mutual coupling will modify the effective excitations, causing the actual currents to be complex and shifting the phase centers. Illustrations of this effect are presented later. In any case, this chapter is not concerned with definitions of phase center position, but with the proper choice of PRI to maximize clutter cancellation performance. With excitations distorted by array mutual coupling, it is not easy to predict the best PRI for a given set of antenna excitations, and the examples that follow will show that a departure from the geometrical spacing can sometimes improve clutter-cancellation performance significantly. In any case, if the best spacing is known (from theoretical analysis and/or testing), the radar PRI can be adjusted to achieve the desired effective phase center displacement.

For a nadir-pointed SBR, as depicted in Figure 2.9, in (2.12) the range of integration over the Earth's surface will be $0 < \phi < 2\pi$ and $0 < \theta < \theta_{max}$, where

$$\theta_{max} = \sin^{-1}(\frac{R_e}{R_e + h}) \qquad (2.16)$$

where R_e is the radius of the Earth (this is taken to be 6371 km) and h is the altitude of the radar. From (2.16), for a low-altitude SBR with $h = 1000$ km, $\theta_{max} = 60°$.

The DPCA arrangement with two phase centers may be viewed as an adaptive nulling system with two antenna channels. Note: In Chapter 5, multiple auxiliary channels are included to suppress jamming as well as clutter. Since the two patterns are expected to be very well matched, and the pulse timing corresponds to the physical separation of their phase centers, the returns from stationary clutter should be nearly equal in the two channels. The target returns, however, will differ by a phase shift that is related to the actual target range rate (Doppler), and this phase determines the ideal steering vector in the two-channel system. However, nearly equal performance is attained when a single steering vector, $[1, 0]^T$, is used. This is a mismatched steering vector, but it suffices for all target range rates of interest and hence simplifies the processing considerably. The performance of this two-channel

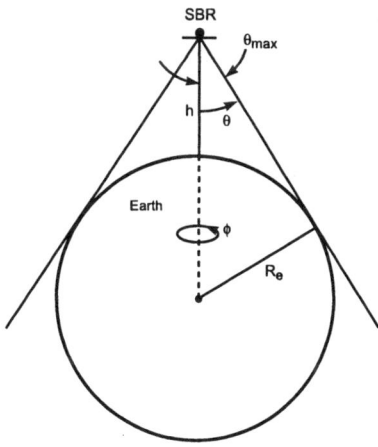

Figure 2.9 Geometry for nadir-pointed space-based radar antenna.

nulling system depends on the true target Doppler (range rate) – being very poor for targets with nearly zero doppler, since these are lost in the clutter return. The optimum target range rate is one-half the radar blind speed, and detection performance is best for this case.

The output SNR of this DPCA adaptive nulling system is, in any case, inversely proportional to the quantity

$$C = 1 - \frac{|M_{12}|^2}{M_{11}M_{22}} \tag{2.17}$$

which plays the role of a cancellation factor, reducing the clutter power perceived by the radar. In the calculated results presented in this chapter, the cancellation is expressed as a positive decibel value by taking $10\log_{10}(1/C)$. A derivation of (2.17) is given in the next section.

2.3.3 DERIVATION OF CLUTTER CANCELLATION FACTOR

Consider an N-channel adaptive nulling signal processor that uses a quiescent steering weight vector w_q prior to generating an adaptive weight vector w_a for suppression of clutter and detection of targets. The effect of jamming on DPCA performance is ignored in this chapter, but it will be addressed in Chapter 5. It is assumed that w_q is normalized, such that $w_q^\dagger w_q = 1$, where the † symbol means complex conjugate transpose. Let it be assumed at first that the true channel covariance matrix M is known and that it consists of receiver noise and radar clutter (backscatter from the Earth's surface). Let the N-channel output vector prior to weighting be denoted x. For a given weight vector w, the weighted output can be expressed as [7, 15]

$$y = w^\dagger x \tag{2.18}$$

Detection will be based on the magnitude of the adaptively weighted quantity

$$y_a = w_a^\dagger x \tag{2.19}$$

where

$$w_a = M^{-1} w_q \tag{2.20}$$

Now, let s denote the desired signal (target) vector and let n be the noise-only channel vector. It is assumed that $E(nn^\dagger) = M$ and $E(n) = 0$, where E means mathematical expectation. The desired signal vector can be written as

$$s = Ap \tag{2.21}$$

where A is a complex amplitude parameter and p is a normalized vector that contains the relative signal phase shift between channels. It follows that the desired signal expected value or mean is $E(s) = Ap$.

The channel output vector is equal to the superposition of signal and noise as

$$x = s + n \tag{2.22}$$

From the above it is clear that the expected value of the signal plus noise vector is given by

$$E_{s+n}(x) = Ap \tag{2.23}$$

The signal-to-noise ratio is defined as the ratio of the output power of the signal alone to the output power of the noise alone. It follows then that the weighted channel output signal-to-noise ratio can be expressed as

$$SNR \equiv \frac{|E_{s+n}(y) - E_n(y)|^2}{E_n|y|^2} \tag{2.24}$$

Now, in (2.24) $E_n(y) = 0$ and

$$E_{s+n}(y) = w^\dagger E_{s+n}(x) \tag{2.25}$$

where the weight vector w is equal to either w_q before adaption or w_a after adaption. Before adaption, the expected value of y_q for signal plus noise is

$$E_{s+n}(y_q) = A w_q^\dagger p \tag{2.26}$$

The quiescent output power due to noise is expressed as

$$E_n|y_q|^2 = w_q^\dagger M w_q \tag{2.27}$$

Substituting (2.26) and (2.27) into (2.24) yields, for the quiescent signal-to-noise ratio,

$$SNR_q = |A|^2 \frac{|w_q^\dagger p|^2}{w_q^\dagger M w_q} \tag{2.28}$$

To find the adapted SNR, first substitute (2.20) and (2.23) in (2.25), which produces

$$E_{s+n}(y_a) = A w_q^\dagger M^{-1} p \tag{2.29}$$

Next, the output power due to clutter and receiver noise is given by

$$E_n|y_a|^2 = w_a^\dagger M w_a \tag{2.30}$$

Substituting (2.20) in (2.30) yields

$$E_n|y_a|^2 = w_q^\dagger M^{-1} w_q \tag{2.31}$$

Equations (2.29) and (2.31) are substituted into (2.24) with the result

$$SNR_a = |A|^2 \frac{|w_q^\dagger M^{-1} p|^2}{w_q^\dagger M^{-1} w_q} \qquad (2.32)$$

In practice, the covariance matrix is not known and is usually estimated from some set of sample vectors which have the same noise statistics as x, but that are free of signal components. This estimation procedure causes a loss in output SNR which was first studied by Reed, Mallet, and Brennan [15] for the matched case in which $p = w_q$, and by Boroson [16] for the general mismatched case. In either case, if the covariance estimate is based on a sufficiently large number of samples, the loss in SNR is small, and it will be ignored in the remainder of this discussion.

In the two-phase center DPCA case, there are only two channels ($N = 2$), and a typical signal vector has form $s = Ap$, where

$$p = \frac{1}{\sqrt{2}} \begin{bmatrix} 1 \\ e^{i\psi} \end{bmatrix} \qquad (2.33)$$

This equation for p is a normalized vector, and the phase angle ψ corresponds to the Doppler shift associated with the range rate of the target itself. Let the covariance matrix of the clutter-plus-receiver noise output of the two channels be

$$M = \begin{bmatrix} M_{11} & M_{12} \\ M_{21} & M_{22} \end{bmatrix} \qquad (2.34)$$

As mentioned earlier, instead of steering for a variety of values of ψ, it has been found to be sufficient to use a single "mismatched" steering vector,

$$w_q \equiv \begin{bmatrix} 1 \\ 0 \end{bmatrix} \qquad (2.35)$$

Substituting (2.33), (2.34), and (2.35) into (2.28) yields

$$SNR_q = \frac{|A|^2}{2M_{11}} \qquad (2.36)$$

which is the SNR for channel 1 alone. The clutter returns will usually dominate the noise, and since they will have very nearly the same power in the two channels (if the latter are well matched), then the SNR in channel 2 will be essentially the same as that of channel 1.

Next, to compute the adapted SNR, the inverse of M is needed and is given by the adjoint of M divided by the determinant of M, or,

$$M^{-1} = \frac{1}{\Delta_M} \begin{bmatrix} M_{22} & -M_{12} \\ -M_{21} & M_{11} \end{bmatrix} \qquad (2.37)$$

where
$$\Delta_M = M_{11}M_{22} - M_{12}M_{21} \qquad (2.38)$$
is the determinant of M. Substituting (2.33), (2.35), and (2.37) into (2.32) produces the signal-to-noise ratio after adaption,
$$SNR_a = \frac{|A|^2}{2\Delta_M} \frac{|M_{22} - M_{12}e^{i\psi}|^2}{M_{22}} \qquad (2.39)$$
This expression can be rewritten in the form
$$SNR_a = \frac{|A|^2}{2CM_{11}}|1 - \frac{M_{12}}{M_{22}}e^{i\psi}|^2 \qquad (2.40)$$
where C, the clutter cancellation factor, is defined by (2.17). For strong clutter, M_{12} will be very nearly equal to M_{22}, and the factor
$$|1 - \frac{M_{12}}{M_{22}}e^{i\psi}|^2 \approx |1 - e^{i\psi}|^2 = 4\sin^2(\psi/2) \qquad (2.41)$$
then describes the variation of SNR with true target Doppler. For targets with very small range rates, ψ will be nearly zero, and the output SNR itself will be very small. These are targets that are lost in the clutter return. When $\psi = \pi$, the target range rate corresponds to the "optimum speed," in the terminology of conventional MTI radar, and the second factor in the SNR formula (2.40) will represent approximately 6 dB of gain. In this case, the target returns from the two phase centers are combining coherently, and the effective gain of the full antenna is restored.

The other factor $|A|^2/(2CM_{11})$ in (2.40) is just like the SNR on channel 1 (2.36), but with the clutter-plus-noise power M_{11} reduced by the "cancellation factor" C. This is the quantity in terms of which the DPCA performance is evaluated in the body of this chapter.

2.4 RESULTS

2.4.1 INTRODUCTION

In general, two-dimensional radiation patterns are required to completely evaluate the DPCA performance of an antenna (see (2.13)). However, in this chapter only one-dimensional principal plane pattern cuts are used to calculate the DPCA performance, and (2.13) is replaced by the corresponding one-dimensional integral. Simulations not shown here indicate that the one-dimensional pattern data yields clutter cancellation results that are typically

3-4 dB higher than full two-dimensional data. Further, it is assumed here that the clutter is incident only from angles less than 60° from broadside. This is the geometry that would occur with a low altitude SBR. To properly use the data that follows, it is desirable that the mutual-coupling limited (upper-bound) clutter cancellation be at least 5 to 10 dB higher than the design goal. This upper bound is chosen because of the assumption of one-dimensional pattern data.

2.4.2 SUBSCALE 96-ELEMENT PLANAR PHASED ARRAY

In this section, theoretical predictions of the upper-bound DPCA clutter cancellation for a subscale 96-element SBR phased array antenna are made. A study discussed in Chapter 11 has investigated measured DPCA performance for an experimental L-band 96-element test array. A photograph of the 8-row by 12-column (96 elements) array is shown in Figure 2.10. The experimental array consists of monopole antenna elements arranged in a hexagonal lattice with 12.7-cm spacing as depicted in Figure 2.11(a). The monopole length is $l = 5.842$ cm and the monopole diameter is $d = 3.175$ mm. The 96-element receive portion of the array is surrounded by two guard bands of elements (the total number of guard elements is also 96). The array element spacing was chosen to allow a full conical scan to 60° from broadside. The operating frequency range for this array is 1.2-1.4 GHz with center frequency 1.3 GHz. The element spacing at center frequency is, thus, 0.55λ. The array amplitude and phase weighting was achieved using 6-bit phase shifters (64 phase states with 5.625° phase steps) and 7-step (coarse) attenuators (0 dB, 1.3 dB, 2.8 dB, 4.6 dB, 6.8 dB, 10 dB, and >55 dB attenuation states). Two independent beamformers (coherent power combiners) were used to implement two-phase center DPCA, as shown in Figure 2.3.

As mentioned earlier, with hardware it is often impractical to investigate the effects of varying the antenna array parameters, such as type of radiating element and number of guard bands. With software, variation of parameters

Figure 2.10 Photograph of a 96-element displaced phase center antenna (DPCA) test array.

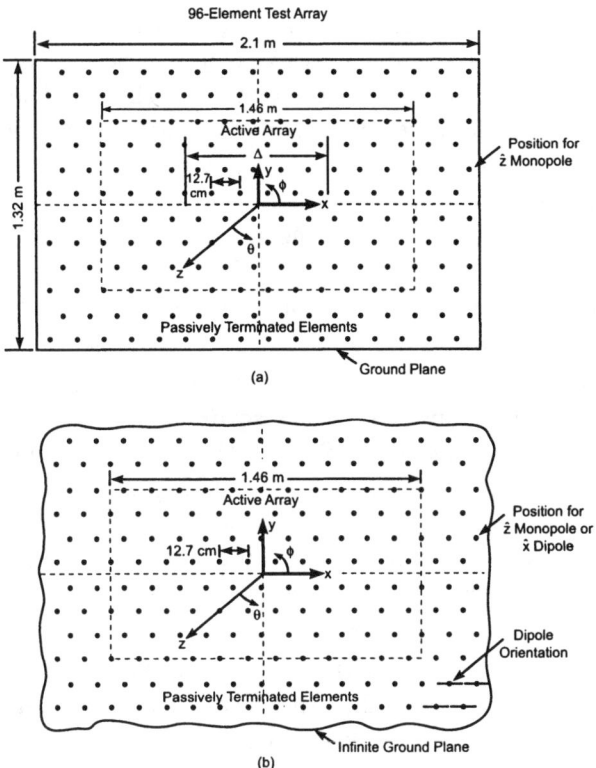

Figure 2.11 96-element DPCA test array layout: (a) experimental monopole array and (b) theoretical array (monopoles or dipoles).

is done relatively easily. In this section, comparisons of method of moments predictions and published measurements will be made where possible. Only a monopole array has been fabricated and tested here, but in the theoretical simulations both monopoles and dipoles have been analyzed. A layout for the theoretical arrays is shown in Figure 2.11(b). Notice that the only difference between the experimental and theoretical monopole arrays is in the ground plane size – the experimental ground plane of course is finite and the theoretical ground plane has been assumed infinite for convenience. A finite ground plane could be included in the theory at the expense of more computation time. The dipoles are assumed to have the same wire diameter as the monopoles. The dipole half-length is $l = 5.588$ cm and the spacing above ground plane is $h = 5.334$ cm.

Consider first the theoretical radiation patterns of the monopole array when steered to $\theta = 50°$. Figure 2.12 shows the transmit pattern using a -10 dB cosine tapered (no pedestal) illumination. Figure 2.13 shows the receive patterns, amplitude, and phase, respectively, for the case where the phase center separation is 3 columns (38.1 cm). (The receive illumination was also assumed to be a -10 dB cosine taper.) The broadside null of the monopole is apparent in these patterns. Notice that the amplitude and phase of the two phase centers track very closely when $|\theta| < 60°$. For $|\theta| > 60°$ the two phase centers become mismatched. The importance of this is that the clutter cancellation capability is dependent on the selection of cutoff angle in the evaluation of (2.12). Figure 2.13 clearly shows that the effect of mutual coupling is to create a mismatch between the two phase centers. Similar results are found for dipole elements.

Consider now Figure 2.14, which shows the clutter cancellation of monopole and dipole arrays for various scan angles ($\leq 55°$) as a function of phase center displacement in columns. Notice that for both monopoles and dipoles, the cancellation tends to slowly degrade as the phase center displacement increases. For this particular array design (8×12 elements, hexagonal lattice, two guard bands), it is observed that dipoles have a higher value of clutter cancellation compared to monopoles.

For the monopole array 40° scan case, experimental data for the 96-element test array were available. These data are also plotted in Figure 2.14(a) using the symbol •. Notice that the experimental data are about 10 dB below the theoretical predictions. This result is expected because the theory does not include any module errors or finite ground plane effects – only array mutual coupling effects. However, the variation as a function of phase center displacement is in good agreement. In other words, the theory is predicting the correct trend.

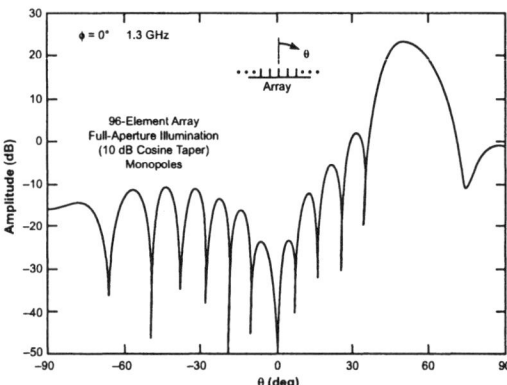

Figure 2.12 Calculated transmit radiation pattern using -10 dB cosine taper for the 96-element monopole test array. Scan angle is $\theta = 50°$.

In Figure 2.15, the same theoretical results are replotted as a function of scan angle for both monopoles and dipoles with three- and six-column phase center displacements. For this size (8 × 12) monopole array, it can be concluded that the upper-bound cancellation is 50 dB at 40° scan, but only 30 dB at 55° scan. For a dipole array, the upper-bound cancellation is 48 dB at 40° scan and 40 dB at 55° scan. Later in this chapter, it will be shown that for large arrays there is little difference in clutter cancellation between monopole and dipole array elements.

One of the potential effects of mutual coupling is to shift the phase centers from their assumed positions, as mentioned in Section 2.2. The quantity Δ_x defined in (2.14) can be thought of as a tuning parameter by which the radar PRI is optimized for the given excitations of the receive antennas.

To demonstrate this phase-center shifting (tuning) effect, consider Figures 2.16 and 2.17, which depict clutter cancellation as a function of change in apparent phase center displacement for monopoles and dipoles. The

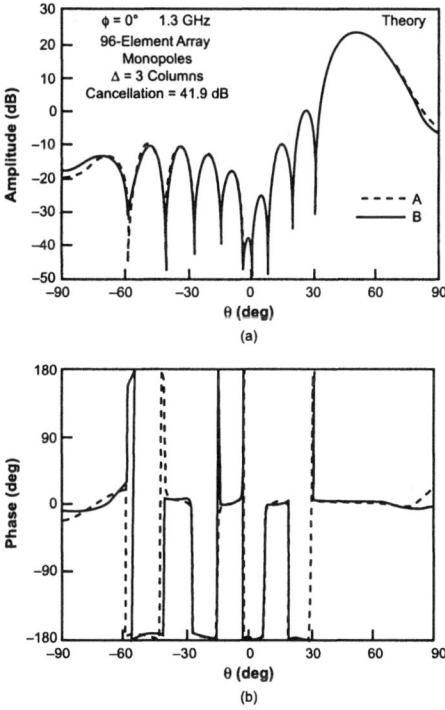

Figure 2.13 Receive array radiation patterns using −10 dB cosine taper for the 96-element monopole test array. Phase center separation is three columns. Scan angle is $\theta = 50°$. (a) Amplitude and (b) phase.

desired phase center displacement was chosen to be three columns (1.65λ), and scan angles of 40° and 55° were examined. Figure 2.16 is for the case where there are no guard bands, and Figure 2.17 is for the case of two guard bands. Notice that the phase center displacement difference Δ_x (in wavelengths) is larger when there are no guard bands. For example, the difference is as large as 0.125λ for the 55° scan angle with monopoles. For the same monopole array case but now with two guard bands (Figure 2.17(a)) the displacement difference is only −0.025λ. Thus, one important feature of the guard bands is to stabilize the location of the phase center. The other benefit of guard bands is to increase the clutter cancellation capability, as demonstrated in Figure 2.18. Here, the number of guard bands was varied from zero to four. The improvement in performance is greater than 10 dB for the 40° scan case when two or more guard bands are used, compared to zero guard bands. It is interesting to note that the clutter cancellation does not improve substantially (only a few decibels) for the 55° scan with monopoles. In all subsequent

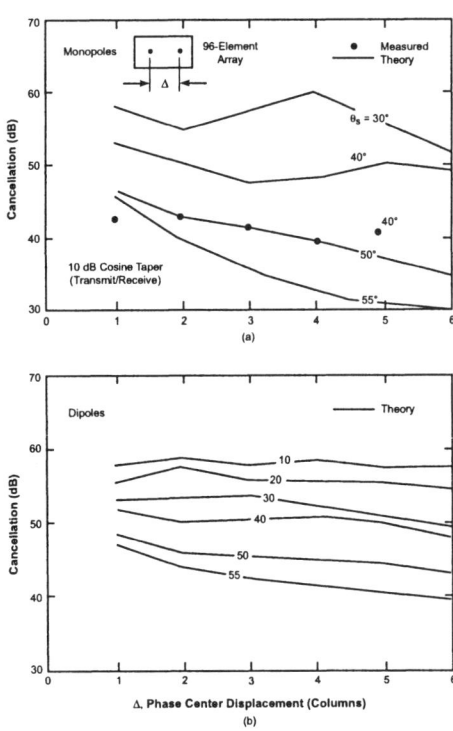

Figure 2.14 Calculated clutter cancellation as a function of phase center displacement in columns for the 96-element test array. (a) Monopole elements (measured data shown as solid dots) and (b) dipole elements.

performance results shown in this chapter, it has been assumed that the PRI is matched to the geometrical phase center spacing of the antennas. In those cases (the majority) where guard bands are present, this value is very close to the optimum choice. In the others, it may be assumed that performance could be improved somewhat by a suitable adjustment of the PRI.

2.4.3 SUBSCALE AND FULL-SCALE PLANAR PHASED ARRAYS

The previous section dealt with a specific subscale phased array antenna design – a 96-element planar array. In this section, the effect of the array size on DPCA performance will be investigated. Array sizes up to 16 meters are considered. The designation subscale array will be used arbitrarily to refer to antennas less than 12 meters in length. Of primary interest is the effect of array length along the dimension of phase center displacement. The number of active rows is fixed at eight, and the number of active columns is made variable, as depicted in Figure 2.19. With eight active rows and two guard bands, for a 16-meter aperture there are a total of 148 columns by 12 rows (1776 unknowns) in the square lattice case. For the same aperture size, with a hexagonal lattice, the number of columns is 132 and the number of rows is 12 (1584 unknowns). Both monopole and dipole radiators are evaluated here. A low altitude SBR system is assumed as in the previous sections. The array element spacing is selected to provide ±60° scan coverage. At center frequency, 1.3 GHz, with a square lattice the element spacing is 0.473λ and with a hexagonal lattice the spacing is 0.55λ. The monopole length is 6.35 cm

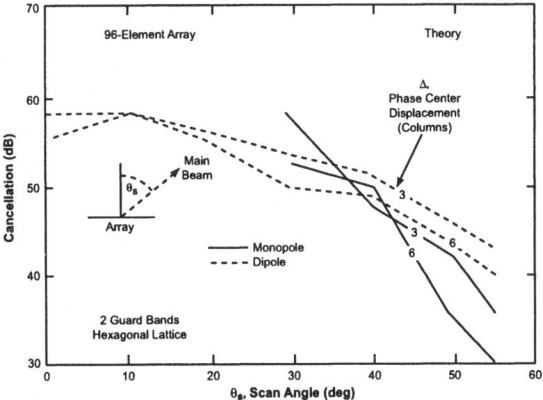

Figure 2.15 Calculated clutter cancellation as a function of scan angle using monopole and dipole elements with three-and six-column phase center separation for the 96-element test array.

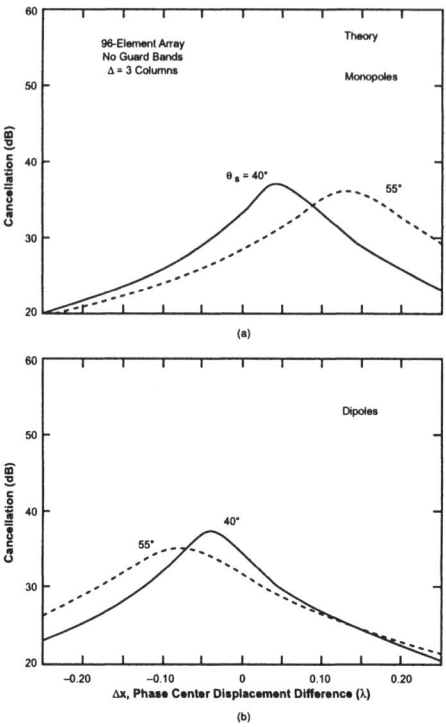

Figure 2.16 Calculated clutter cancellation versus phase-center displacement difference in wavelengths (with no guard bands) for the 96-element DPCA array shown in Figure 2.11(b). The ideal (desired) phase center displacement is three columns. (a) Monopole elements and (b) dipole elements.

for the square lattice and 5.842 cm for the hexagonal lattice. The dipole half-length is 5.334 cm in the square lattice and 5.588 cm in the hexagonal lattice. The dipole spacing over the ground plane is 5.334 cm.

Consider Figure 2.20(a), which shows the cancellation as a function of number of active columns for a square lattice. Here, the number of active columns has been varied from 12 to 144. For the two scan angles shown, it is clear for larger array size that the monopole offers a slight improvement over dipoles. However, in Figure 2.20(b) (hexagonal lattice) the cancellation is approximately the same for either monopoles or dipoles. Replotting the same results in Figure 2.21, the effect of the array lattice on cancellation is clearly shown.

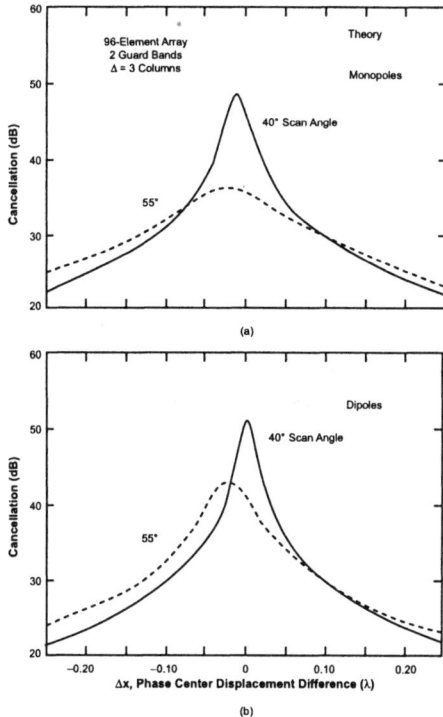

Figure 2.17 Calculated clutter cancellation versus phase center displacement difference in wavelengths (with two guard bands) for the 96-element DPCA array shown in Figure 2.11(b). The ideal (desired) phase center displacement is three columns. (a) Monopole elements and (b) dipole elements.

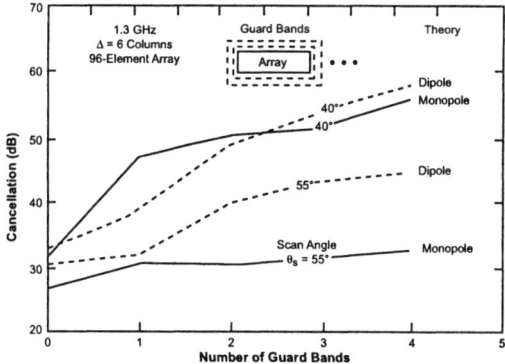

Figure 2.18 Calculated clutter cancellation versus number of guard bands for the 96-element DPCA array shown in Figure 2.11(b). The phase center displacement is six columns.

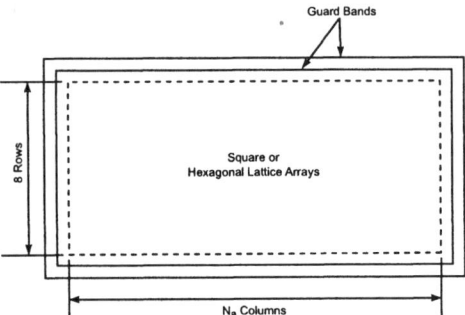

Figure 2.19 Subscale and full-scale array geometry. The number of active rows is fixed at eight. The number of active columns is variable. Two guard bands surround the active array.

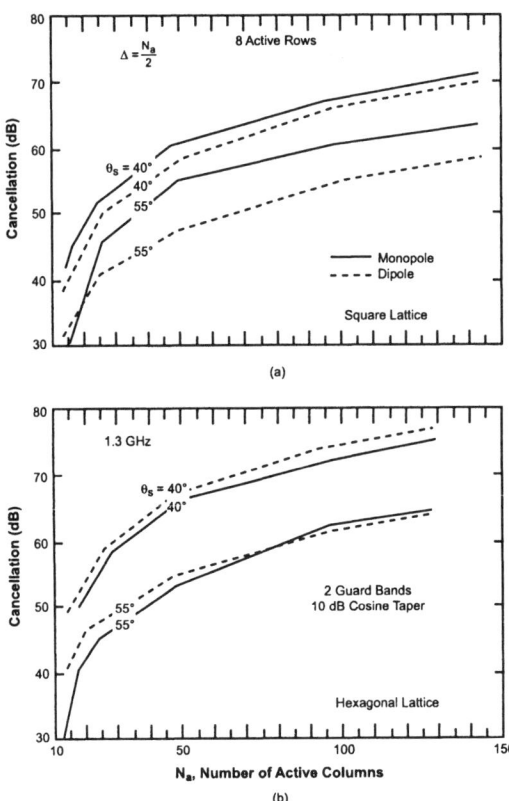

Figure 2.20 Clutter cancellation as a function of number of active array columns for one-half aperture phase center spacing using monopole and dipole antenna elements. (a) Square lattice and (b) hexagonal lattice.

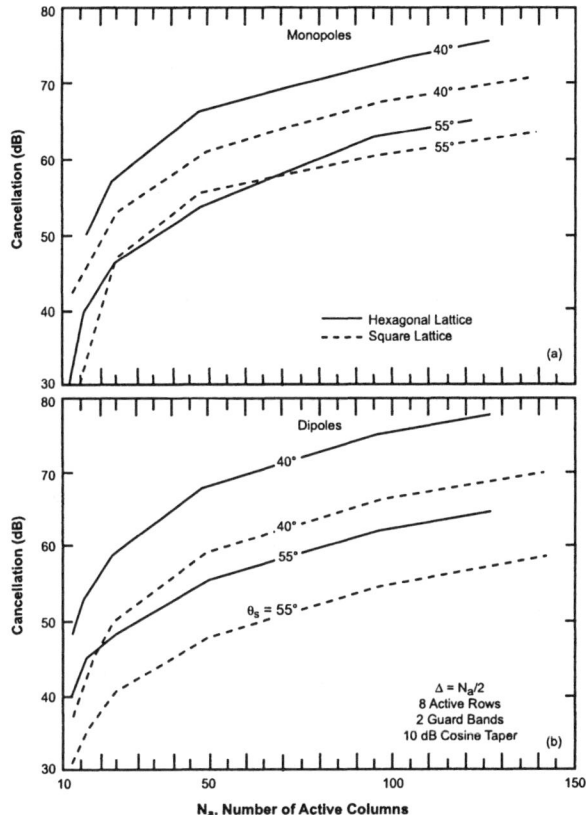

Figure 2.21 Clutter cancellation as a function of number of active array columns for one-half aperture phase center spacing using square and hexagonal array lattices. (a) Monopole elements and (b) dipole elements.

The effect of the lattice is greater for dipoles than for monopoles. In Figure 2.22 the same results are again replotted, this time as a function of array length in meters. From these data it is observed that, for arrays greater than 4m, the cancellation increases typically by about 8 dB when the array length increases by a factor of two. It should be noted that the cancellation shown here does not include the effects of element weighting errors. That is, the high values of cancellation may not actually be achieved in practice. The cancellation will be limited by the random amplitude and phase errors produced by the receive modules.

Now, consider the effect of varying the phase center displacement for a fixed size array consisting of 8 active rows and 128 active columns. The array lattice is assumed to be hexagonal and there are two guard bands. Two scan angles, 40° and 55°, were considered for monopoles and dipoles and the phase

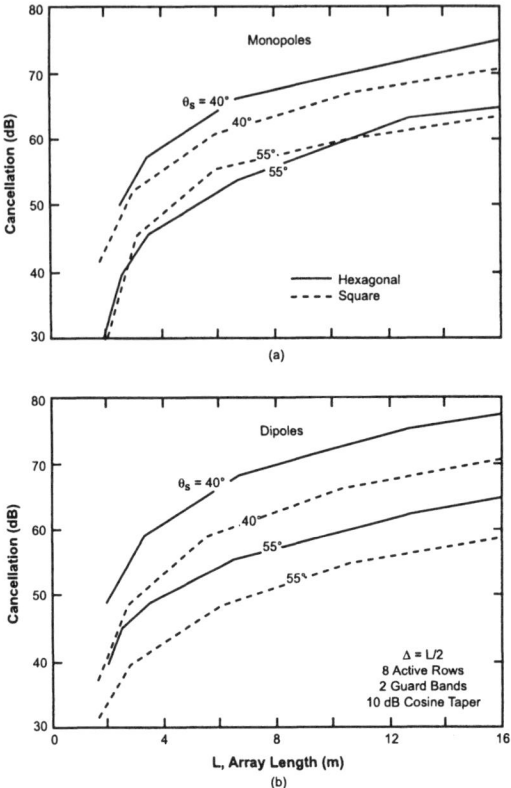

Figure 2.22 Clutter cancellation as a function of array length for one-half aperture phase center spacing using square and hexagonal array lattices. (a) Monopole elements and (b) dipole elements.

center displacement was varied from 8 to 64 columns. The results are shown in Figure 2.23 where it is observed that the clutter cancellation for this large array is insensitive to phase center displacement. To show that these results do not depend on the number of active rows, the number of columns was fixed at 128 and the number of active rows was varied from 4 to 16. The array lattice is hexagonal, and there are two guard bands. The phase center displacement was assumed to be 64 columns. Figure 2.24 clearly shows that there is little change in the clutter cancellation as the number of active rows is varied.

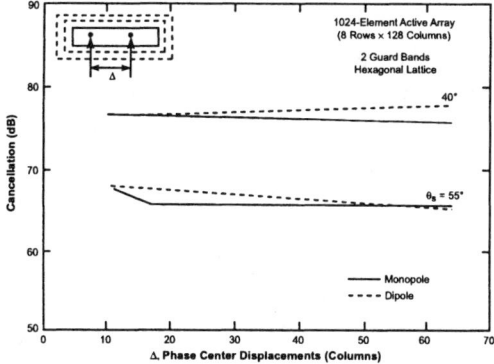

Figure 2.23 Clutter cancellation versus phase center displacement in columns for an 8 active row by 128 active column DPCA array.

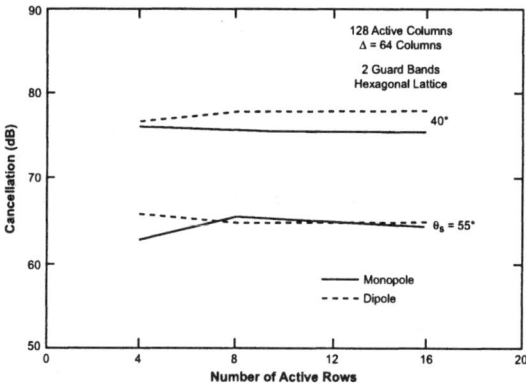

Figure 2.24 Clutter cancellation versus number of active rows in an array having 128 active columns.

2.5 SUMMARY

This chapter has investigated the subject of array mutual coupling effects on the displaced phase center antenna clutter cancellation capability of corporate-fed phased array antennas. Mutual coupling effects have been modeled using the method of moments for monopole and dipole phased array antennas. Finite array edge effects and polarization are inherently included in the analysis. The assumption of an infinite ground plane was made in the theoretical formulation, and finite ground plane size would be expected to have some degradation on DPCA clutter cancellation. The effects of transmit/receive module random amplitude and phase errors have been ignored in this analysis. The theoretical results presented are the upper-bound clutter cancellation limited by array mutual coupling alone.

In Section 2.3 the theoretical formulation was discussed. A derivation of the factor used to quantify clutter cancellation was given.

A variety of subscale and full-scale SBR antennas have been analyzed in Section 2.4. One such array was a 96-element planar array of monopoles for which measured DPCA clutter cancellation capability data were available. The theoretical calculations were shown to be consistent with measurements.

Based on the results presented in this chapter, it has been shown that the clutter cancellation capability of an array is sensitive to many parameters. These parameters include the scan angle, phase center displacement, array size, array lattice, number of passively terminated element guard bands, and radiating element type. Some conclusions that can be made are as follows: The clutter cancellation tends to increase as the array size or number of guard bands increases. For large arrays, a hexagonal lattice array provides better cancellation than does a square lattice array. For smaller arrays, the best cancellation that can be achieved for a given lattice depends on the array element – a straight dipole element was shown to have better cancellation that a monopole element. The array element type strongly affects the amount of clutter cancellation achieved in subscale arrays.

Further research can be conducted to better understand how the mutual coupling in an array affects DPCA performance. For example, in the present moment method model one unknown current function per element of the array was used – it would be desirable to be able to vary the number of unknowns per element of the array to check the convergence of the clutter cancellation. It would be desirable to compute the clutter cancellation using full two-dimensional radiation patterns. Other types of array radiating elements such as microstrip patches and waveguides could be analyzed. The effect of a finite size ground plane and/or other surrounding structures would be important to include in the analysis. Also, the effect of transmit/receive module errors could be included in the mutual coupling results. Three or more displaced phase centers could also be analyzed. These added features would likely provide a clutter cancellation prediction capability that would more closely simulate a practical DPCA radar system.

2.6 PROBLEM SET

2.1 Based on the results presented in Figure 2.20(a), assuming a square array lattice determine the minimum number of active columns, N_a, required to achieve 50-dB and 60-dB clutter cancellation for both the monopole and dipole array element cases. Based on these results for arrays with a square lattice, for clutter cancellation in the range of 50 dB to 60 dB, which array element would provide the greatest clutter cancellation for a given array size? Repeat this exercise for arrays with a hexagonal lattice using the results presented in Figure 2.20(b).

References

[1] Stone, M.L., and W.J. Ince, "Air-to-Ground MTI Radar Using a Displaced Phase Center Phased Array," *The Record of the IEEE 1980 International Radar Conference*, Arlington, Virginia, April 28-30, 1980, pp. 225-230.

[2] Nichols, B.E., (ed.), "Multiple-Antenna Surveillance Radar (MASR) Experiment, Phase I: Concept Validation System," Project Report TST-16, *Lincoln Laboratory, Massachusetts Institute of Technology*, January 13, 1978, DDC AD-B025363-L.

[3] Kelly E.J., and G. N. Tsandoulas, "A Displaced Phase Center Antenna Concept for Space Based Radar Applications," *IEEE Eascon*, September 1983, pp. 141-148.

[4] Tsandoulas, G.N., "Space-Based Radar," *Science*, July 17, 1987, pp. 257-262.

[5] Muehe, C.E., and M. Labitt, "Displaced-Phase-Center Antenna Technique," *Lincoln Laboratory Journal*, Vol. 12, No. 2, 2000, pp. 281-296.

[6] Fenn, A.J., F. G. Willwerth, and H. M. Aumann, "Displaced Phase Center Antenna Near Field Measurements for Space Based Radar Applications," *Proceedings of the Phased Arrays 1985 Symposium*, RADC-TR-85-171 In-House Report, August 1985, ADA-169316, pp. 303-318.

[7] Fenn, A.J., and E.J. Kelly, "Theoretical Effects of Array Mutual Coupling on Clutter Cancellation in Displaced Phase Center Antennas," *Technical Report 1065*, Lincoln Laboratory, Massachusetts Institute of Technology, 2000.

[8] Skolnik, M.L, *Radar Handbook*, 2nd ed., New York: McGraw-Hill, 1990, p. 16.10.

[9] Shnitkin, H., "A Unique Joint STARS Phased-Array Antenna," *Microwave Journal*, Vol. 34, No. 1, 1991, pp. 131-141.

[10] Lacomme, P., J.P. Hardange, J.C. Marchais, and E. Normant, *Air and Spaceborne Radar Systems: An Introduction*, New York: William Andrew Publishing, 2001, pp. 173-176.

[11] Fenn, A.J., "Theoretical and Experimental Study of Monopole Phased Array Antennas," *IEEE Trans. Antennas Propagat.*, Vol. AP-34, No. 10, October 1985, pp. 1118-1126.

[12] Fenn, A.J., "Element Gain Pattern Prediction for Finite Arrays of V-Dipole Antennas over Ground Plane," *IEEE Trans. Antennas Propagat.*, Vol. AP-36, No. 11, November 1988, pp. 1629-1633.

[13] Stutzman, W.L, and G.A. Thiele, *Antenna Theory and Design*, 2nd ed., New York: Wiley, 1998.

[14] Hansen, R.C., "Formulation of Echelon Dipole Mutual Impedance for Computer," *IEEE Trans. Antennas Propagat.*, Vol. AP-20, No. 6, November 1972, pp. 780-781.

[15] Reed, I.S., J.D. Mallet, and L.E. Brennan, "Rapid Convergence Rate in Adaptive Arrays," *IEEE Trans. Aerospace and Electronic Systems*, Vol. AES-10, No. 6, November 1974, pp. 853-863.

[16] Boroson, D.M., "Sample Size Considerations for Adaptive Arrays," *IEEE Trans. Aeorospace and Electronic Systems*, Vol. AES-16, No. 4, July 1980, pp. 446-451.

3
Focused Near-Field Technique for Evaluating Adaptive Phased Arrays

3.1 INTRODUCTION

Electrically large phased array antennas having partially adaptive or fully adaptive nulling capability are often suggested for radar or communications applications. The adaptive nulling performance of these antennas is principally tested using conventional far-field antenna ranges with far-field (or plane wave) interferers. At microwave frequencies, this can lead to significant far-field range distances which require that tests be made outdoors. For example, given a 20m aperture at 1 GHz and 10 GHz, the usual far-field range criterion $R = 2D^2/\lambda$ yields a far-field test distance of approximately 2.7 km and 27 km, respectively. Multiple, widely separated interferers make the far-field range design more difficult. Near-field testing, suitable for indoor measurements, is desirable as has been demonstrated in the case of near-field scanning [1] for far-field pattern determination and compact range-reflector techniques [2] for far-field radiation pattern and radar cross-section determination.

Conventional near-field scanning (see Chapters 11 and 12) does not appear practical in the adaptive nulling antenna situation, as a mathematical transformation is usually required to determine the far-field performance. Evaluation of adaptive nulling performance requires a real-time interference wavefront (or multiple wavefronts) impinging on the adaptive antenna. Furthermore, near-field scanning reported in the literature is strictly single tone and is not compatible with adaptive nulling tests that often require the presence of a wide-bandwidth noise interferer (jammer). A compact range reflector (a parabolic reflector that converts a spherical wave from a feed horn into a plane wave) is applicable to near-field adaptive nulling as it generates a real-time wavefront. However, the compact range reflector does not easily lend itself to multiple interferers widely separated in angle. Thus, an alternate approach to adaptive array near-field testing needs to be considered.

If the requirement for plane wave test conditions is relaxed and spherical wave incidence is allowed, for a near-field focused phased array antenna, near-field testing that is nearly equivalent to far-field testing is possible. In this chapter, it is shown, by example, that at *one to two aperture diameters range*, J near-field interferers (where J is a number typically less than the number of adaptive array channels) can be equivalent to J far-field interferers. The contrast between plane wave incidence and spherical wave incidence is depicted in Figure 3.1. The amount of wavefront dispersion observed by the adaptive array is a function of the nulling channel bandwidth, array length, and angle of incidence. Interference wavefront dispersion is an effect that can limit the depth of null (or cancellation) achieved by an adaptive antenna [3].

Basic dispersion models for spherical wave incidence and plane wave incidence can be made by considering only the wavefront dispersion observed by the end points of an adaptive array. This calculation is useful in gaining some initial insight into how near-field (NF) nulling will relate to far-field (FF) nulling. Consider first, a plane wave arriving from infinity and an array of length L as shown in Figure 3.1(a). The dispersion for this case is denoted γ_{FF} and is computed according to the product of bandwidth and time delay as

$$\gamma_{FF} = \frac{BL}{c}\beta_{FF} \qquad (3.1)$$

where B is the nulling bandwidth and c is the speed of light. In (3.1) the quantity β_{FF} will be referred to as the far-field dispersion multiplier, which is simply given by

$$\beta_{FF} = \cos\theta_i \qquad (3.2)$$

where θ_i is the angle of incidence with respect to the axis of the array. Note that the dispersion is maximum for endfire incidence ($\theta_i = 0$) and is zero for broadside incidence ($\theta_i = \pi/2$). Next, consider the same array and now a point source, at range $r = r_i$ and angle $\theta = \theta_i$, which produces a spherical wavefront as depicted in Figure 3.1(b). The near-field dispersion multiplier, denoted γ_{NF}, is given by

$$\gamma_{NF} = \frac{BL}{c}\beta_{NF} \qquad (3.3)$$

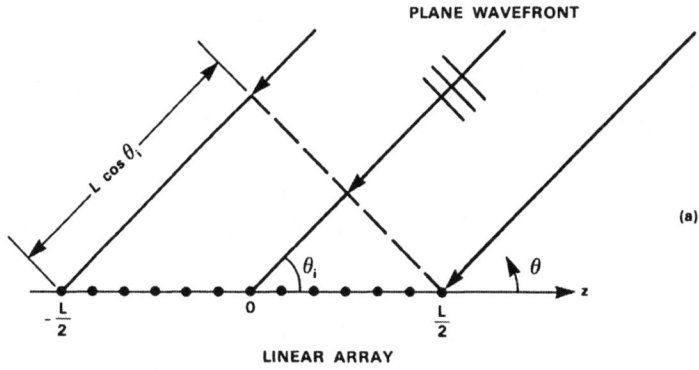
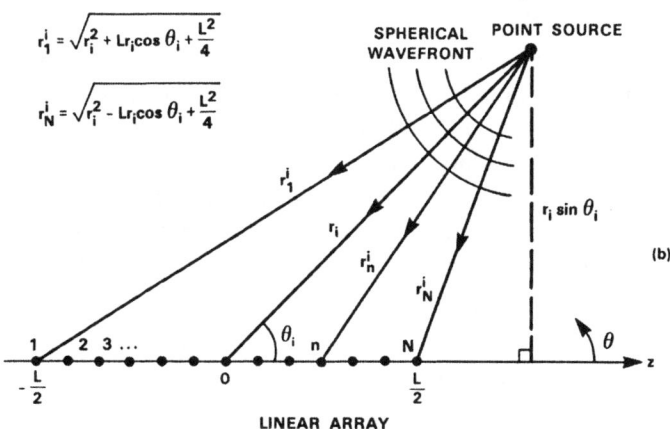

Figure 3.1 Geometry for far-field and near-field interference for a linear array. (a) Plane wavefront and (b) spherical wavefront. © 1990 IEEE [8].

where the quantity β_{NF} denotes the near-field dispersion multiplier which, from the difference between the path lengths r_1^i and r_N^i, is expressed as

$$\beta_{NF} = \sqrt{\alpha^2 + \alpha \cos\theta_i + \frac{1}{4}} - \sqrt{\alpha^2 - \alpha \cos\theta_i + \frac{1}{4}} \qquad (3.4)$$

where

$$\alpha = \frac{F}{L} \qquad (3.5)$$

and where F is the focal length of the array such that $r_i = F$. In comparing (3.1) and (3.3) it is seen that the far-field and near-field dispersions differ only by their respective dispersion multipliers β_{FF} and β_{NF}. If near-field nulling is equivalent to far-field nulling then (3.4) must be equal to (3.2).

Figure 3.2 shows a plot of β_{FF} and β_{NF} versus angle of incidence for values of $\alpha = 0.2$ to 2 (that is, focal lengths $0.2L$ to $2L$). From this figure it is seen that the near-field dispersion approaches the value of the far-field dispersion for source range distances greater than approximately one aperture diameter ($\alpha \geq 1$). Clearly, at source range distances such that $\alpha \leq 0.5$ (one-half aperture diameter) the near-field dispersion is significantly different from the far-field dispersion. Thus, for this simple dispersion model it is expected that near-field adaptive nulling will be similar to far-field nulling at source range distances greater than one aperture diameter.

A more accurate dispersion model is essentially the characteristics of the interference covariance matrix – namely its eigenvalues or degrees of freedom [3,4] as discussed in Chapter 1. The covariance matrix contains all wavefront dispersion presented to the adaptive channels, not just the contributions from the array edge as modeled above in (3.2) and (3.4). The interference covariance matrix eigenvalues can be used to quantify and to compare the dispersion present for plane wave and spherical wave incidence. For near-field adaptive nulling to be equivalent to far-field adaptive nulling, it is assumed that the NF/FF interference covariance matrix eigenvalues must be equivalent. Additionally, it is assumed that the NF/FF adaptive array weights, cancellation of interference power, and radiation patterns must also be equivalent.

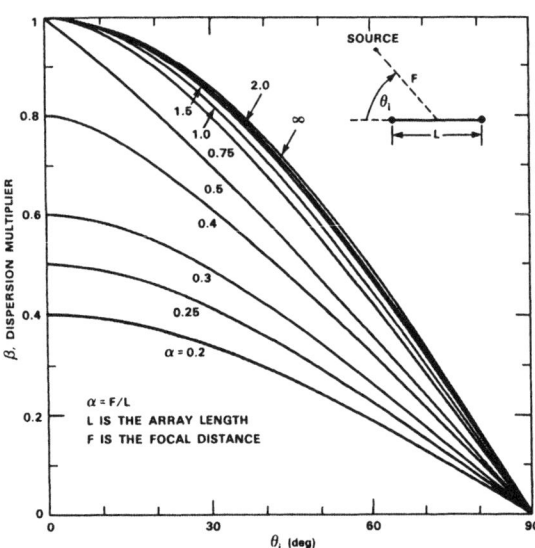

Figure 3.2 Dispersion multiplier as a function of interferer incidence angle for various interferer range distances. © 1990 IEEE [8].

An example of near-field adaptive nulling has been given by Hudson [5]. A two-element array with a single near-field continuous wave (CW) interferer was investigated. It was shown that a null could be formed at a near-field point.

Investigations of NF/FF adaptive nulling for linear arrays of isotropic receive elements have been performed for interference and clutter by this author [6-9]. Near-field adaptive nulling results for a single interferer at a range of $1.7L$ for sidelobe canceller [6, 8] and fully adaptive [7, 8] arrays have been presented. The effects of nulling bandwidth were taken into account. Comparisons with far-field adaptive nulling indicated an excellent NF/FF equivalence. A detailed analysis of near-field adaptive nulling in the range of one to two aperture diameters has been given for a sidelobe canceller antenna [8]. Application of this near-field technique to testing main beam clutter cancellation is also documented [9, 10]. The purpose of the present chapter is to review the focused near-field nulling technique for fully adaptive arrays and sidelobe canceller arrays. In this chapter, the arrays analyzed have isotropic receive antenna elements and the effects of array mutual coupling and polarization are ignored. Chapters 4 and 5 will investigate the situation where mutual coupling and polarization are included in the near-field nulling analysis.

This chapter is organized as follows: In Section 3.2, a theory for analyzing and comparing both near-field and far-field adaptive nulling is presented. General adaptive nulling concepts are addressed first. Near-field focusing and the near-field nulling concept are then described. The theory is developed for the situation of a linear array of isotropic receive elements and isotropically radiating interference sources. Equations for the NF/FF covariance matrices and radiation patterns of fully adaptive and sidelobe canceller adaptive arrays are given. A discussion of the boundaries of the near, Fresnel, and far zones is included. The emphasis of this chapter is on near-field nulling, although the theory is equally valid in the Fresnel zone. In Section 3.4, results are presented that show that, at one to two aperture diameters range, a fully adaptive array responds in the same manner to near-field sources as it does to far-field sources. Results for a sidelobe canceller array are briefly addressed in Section 3.4.

3.2 THEORY

3.2.1 ADAPTIVE NULLING FORMULATION

Consider an N-element linear array of isotropic point receiving antennas. Let an interference wavefront be impressed across the array, which results in a set of induced terminal voltages denoted as $v_1, v_2, \cdots, v_N$. As shown in

Figure 3.3, two types of arrays will be considered – fully adaptive arrays and sidelobe canceller arrays. The number of adaptive channels is denoted as M. For the fully adaptive array $M = N$, and for the sidelobe canceller $M = 1 + N_{aux}$ where N_{aux} is the number of auxiliary channels. In this chapter, ideal (error-free) adaptive array weights are assumed with $w = (w_1, w_2, \cdots, w_M)^T$ denoting the adaptive channel weight vector and $W = (W_1, W_2, \cdots, W_N)^T$ denoting the phased array element weight vector in the sidelobe canceller case (superscript T means transpose). The fundamental quantities required in fully characterizing the incident field, for adaptive nulling purposes, are the adaptive channel cross correlations.

The cross correlation R_{mn} of the received voltages in the mth and nth adaptive channels is given by

$$R_{mn} = E(v_m v_n^*) \tag{3.6}$$

where $*$ means complex conjugate and $E(\cdot)$ means mathematical expectation. Since v_m and v_n represent voltages of the same waveform, but at different times, R_{mn} is also referred to as an autocorrelation function.

In the frequency domain, (3.6) can be expressed as the frequency average

$$R_{mn} = \frac{1}{B} \int_{f_1}^{f_2} v_m(f) v_n^*(f) \, df \tag{3.7}$$

where $B = f_2 - f_1$ is the nulling bandwidth. It should be noted that $v_m(f)$ takes into account the wavefront shape, which can be spherical in the near-field case or planar in the far-field case.

Let the channel or interference covariance matrix for the ith source be denoted by R_i. If there are J uncorrelated broadband interference sources, then the J-source covariance matrix is the sum of the covariance matrices for the individual sources, that is,

$$R = \sum_{i=1}^{J} R_i + I \tag{3.8}$$

where I is the identity matrix used to represent the thermal noise level of the receiver.

Prior to generating an adaptive null, the adaptive channel weight vector, w, is chosen to synthesize a desired quiescent radiation pattern. When interference is present, the optimum set of weights, denoted w_a, to form an adaptive null is computed by [10]

$$w_a = R^{-1} \cdot w_q \tag{3.9}$$

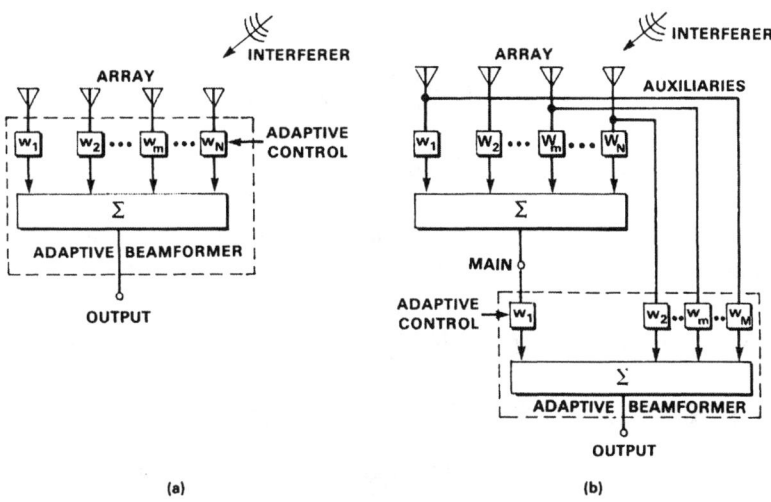

Figure 3.3 Adaptive beamformer arrangements. (a) Fully adaptive array and (b) partially adaptive array (sidelobe canceller array). © 1990 IEEE [8].

where $^{-1}$ means inverse and w_q is the quiescent weight vector.

The output power at the adaptive array summing junction is given by

$$p = w^\dagger \cdot R \cdot w \qquad (3.10)$$

where † means complex conjugate transpose. The interference-to-noise ratio, denoted INR, is computed as the ratio of the output power (defined in (3.10)) with the interferer present to the output power with only receiver noise present, that is,

$$\text{INR} = \frac{w^\dagger \cdot R \cdot w}{w^\dagger \cdot w} \qquad (3.11)$$

The adaptive array cancellation ratio, denoted C, is defined here as the ratio of interference output power before adaption to the interference output power after adaption, that is,

$$C = \frac{p_q}{p_a} \qquad (3.12)$$

Substituting (3.10) in (3.12) yields

$$C = \frac{w_q{}^\dagger R w_q}{w_a{}^\dagger R w_a} \qquad (3.13)$$

Next, the covariance matrix defined by the elements in (3.7) is Hermitian (that is, $R = R^\dagger$), which, by the spectral theorem, can be decomposed in eigenspace as [12]

$$R = \sum_{k=1}^{M} \lambda_k e_k e_k^\dagger \qquad (3.14)$$

where $\lambda_k, k = 1, 2, \cdots, M$ are the eigenvalues of R, and $e_k, k = 1, 2, \cdots, M$ are the associated eigenvectors of R. The covariance matrix eigenvalues $(\lambda_1, \lambda_2, \cdots, \lambda_M)$ are a convenient quantitative measure of the utilization of the adaptive array degrees of freedom as discussed in Chapter 1.

3.3 NEAR-FIELD FORMULATION

3.3.1 FOCUSED NEAR-FIELD NULLING CONCEPT

To properly utilize the near-field nulling technique described here, it is assumed that the quiescent near-field radiation pattern of the array should have the same characteristics as the quiescent far-field radiation pattern of the array. This means typically that a main beam and sidelobes should be formed. To produce an array near-field pattern that is approximately equal to the far-field pattern, phase focusing can be used [13]. Consider Figure 3.4, which shows a CW calibration source (typically operating at center frequency f_c) located at a desired focal point of the array. The array can maximize the signal received from the calibration source by adjusting its phase shifters such that the spherical wavefront phase variation is removed. One way to do this is to choose a reference path length as the distance from the focal point to the center of the array. This distance is denoted r_F, and the distance from the focal point to the nth array element is denoted r_n^F. Ignoring the time-harmonic quantity $e^{j\omega t}$, where $\omega = 2\pi f$ is the radian frequency, the voltage received at the nth array element, due to the CW source at the focal point, is expressed as

$$v_n^F = \frac{e^{j2\pi(r_F - r_n^F)/\lambda_c}}{r_n^F} \qquad (3.15)$$

where the wavelength $\lambda_c = c/f_c$ where c is the speed of light. To maximize the received voltage at the array output, it is necessary to add Ψ_n radians to the nth element, that is,

$$\Psi_n = 2\pi(r_n^F - r_F)/\lambda_c \qquad (3.16)$$

The resulting radiation pattern at range r_F looks much like an ordinary far-field pattern, as will be shown in Section 3.4. A main beam will be pointed at the array focal point. Sidelobes will exist at angles away from the main beam. An interferer can then be placed on a near-field sidelobe, at range $r_i = r_F$, as depicted conceptually in Figure 3.4.

3.3.2 COVARIANCE MATRIX FOR NEAR-FIELD INTERFERENCE

The near-field covariance matrix elements are derived in the following manner [8]: Refer to Figure 3.1(b), which shows the near-field geometry under consideration. An interference source, whose power spectral density is assumed to be bandlimited white noise, is represented by the index i in a set of J sources. Let it be assumed that the power received at the center element of the array from the ith source, denoted P_i, is measured relative to thermal noise. The voltage received by the nth array element is then given by

$$v_n^i = \sqrt{P_i}\frac{e^{-j2\pi f r_n^i/c}}{r_n^i} \tag{3.17}$$

where r_n^i is the distance between the nth array element and the interferer and c is the speed of light. Substituting (3.17) in (3.7) and integrating gives the correlation between either two channels in a fully adaptive array or two auxiliary channels in a sidelobe canceller array as

$$R_{mn}^{NF} = P_i \frac{r_i^2}{r_m^i r_n^i} e^{-j2\pi \tau_{mn} f_c} \frac{sin(\pi B \tau_{mn})}{\pi B \tau_{mn}} \tag{3.18}$$

where r_i is the range from the array center to the interferer and

$$\tau_{mn} = (r_m^i - r_n^i)/c \tag{3.19}$$

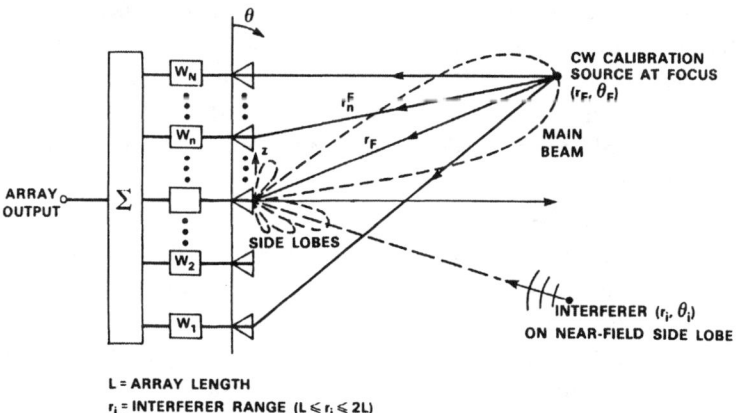

Figure 3.4 Adaptive phased array antenna near-field focusing concept. A CW radiating source is used to illuminate the array with a calibration signal, and phase corrections are applied at the array weights to focus the antenna. The array then forms a main beam at the location of the focal point, and an interference source is placed at the location of a near-field sidelobe. © 1990 IEEE [8].

is the interferer time differential factor between elements m and n. In (3.18), r_i^2 has been included as a convenient normalization factor.

Next, for a sidelobe canceller array, the main channel received voltage is expressed as

$$v_{main} = \sqrt{P_i} \sum_{m=1}^{N} W_m^* \frac{e^{-j2\pi f r_m^i/c}}{r_m^i} \qquad (3.20)$$

Substituting (3.17) and (3.20) in (3.7) and integrating yields the cross correlation between the main channel and the nth auxiliary channel

$$R_{main,aux_n}^{NF} = P_i r_i^2 \sum_{m=1}^{N} W_m^* \frac{e^{-j2\pi \tau_{mn} f_c}}{r_m^i r_n^i} \frac{\sin(\pi B \tau_{mn})}{\pi B \tau_{mn}} \qquad (3.21)$$

The main channel cross correlation is found using (3.20) in (3.7) as

$$R_{main,main}^{NF} = P_i r_i^2 \sum_{m=1}^{N} \sum_{n=1}^{N} W_m^* W_n \frac{e^{-j2\pi \tau_{mn} f_c}}{r_m^i r_n^i} \frac{\sin(\pi B \tau_{mn})}{\pi B \tau_{mn}} \qquad (3.22)$$

3.3.3 ARRAY RADIATION PATTERN IN THE FOCUSED NEAR FIELD

In this section, equations used to compute antenna near-field focused directivity patterns are given [8]. These patterns are in contrast to the near-field pattern of a far-field focused array. In the near-field region, taken here as one to two aperture diameters, if the antenna is near-field focused, the pattern will be approximately the same as a far-field focused/far-field observed pattern. This pattern equivalence is assumed desirable for near-field nulling, because the quiescent near-field conditions can be made to look the same as the quiescent far-field conditions.

Let r_o be the observation range relative to the midpoint of the array antenna and let z_n' be the location of the nth array element. Next, let r_n' be the distance from the observation point to the nth array element. The near-field radiation pattern can be calculated by superposition of the spherical waves received at the array elements. These received signals are appropriately weighted by either the adaptive array weights in the case of a fully adaptive array, or by the array weights and auxiliary channel weights in the case of a sidelobe canceller array.

The antenna directivity pattern D is given by the ratio of the radiation intensity, denoted $U(\theta, \phi)$, to the average radiation intensity, U_{ave}, or denoted

$$D(\theta, \phi) = \frac{U(\theta, \phi)}{U_{ave}} \qquad (3.23)$$

where
$$U_{\text{ave}} = \frac{1}{4\pi} \int_{\theta=0}^{\pi} \int_{\phi=0}^{2\pi} U(\theta, \phi) \sin\theta\, d\theta\, d\phi \tag{3.24}$$

The radiation intensity in (3.23) and (3.24) is defined here by

$$U(\theta, \phi) = |P(\theta, \phi)|^2 = P(\theta, \phi)P^*(\theta, \phi) \tag{3.25}$$

where $P(\theta, \phi)$ is the radiation pattern, and * denotes conjugate.

For a fully adaptive linear array positioned along the z-axis, the near-field received voltage pattern at range r_o is independent of ϕ and is given in exact form by the following summation of spherical waves,

$$P(\theta) = r_o \sum_{n=1}^{N} w_n^* \frac{e^{-jk(r_n' - r_o)}}{r_n'} \tag{3.26}$$

where $k = 2\pi/\lambda$ is the phase constant (angular wavenumber), and w_n is the complex weight (either quiescent or adaptive) for the nth array element. In spherical coordinates, r_n' is determined from the law of cosines as

$$r_n' = \sqrt{r_o^2 + (z_n')^2 - 2r_o z_n' \cos\theta} \tag{3.27}$$

The numerator in (3.23) controls the shape of the directivity pattern and can be computed in exact form from (3.25) and (3.26) as the following double summation,

$$U(\theta) = r_o^2 \sum_{m=1}^{N} \sum_{n=1}^{N} w_m w_n^* \frac{e^{jk(r_m' - r_o)} e^{-jk(r_n' - r_o)}}{r_m' r_n'} \tag{3.28}$$

or, equivalently, in exact form as the following single summation

$$U(\theta) = r_o^2 |\sum_{n=1}^{N} w_n^* \frac{e^{-jk(r_n' - r_o)}}{r_n'}|^2 \tag{3.29}$$

The denominator U_{ave} in (3.23) can be computed in approximate form by making the following assumptions. It is now assumed that the radial distance from the focused antenna under test is at least one to two aperture diameters or greater. This assumption means that r_m' and r_n' are not too dissimilar, such that for amplitude comparison (3.28) becomes

$$U(\theta) \approx \sum_{m=1}^{N} \sum_{n=1}^{N} w_m w_n^* e^{jk(r_m' - r_n')} \tag{3.30}$$

Next, (3.27) is expanded using the Binomial Theorem, such that

$$r'_n = r_o - z'_n \cos\theta + \frac{(z'_n)^2 \sin^2\theta}{2r} + \frac{(z'_n)^3 \cos\theta \sin^2\theta}{2r^2} + \cdots \quad (3.31)$$

Since near-field phase focused arrays are of primary interest here, the squared and cubic terms of (3.31) can be neglected, since the focused near-field radiation pattern is known to be approximately the same as a conventional far-field radiation pattern. Thus, to compute the average directivity, the approximation

$$r'_n \approx r_o - z'_n \cos\theta \quad (3.32)$$

is used for an array focused in the near field. Substituting (3.32) into (3.30) yields

$$U(\theta) \approx \sum_{m=1}^{N}\sum_{n=1}^{N} w_m w_n^* e^{jk(z'_n - z'_m)\cos\theta} \quad (3.33)$$

Substituting (3.33) into (3.24) yields

$$U_{\text{ave}} = \frac{1}{4\pi}\sum_{m=1}^{N}\sum_{n=1}^{N} w_m w_n^* \int_{\theta=0}^{\pi}\int_{\phi=0}^{2\pi} e^{jk(z'_n - z'_m)\cos\theta} \sin\theta\, d\theta\, d\phi \quad (3.34)$$

For the case of one-half wavelength array element spacing, (3.34) reduces to

$$U_{\text{ave}} = \sum_{n=1}^{N} |w_n|^2 \quad (3.35)$$

For a fully adaptive (FA) linear array, assuming $\lambda/2$ element spacing and using (3.29) and (3.35), the near-field directivity pattern is given by

$$D_{FA}^{NF}(r_o, \theta) \approx \frac{r_o^2 |\sum_{n=1}^{N} w_n^* \frac{e^{-jk(r'_n - r_o)}}{r'_n}|^2}{\sum_{n=1}^{N} |w_n|^2} \quad (3.36)$$

The radiation pattern of the sidelobe canceller adaptive array is equal to the sum of the main channel radiation pattern and the weighted auxiliary channels radiation pattern. For a sidelobe canceller (SLC) array, the near-field directivity pattern is found to be

$$D_{SLC}^{NF}(r_o, \theta) \approx$$

$$\frac{r_o^2 |w_1^* \sum_n W_n^* \frac{e^{-jk(r'_n - r_o)}}{r'_n} + \sum_m w_{aux_m}^* \frac{e^{-jk(r'_{aux_m} - r_o)}}{r'_{aux_m}}|^2}{|w_1|^2(\sum_n |W_n|^2 + 2\text{Re}(w_1 \sum_m W_{aux_m} w_{aux_m}^*) + \sum_m |w_{aux_m}|^2)} \quad (3.37)$$

where w_1 is the main channel weight and $Re(\cdot)$ means real part, and where the summation on m and n is from 1 to N_{aux} and from 1 to N, respectively. It should be noted that the shape of the near-field patterns given by (3.36) and (3.37) is controlled by the quantity in the numerator (radiation intensity) and is exact. The denominator in (3.36) and (3.37) is the average radiation intensity, which controls only the directivity level and has been computed by assuming that the NF average radiation intensity is equal to the FF average radiation intensity. This is equivalent to neglecting the squared and cubic phase terms in the integration of the NF radiation intensity.

For convenience, this chapter will assume that the maximum pattern gain is equal to the maximum directivity, which is appropriate for isotropic receive antenna elements. All of the array radiation pattern plots given in the results section (Section 3.4) will be displayed in terms of gain in dBi (decibels with respect to isotropic).

3.3.4 FAR-FIELD FORMULATION

3.3.4.1 Covariance Matrix

For a far-field interferer, the covariance matrix elements are readily determined using the near-field results presented in the previous section. In the far field (range is infinity), the interferer distance from the nth element is given exactly by

$$r_n^i = r_o - z_n' \cos \theta_i \tag{3.38}$$

Substituting (3.38) in (3.19) yields the far-field interferer time differential factor

$$\tau_{mn}^{FF} = \frac{(z_n' - z_m')\cos \theta_i}{c} \tag{3.39}$$

Substituting τ_{mn}^{FF} for τ_{mn} in (3.18) and letting r tend to infinity gives the far-field interference correlation between two auxiliary channels, as

$$R_{mn}^{FF} = P_i e^{-j2\pi \tau_{mn}^{FF} f_c} \frac{sin(\pi B \tau_{mn}^{FF})}{\pi B \tau_{mn}^{FF}} \tag{3.40}$$

Similarly, in the far field (3.21) and (3.22) reduce to

$$R_{main,aux_n}^{FF} = P_i \sum_{m=1}^{N} W_m^* e^{-j2\pi \tau_{mn}^{FF} f_c} \frac{sin(\pi B \tau_{mn}^{FF})}{\pi B \tau_{mn}^{FF}} \tag{3.41}$$

and

$$R_{main,main}^{FF} = P_i \sum_{m=1}^{N} \sum_{n=1}^{N} W_m^* W_n e^{-j2\pi \tau_{mn}^{FF} f_c} \frac{sin(\pi B \tau_{mn}^{FF})}{\pi B \tau_{mn}^{FF}} \tag{3.42}$$

3.3.5 ARRAY RADIATION PATTERN

In the far field the observation distance from the nth element is given by

$$r'_n = r_o - z'_n \cos\theta \qquad (3.43)$$

For a fully adaptive array, the far-field directivity pattern, assuming $\lambda/2$ element spacing, is found by substituting (3.43) in (3.36) and letting the range tend to infinity, with the result

$$D_{FA}^{FF}(\theta) = \frac{|\sum_{n=1}^{N} w_n^* e^{jkz'_n \cos\theta}|^2}{\sum_{n=1}^{N} |w_n|^2} \qquad (3.44)$$

Similarly, for a sidelobe canceller array, the far-field directivity pattern is found by substituting (3.36) in (3.37) with the result

$$D_{SLC}^{FF}(\theta) =$$
$$\frac{|w_1^* \sum_n W_n^* e^{jkz'_n \cos\theta} + \sum_m w_{aux_m}^* e^{jkz'_{aux_m} \cos\theta}|^2}{|w_1|^2 (\sum_n |W_n|^2 + 2Re(w_1 \sum_m W_{aux_m} w_{aux_m}^*) + \sum_m |w_{aux_m}|^2)} \qquad (3.45)$$

where the summation on m and n is from 1 to N_{aux} and from 1 to N, respectively.

3.3.6 NEAR-FIELD BOUNDARY

The hemispherical volume in front of an antenna of length L can be categorized by three regions: near, Fresnel, and far zones. A simple equation for computing the maximum range of the near zone (or beginning of the Fresnel zone) is [14]

$$r_{max} = 0.62 \sqrt{\frac{L^3}{\lambda}} \qquad (3.46)$$

Normalizing this distance by the aperture length yields the near zone range inequality

$$\frac{r}{L} \leq 0.62 \sqrt{\frac{L}{\lambda}} \qquad (3.47)$$

For a given r/L, the minimum aperture length, L_{min}, which satisfies (3.47) is expressed as

$$L_{min} = 2.6 \left(\frac{r}{L}\right)^2 \lambda \qquad (3.48)$$

In this chapter, the maximum finite range considered is two times the aperture length, that is, $r/L = 2$. Substituting $r/L = 2$ in (3.48) yields $L_{min} = 10.4\lambda$. The adaptive 32-element linear array example considered in this chapter has $L = 15.5\lambda$, which means that the near-field criterion is met.

3.4 RESULTS

3.4.1 FOCUSED LINEAR ARRAY QUIESCENT CONDITIONS

Consider a 32-element linear array with one-half wavelength element spacing at center frequency 1.3 GHz. The antenna length for this array of isotropic elements is 3.58m, which is similar to the experimental array tested in Chapter 6. Focal lengths of L, $1.5L$, and $2L$ will be examined. A scan angle of 30 degrees from broadside and a Chebyshev illumination that generates -40 dB uniform far-field sidelobes are assumed. The amplitude of the array NF/FF quiescent weights (w_q) is independent of range, as shown in Figure 3.5, because "phase-only" focusing is assumed. After positioning a CW radiation source at the desired location in the near field, the near-field phase focusing relation given by (3.16) is used to generate the relative phase commands at each array element. The resulting near-field focused/near-field observation quiescent patterns are shown in Figure 3.6. Figure 3.6(a) shows the near-field result (dashed curve) obtained at $F/L = 1$. Included in this figure is the conventional far-field pattern (solid curve) observed at infinite range under the condition of focusing at infinite range. Figures 3.6(b) and 3.6(c) give the corresponding results at $F/L = 1.5$ and $F/L = 2$, respectively. It can be noted that the near-field sidelobe envelope behaves much like the far-field sidelobe envelope, except in the vicinity of the main beam and at endfire. Also, notice that the near-field nulls and far-field nulls are not aligned. As the near-field range increases from L to $2L$, it is observed that the near-field pattern behaves more like the far-field pattern.

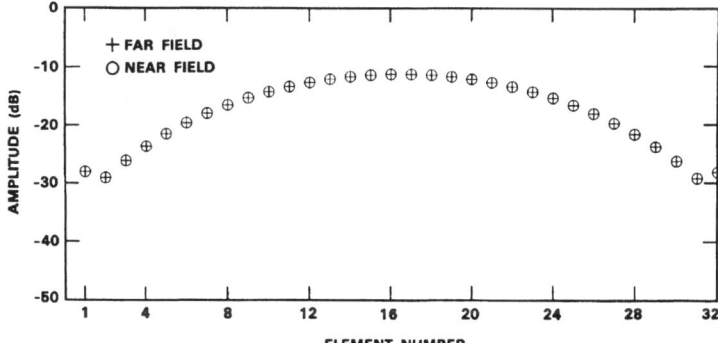

Figure 3.5 Amplitude of near-field and far-field weights, for a 32-element linear array, before nulling. Illumination is for -40 dB Chebyshev array factor. Array element spacing is $\lambda/2$. © 1990 IEEE [8].

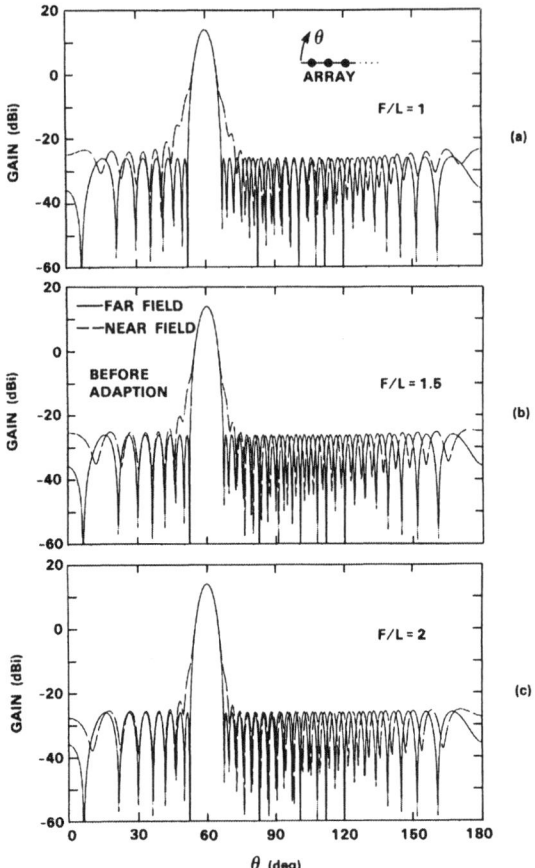

Figure 3.6 Focused 32-element linear array near-field/far-field radiation patterns before nulling. Scan angle is $\theta = 60°$ with -40 dB Chebyshev illumination. Array element spacing is $\lambda/2$. Dashed curve is for near-field focusing with near-field observation. Solid curve is for far-field ($r = \infty$) focusing with far-field observation. (a) $F/L = 1$, (b) $F/L = 1.5$, and (c) $F/L = 2$. © 1990 IEEE [8].

3.4.2 FULLY ADAPTIVE ARRAY BEHAVIOR

In this section the adaptive nulling characteristics of a fully adaptive 32-element linear array of isotropic elements with one-half wavelength spacing are investigated. The covariance matrix size is 32×32 in this case, and so there are 32 eigenvalues or degrees of freedom. The array quiescent conditions are the same as those described in the previous section. The quiescent radiation patterns were shown in Figure 3.6. Near-field ranges of $F/L = 1$,

1.5, and 2 are examined. In all near-field examples, the interference source range and focal range are equal. For the example array size, the actual near-field test distances are 3.58m ($F/L = 1$), 5.36m ($F/L = 1.5$), and 7.15m ($F/L = 2$). The far-field test distance is assumed to be at range $F/L = \infty$.

3.4.2.1 Single Interferer

Consider the case of one interference source ($J=1$). Let an interferer, with power 50 dB above receiver noise at the array output prior to adaption, be located at $\theta = 33°$ both for finite range. Note that the interferer lies on the fifth sidelobe to the left of the main beam. Initially, two nulling bandwidths will be considered: $B = 1$ MHz (narrowband, $BL\cos\theta/c = 0.01$) and $B = 100$ MHz (wideband, $BL\cos\theta/c = 1.0$).

For $B = 1$ MHz, the adaptive array radiation patterns, at center frequency 1.3 GHz, are shown in Figure 3.7. Figure 3.7(a) is for a near-field range of one aperture diameter ($F/L = 1$). Figures 3.7(b) and 3.7(c) are for near-field ranges of $1.5L$ and $2L$, respectively. For each case, the adaptive cancellation ratio was computed to be 50 dB. This is expected because ideal nulling weights are used, and there are 32 degrees of freedom available for canceling the interference. The consumption of adaptive array degrees of freedom is depicted in Figure 3.8. It is seen that 2 eigenvalues are significantly above the receiver noise level. The remaining eigenvalues are at the receiver noise level (0 dB); thus, only 2 degrees of freedom are engaged. The near-field eigenvalues are in good agreement with the corresponding far-field eigenvalues, indicating that the degrees of freedom are consumed the same. The amplitude of the adaptive array NF/FF weights (w_a) is given in Figure 3.9, and good agreement is evident. To show the sensitivity of a near-field null to range and angle, a two-dimensional contour radiation pattern is given in Figure 3.10. The pattern in Figure 3.10 has been computed for the $F/L = 1$ case.

Next, for $B = 100$ MHz, the adaptive array radiation patterns are shown in Figure 3.11. Again, for each near-field range considered, the adaptive array cancellation ratio was computed to be 50 dB. Since the bandwidth has increased, more degrees of freedom are consumed, but there are still ample degrees of freedom remaining. In Figure 3.12, it is seen that 6 eigenvalues are above the receiver noise level. Equivalently, 6 degrees of freedom are consumed – this is true both for the near-field and far-field degrees of freedom. The amplitude of the adaptive array NF/FF weights is depicted in Figure 3.13. Only minor differences between the NF and FF adaptive weights are observed.

To compare near-field and far-field consumption of the adaptive array degrees of freedom as a function of nulling bandwidth, the dominant

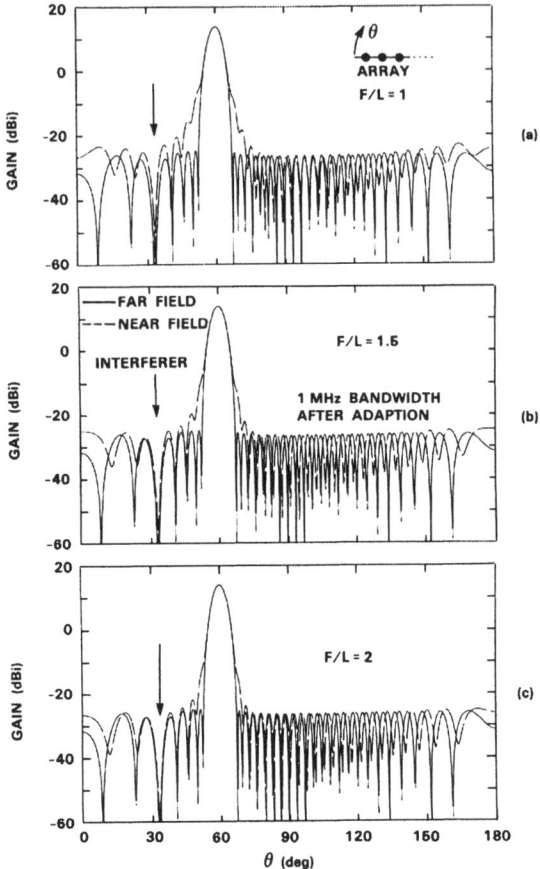

Figure 3.7 Focused 32-element fully adaptive linear array near-field/far-field radiation patterns after adaption. Nulling bandwidth is 1 MHz (narrowband case). Interferer is located at $\theta = 33°$. Dashed curve is for near-field focusing/observation/interference. Solid curve is for far-field ($r = \infty$) focusing/observation/interference. (a) $F/L = 1$, (b) $F/L = 1.5$, and (c) $F/L = 2$. © 1990 IEEE [8].

eigenvalues of the interference covariance matrix are presented in Figure 3.14 in the range ($1 \leq F/L \leq 2$). Figure 3.14(a) shows near-field and far-field eigenvalues ($\lambda_1, \lambda_2, \cdots, \lambda_7$) for $F/L = 1$. Figures 3.14(b) and 3.14(c) show the corresponding NF and FF eigenvalues for $F/L = 1.5$ and $F/L = 2$, respectively. Notice that each of the near-field and far-field eigenvalues are in good agreement, that is, $\lambda_1^{NF} \approx \lambda_1^{FF}, \lambda_2^{NF} \approx \lambda_2^{FF}, \cdots, \lambda_7^{NF} \approx \lambda_7^{FF}$. It is observed that the near-field eigenvalues are not particularly sensitive to range.

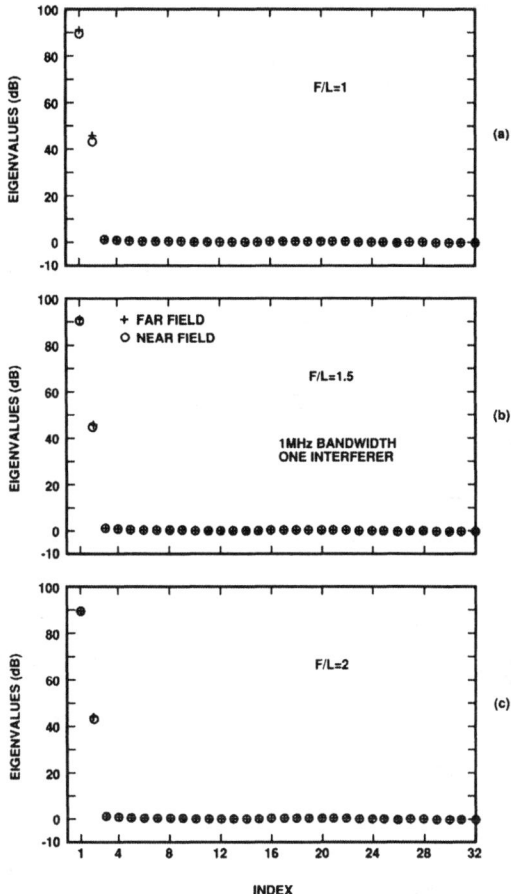

Figure 3.8 Covariance matrix eigenvalues for focused 32-element fully adaptive linear array. One near-field/far-field interferer is located at $\theta = 33°$. Nulling bandwidth is 1 MHz (narrowband case). (a) $F/L = 1$, (b) $F/L = 1.5$, and (c) $F/L = 2$. © 1990 IEEE [8].

3.4.2.2 Multiple Interferers

To demonstrate the validity of the NF/FF adaptive nulling equivalence for multiple sources, consider the previous array case ($N = 32$) and now where there is a large number of interferers, say $J = 31$. Let the interferers be uncorrelated and uniformly distributed across the field of view with 5-degree spacing covering $5° \leq \theta \leq 170°$, excluding the main beam region. The quiescent weights and quiescent radiation patterns were shown in Figures 3.5 and 3.6, respectively. The nulling bandwidth is assumed to be 1 MHz. The

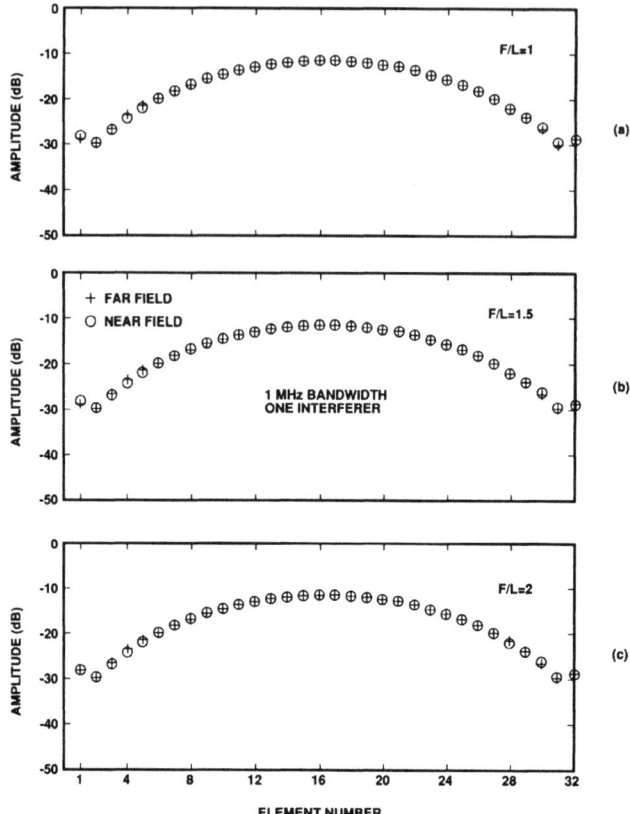

Figure 3.9 Amplitude of focused 32-element linear array weights after nulling. One near-field/far-field interferer is located at $\theta = 33°$. Nulling bandwidth is 1 MHz (narrowband case). (a) $F/L = 1$, (b) $F/L = 1.5$, and (c) $F/L = 2$. © 1990 IEEE [8].

total amount of interference power before adaption is adjusted to be 50 dB above noise at the array output.

The array NF/FF adaptive radiation patterns, at center frequency 1.3 GHz, are shown in Figure 3.15. It is observed that nulls are formed at the interferer positions for both NF and FF patterns. The NF/FF main beam and sidelobe characteristics are approximately equal. The cancellation ratio was computed to be (48.7 dB, 48.4 dB, 48.4 dB, and 48.4 dB) for ranges ($F/L = 1$, 1.5, 2, and ∞), respectively. Note: Complete cancellation, that is, $C = 50$ dB, has not been achieved here because the interference is stressing the adaptive array degrees of freedom. This is observed in Figure 3.16, which shows that a majority of the covariance matrix eigenvalues are significantly above receiver

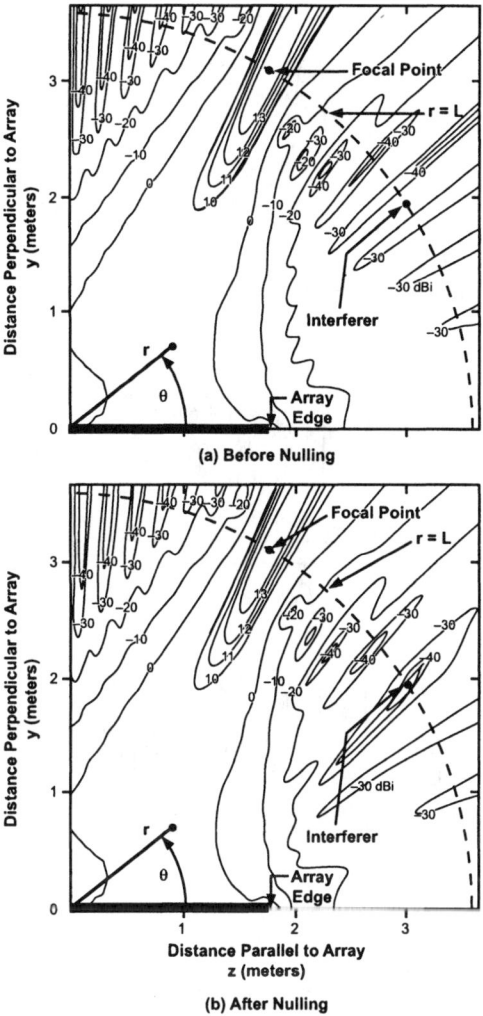

Figure 3.10 Two-dimensional near-field gain distribution, in the vicinity of a near-field interferer, for the focused 32-element linear array. The single interferer is at range $F/L = 1$ ($r_i = 3.58$m) and angle $\theta_i = 33°$. Nulling bandwidth is 1 MHz (narrowband case. (a) Before adaptive nulling and (b) after adaptive nulling. © 1990 IEEE [8].

noise. Note that there is good agreement between the NF and FF eigenvalues. Finally, the adaptive array weights are shown in Figure 3.17. Although there is some shape difference between NF and FF weights, their dynamic ranges are similar. It is observed that the dynamic range covered by the weights is increasing with decreasing range.

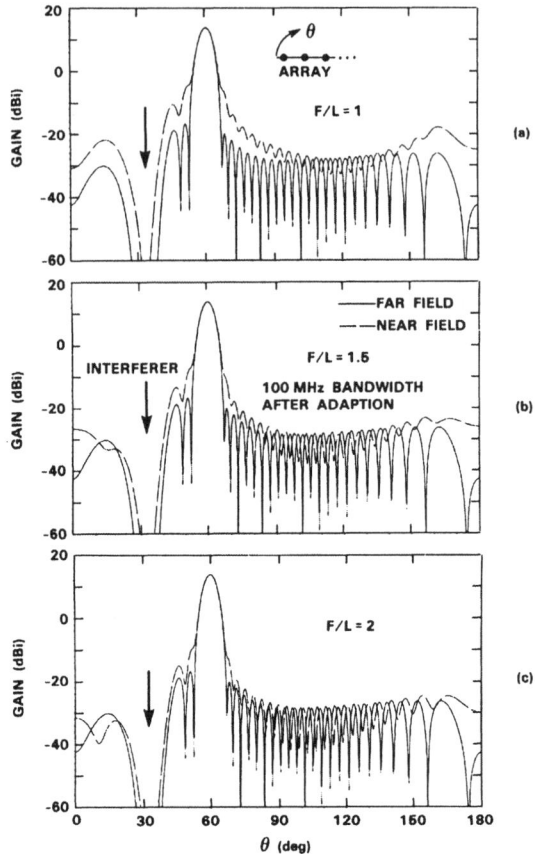

Figure 3.11 Focused 32-element fully adaptive linear array near-field/far-field radiation patterns after adaption. Nulling bandwidth is 100 MHz (wideband case). Interferer is located at $\theta = 33°$. (a) $F/L = 1$, (b) $F/L = 1.5$, and (c) $F/L = 2$. © 1990 IEEE [8].

From these results, it is concluded that 31 near-field (or point source) interferers, arranged equivalently in terms of angle, are equivalent to 31 far-field (or plane wave) interferers. A generalization of this statement would be that J near-field interferers are equivalent to J far-field interferers.

3.4.2.3 Effect of Focal Distance: Sidelobe Canceller Example

The previous sections have shown an equivalence between phase-focused near-field adaptive nulling and conventional far-field adaptive nulling. In theory, it is possible to focus or refocus a phased array at any range distance from the near field to the far field. However, in practice a phased array antenna

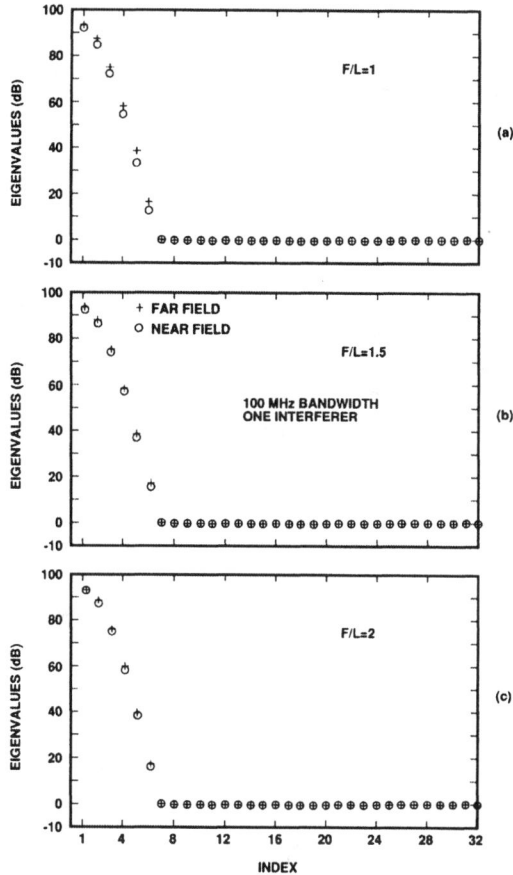

Figure 3.12 Covariance matrix eigenvalues for focused 32-element fully adaptive linear array. One near-field/far-field interferer is located at $\theta = 33°$. Nulling bandwidth is 100 MHz (wideband case). (a) $F/L = 1$, (b) $F/L = 1.5$, and (c) $F/L = 2$. © 1990 IEEE [8].

system may be limited to focusing in the far field, for example, due to the way in which the phased array has been calibrated. A practical phased array system would be expected to operate only under far-field focusing conditions, and the beam steering controller system might not readily allow the array to be refocused into the near-field region. Even if the adaptive antenna cannot be focused in the near field, it is desirable to determine whether some limited near-field adaptive nulling tests could be conducted in an anechoic chamber to verify adaptive array system performance in the presence of jamming. This section briefly investigates the case where the interference source is in the near field and the array is phase focused either at a range distance of infinity or it

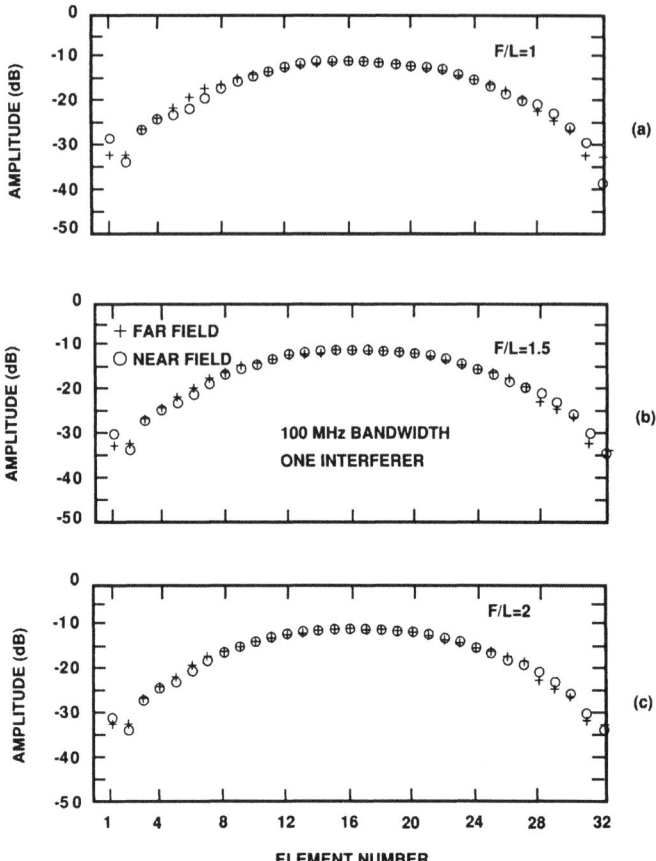

Figure 3.13 Amplitude of focused 32-element linear array weights after nulling. One near-field/far-field interferer is located at $\theta = 33°$. Nulling bandwidth is 100 MHz (wideband case). (a) $F/L = 1$, (b) $F/L = 1.5$, and (c) $F/L = 2$. © 1990 IEEE [8].

is phase focused in the near field.

Consider Figure 3.18, which shows the quiescent near-field radiation directivity pattern for a sidelobe-canceller 32-element array at 1.3 GHz (the same aperture size (3.58m) as the fully adaptive array discussed in the previous section); however, now it is with the array focused and observed at infinity (solid curve) or focused at infinity and observed at a range distance of 3m in the near field (dashed curve). In other words, in Figure 3.18 the solid curve is the conventional far-field gain pattern with the array focused in the far field, and the dashed curve is the case where the array can only be focused at infinity and the test distance is 3m ($r/L = 0.84$). Notice that the

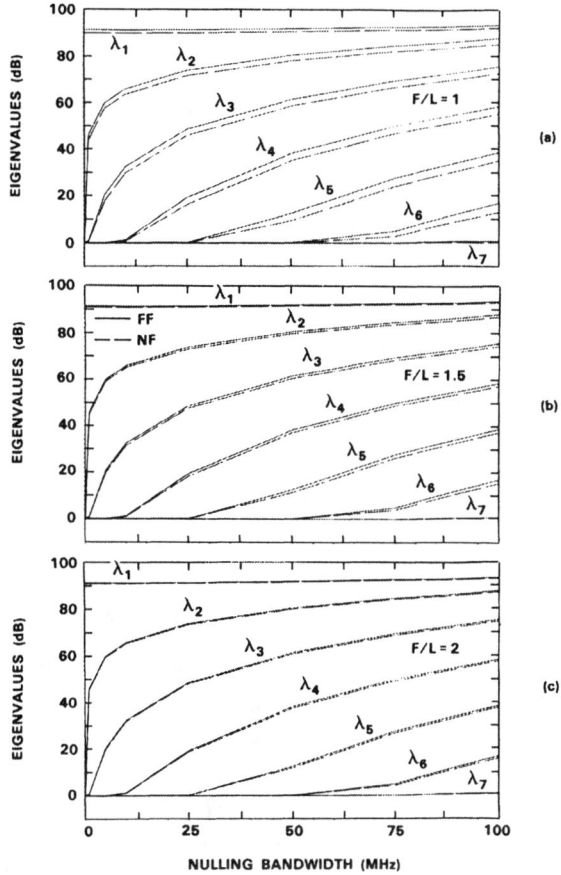

Figure 3.14 Dominant covariance matrix eigenvalues for focused 32-element fully adaptive linear array as a function of nulling bandwidth. One near-field/far-field interferer is located at $\theta = 33°$. (a) $F/L = 1$, (b) $F/L = 1.5$, and (c) $F/L = 2$. © 1990 IEEE [8].

near-field main beam response is substantially broader (due to defocusing) than the conventional far-field main beam response. Also, the near-field main beam peak gain is substantially lower than the far-field peak gain by about 7 dB. It is observed that the wide-angle near-field/far-field sidelobe levels, to the right of the main beam, are in good agreement. Suppose that a single interference source, in the far-field and near-field cases, is located in the sidelobe region at $\theta = 140°$ (50° from broadside). This sidelobe interference source has its power adjusted to produce a main channel output power 30 dB above receiver noise before nulling for both far-field and near-field cases. The nulling bandwidth for this example is assumed to be 1 MHz. The adapted

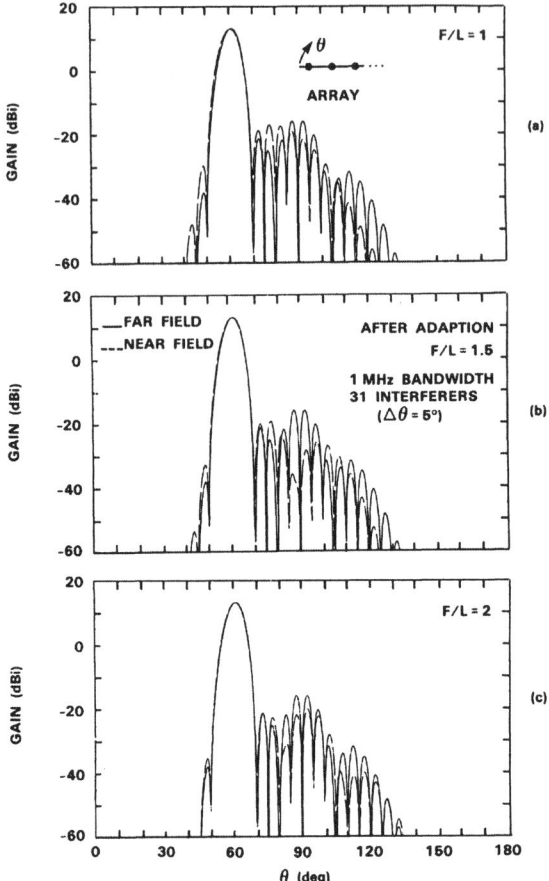

Figure 3.15 Focused 32-element fully adaptive linear array near-field/far-field radiation patterns after adaption. There are 31 interferers with uniform 5° spacing (excluding the main beam region). Nulling bandwidth is 1 MHz. (a) $F/L = 1$, (b) $F/L = 1.5$, and (c) $F/L = 2$. © 1990 IEEE [8].

radiation gain patterns are shown in Figure 3.19 – both patterns have an adaptive null formed at $\theta = 140°$ and the main beams for both cases are not affected by the adaption process. The near-field covariance matrix eigenvalues are computed to be (62.4 dB, 17.6 dB, 0.0 dB) and the far-field eigenvalues are (64.5 dB, 22.9 dB, 0.0 dB), which are very similar. Thus, the degrees of freedom that are being consumed by the two cases are very similar. For both the near-field and far-field cases, the computed adaptive output powers are both 0.0 dB, that is, the interference is suppressed equally for the two cases.

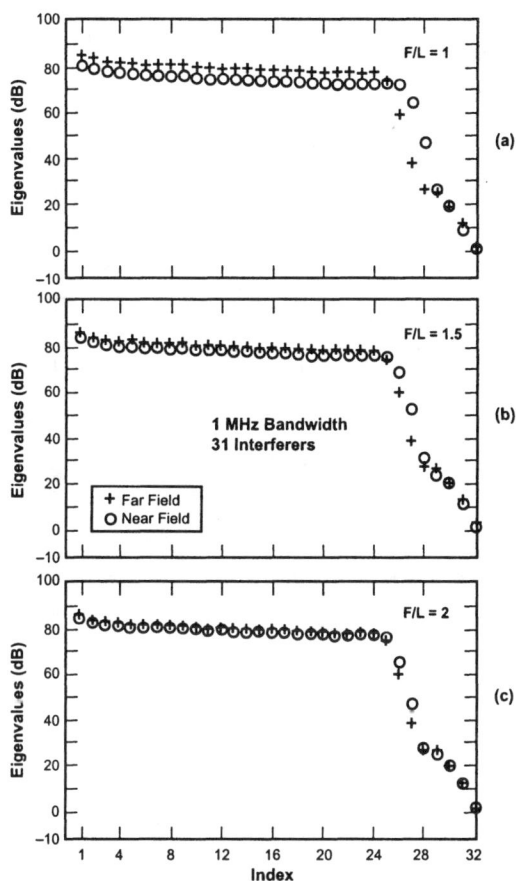

Figure 3.16 Covariance matrix eigenvalues for focused 32-element fully adaptive linear array. There are 31 NF/FF interferers with uniform $5°$ spacing (excluding the main beam region). Nulling bandwidth is 1 MHz. (a) $F/L = 1$, (b) $F/L = 1.5$, and (c) $F/L = 2$. © 1990 IEEE [8].

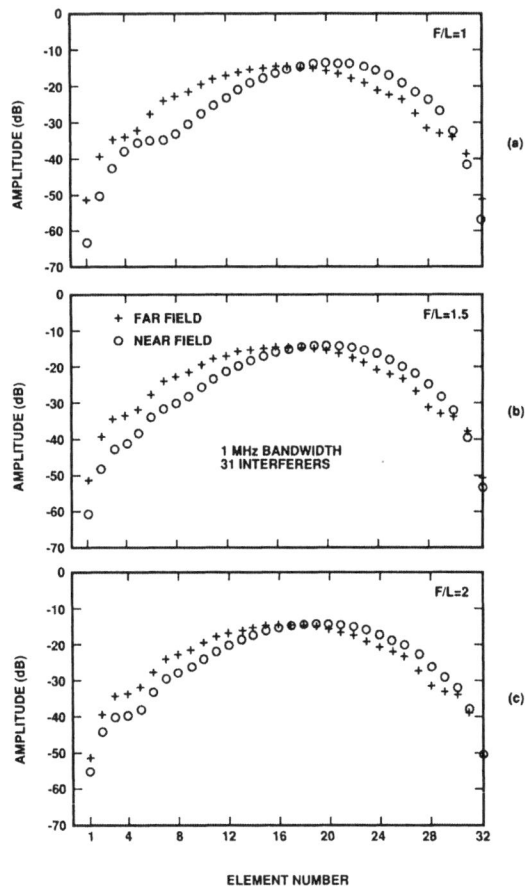

Figure 3.17 Amplitude of focused 32-element fully adaptive linear array weights after adaption. There are 31 NF/FF interferers with uniform $5°$ spacing (excluding the main beam region). Nulling bandwidth is 1 MHz. (a) $F/L = 1$, (b) $F/L = 1.5$, and (c) $F/L = 2$. © 1990 IEEE [8].

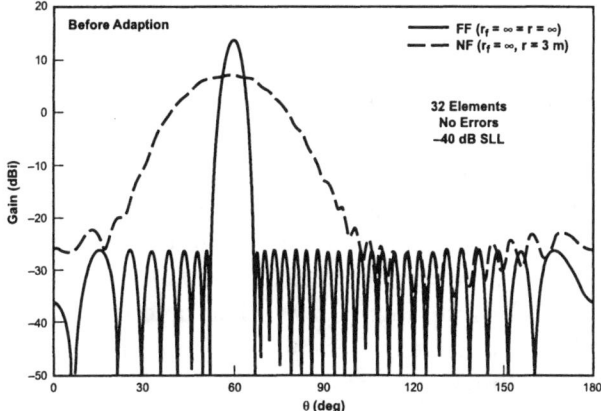

Figure 3.18 Quiescent radiation directivity patterns for a 32-element linear array ($L = 3.58$m) with one-half wavelength spacing at 1.3 GHz, with the array focused and observed at infinity (solid curve) and focused at infinity and observed at a range distance of 3m in the near field (dashed curve). The scan angle is $-30°$ from broadside with a -40 dB Chebyshev illumination for both the near-field and far-field simulations.

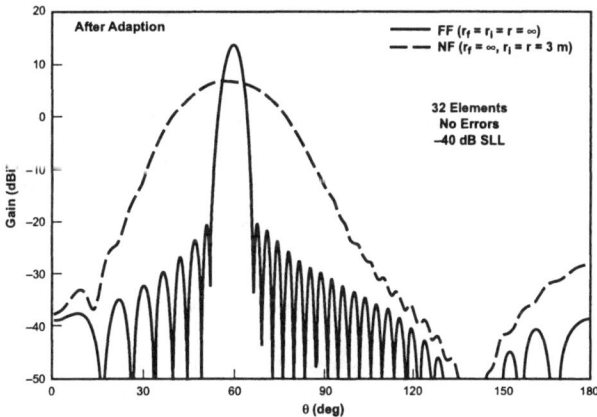

Figure 3.19 Adapted radiation directivity patterns for a 32-element linear array ($L = 3.58$m) with one-half wavelength spacing at 1.3 GHz, with the array focused and observed at infinity (solid curve) and focused at infinity and observed at a range distance of 3m in the near field (dashed curve). The scan angle is $-30°$ from broadside with a -40 dB Chebyshev illumination for both the near-field and far field simulations. A single interferer (sidelobe jammer) is incident at an angle of $50°$ from broadside and the nulling bandwidth is 1 MHz.

Next, let the near-field and far-field interfering source be located at $\theta = 40°$ (50° to the left of broadside), and assume that the nulling bandwidth is 1 MHz. In this case the far-field interferer is located at the angular position of a sidelobe of the far-field focused array pattern as can be observed by referring to the solid curve in Figure 3.18. However, the near-field interferer is in the near-field main beam of the far-field focused array pattern as observed from the dashed curve in Figure 3.18. The interferer source power is adjusted in both the near-field and far-field cases to produce 30 dB main channel output power relative to receiver noise before nulling. The adaptive radiation gain patterns are shown in Figure 3.20. While both the near-field and far-field patterns show that an adaptive null is formed at $\theta = 40°$, the pattern shapes differ very substantially. This difference occurs, obviously, because the main beam has been nulled in the near-field case, whereas in the far-field case a sidelobe was nulled. The near-field covariance matrix eigenvalues are computed to be (35.8 dB, 0.4 dB, 0.0 dB) and the far-field eigenvalues are (61.5 dB, 19.9 dB, 0.0 dB), which differ significantly. Therefore, the degrees of freedom are being consumed in a widely different manner for this example. Nevertheless, the interference is nulled in both cases and the adaptive output powers are both 0.0 dB.

Clearly, the previous examples (Figures 3.18 through 3.20) demonstrate that the quiescent radiation pattern can affect the way in which an array adapts to interference. This effect is shown quantitatively in terms of the radiation patterns and the covariance matrix eigenvalues. Focusing is an important factor in maintaining a strict equivalence between near-field and far-field adaptive nulling. In cases where an adaptive nulling antenna system cannot be refocused in the near field, there could be merit in evaluating the adaptive nulling system performance with near-field sources.

3.4.2.4 Effect of Module Errors: Sidelobe Canceller Example

Finally, consider a sidelobe canceller 32-element linear array with one-half wavelength spacing at 1.3 GHz and with −40 dB SLL in the absence of module illumination errors. Assume now that the actual array illumination is determined by 6-bit modules that have root mean square (rms) amplitude and phase errors of 0.3 dB and 2.0°, respectively. The nulling bandwidth is assumed to be 1 MHz and the end elements of the array (elements number 1 and 32 of this sidelobe cancelling array) are used as the auxiliary nulling channels. The interferer angle of arrival is chosen to be $\theta = 30°$ (−60° from broadside) and the interferer power is adjusted to 50 dB above noise at the array output. Near-field interference ranges of 3m, 4.5m, and 6m are considered here. The quiescent radiation patterns of this array in the near field and far field are shown in Figure 3.21. Although there are module amplitude and phase errors, the near-field and far-field patterns have similar characteristics. The covariance matrix eigenvalues are computed to be (for the range distances of (3m, 4.5m, 6m, infinity)) $\lambda_1 = $ (78.2 dB, 72.1 dB, 75.4 dB, 75.4 dB), $\lambda_2 = $ (33.5 dB, 29.5 dB, 33.6 dB, 34.8 dB), and $\lambda_3 = $ (0 dB, 0 dB, 0 dB, 0 dB). Thus, the adaptive array degrees of freedom are approximately the same for the near-field and far-field interference. Finally, the adaptive array directivity patterns are shown in Figure 3.22 and the single jammer is nulled similarly in all cases (adapted INR is (0.1 dB, 0.0 dB, 0.0 dB, 0.0 dB) for the range distances of (3m, 4.5m, 6m, infinity)). Therefore, even with module errors there is an equivalence between focused near-field nulling and conventional far-field nulling.

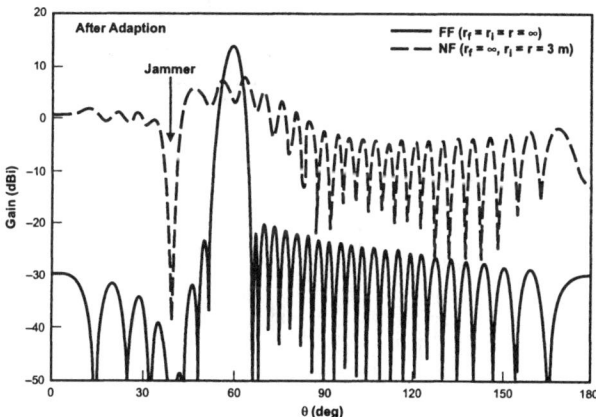

Figure 3.20 Adapted radiation directivity patterns for a 32-element linear array ($L = $ 3.58m) with one-half wavelength spacing at 1.3 GHz, with the array focused and observed at infinity (solid curve) and focused at infinity and observed at a range distance of 3m in the near-field (dashed curve). The scan angle is $-30°$ from broadside with a -40 dB Chebyshev illumination for both the near-field and far-field simulations. A single interferer (sidelobe jammer in the far field and main lobe jammer in the near field) is incident at an angle of 50° from broadside and the nulling bandwidth is 1 MHz.

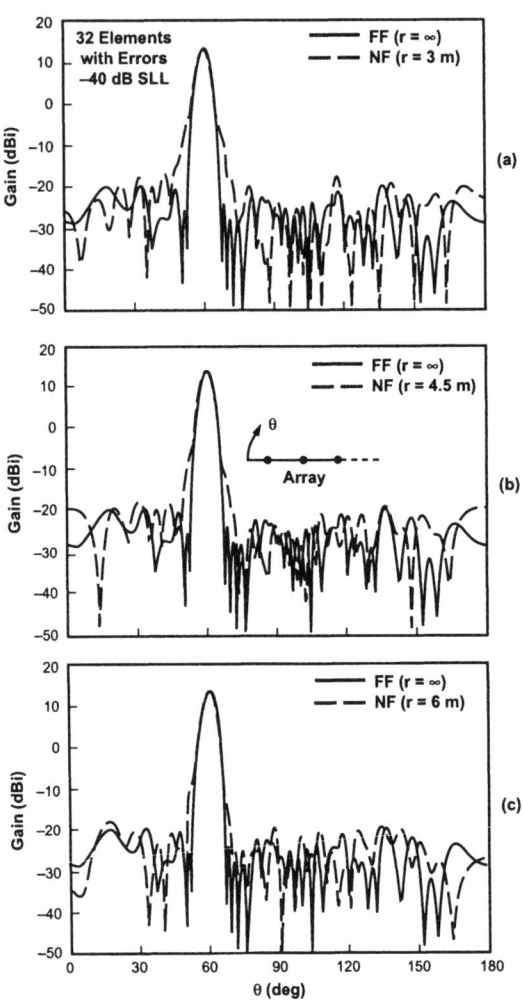

Figure 3.21 Quiescent radiation directivity patterns for a 32-element linear array antenna with one-half wavelength spacing at 1.3 GHz. The scan angle is $-30°$ from broadside with a -40 dB Chebyshev illumination for both the near-field and far-field simulations, and with module errors. (a) $r = 3$m, (b) $r = 4.5$m, and (c) $r = 6$m.

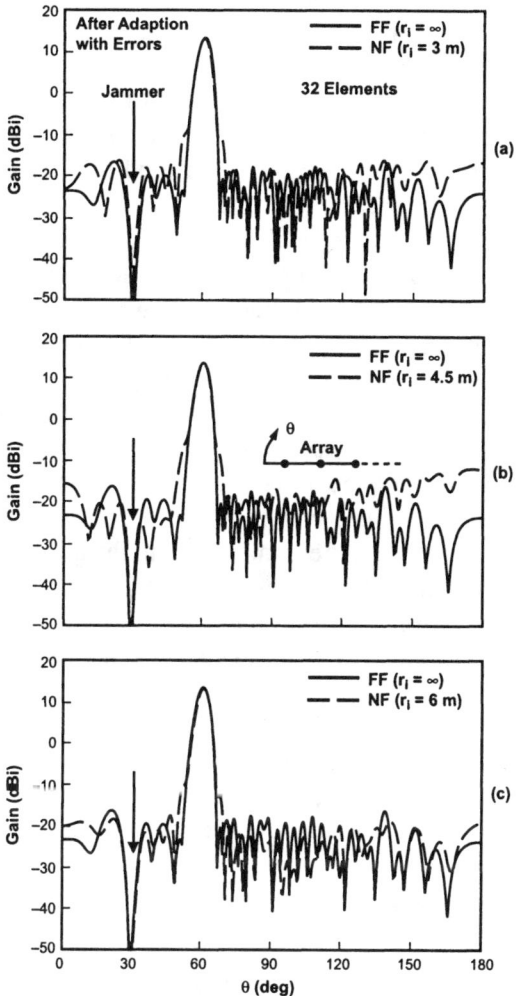

Figure 3.22 Adapted radiation directivity patterns for a 32-element linear array with one-half wavelength spacing at 1.3 GHz. The scan angle is $-30°$ from broadside with a -40 dB Chebyshev illumination for both the near-field and far-field simulations, and with module errors. A single interferer (sidelobe jammer in the far field and main lobe jammer in the near field) is incident at an angle of $60°$ to the left of broadside and the nulling bandwidth is 1 MHz. (a) $r = 3$ m, (b) $r = 4.5$ m, and (c) $r = 6$ m.

3.5 SUMMARY

This chapter has developed an approach to testing adaptive arrays in the near field. A theory for analyzing the behavior of an adaptive array operating in the presence of near-field interference has been developed. The adaptive antenna has been assumed to be a linear array of isotropic receive elements, and the near-field interference is assumed to radiate from an isotropic antenna (point source). Near-field focusing has been used to effectively establish a far-field pattern in the near zone. The near-field range of interest here has been taken to be one to two aperture diameters of the antenna under test. The theoretical section has addressed both sidelobe canceller and fully adaptive arrays. Equations for calculating the adaptive array covariance matrix and antenna radiation patterns, for near-field (spherical wave) and far-field (plane wave) interference were given. Numerical simulations of a fully adaptive linear array and a sidelobe canceller array indicate that the radiation patterns, adapted weights, cancellation, and covariance matrix eigenvalues (degrees of freedom) are effectively the same for near-field and far-field interference. Both single and multiple interference conditions have been analyzed. It has been shown, by example, that J near-field interferers in the presence of a near-field focused adaptive array are equivalent to J far-field interferers in the presence of a far-field focused adaptive array. That is, a one-to-one correspondence can be made between near-field interferers and far-field interferers. The results are expected to be applicable to large planar arrays. Thus, a phased array antenna adaptive nulling system designed for far-field conditions can potentially be evaluated more conveniently using near-field interference sources. In practice, the interference source antennas can be implemented using dipoles or horns. Further exploration of this technique is described in Chapters 4 through 6.

3.6 PROBLEM SET

3.1 Derive (3.18) by substituting (3.17) in (3.7) and integrating.
3.2 Derive (3.21).
3.3 Derive (3.22).
3.4 Verify that (3.35) follows from (3.34) for the case where the array element spacing is $\lambda/2$.
3.5 For a 32-element array of isotropic point receiving antenna elements with $\lambda/2$ spacing, compute the focused near-field and conventional far-field radiation patterns before adaption corresponding to the case shown in Figure 3.6, and after adaption as in Figure 3.7.
3.6 Using (3.18) and (3.40), compute the interference covariance matrix and its eigenvalues corresponding to the example shown in Figure 3.8.

References

[1] Yaghjian, A.D., "An Overview of Near-Field Antenna Measurements," *IEEE Trans. Antennas Propagat.*, Vol. 34, No. 1, 1986, pp. 30-45.

[2] Johnson, R.C., H.A. Ecker, R.A. Moore, "Compact Range Techniques and Measurements," *IEEE Trans. Antennas Propagat.*, Vol. 17, No. 5, 1969, pp. 568-576.

[3] Mayhan, J.T., "Some Techniques for Evaluating the Bandwidth Characteristics of Adaptive Nulling Systems," *IEEE Trans. Antennas Propagat.*, Vol. 27, No. 3, 1979, pp. 363-373.

[4] Fenn, A.J., "Maximizing Jammer Effectiveness for Evaluating the Performance of Adaptive Nulling Array Antennas," *IEEE Trans. Antennas Propagat.*, Vol. 33, No. 10, 1985, pp. 1131-1142.

[5] Hudson, J.E., *Adaptive Array Principles*, New York: Peter Peregrinus LTD, 1981, pp. 24-26.

[6] Fenn, A.J., "Theory and Analysis of Near Field Adaptive Nulling," *1986 IEEE AP-S Symposium Digest, Vol. 2*, IEEE, New York, 1986, pp. 579-582.

[7] Fenn, A.J., "Theory and Analysis of Near Field Adaptive Nulling," *1986 Asilomar Conf. on Signals, Systems and Computers*, Computer Society Press of the IEEE, Washington, D.C., 1986, pp. 105-109.

[8] Fenn, A.J., "Evaluation of Adaptive Phased Array Far-Field Nulling Performance in the Near-Field Region," *IEEE Trans. Antennas Propagat.* Vol. 38, No. 2, 1990, pp. 173-185.

[9] Fenn, A.J., "Theoretical Near Field Clutter and Interference Cancellation for an Adaptive Phased Array Antenna," *1987 IEEE AP-S Symposium Digest, Vol. 1*, IEEE, New York, 1987, pp. 46-49.

[10] Fenn, A.J., "Analysis of Phase-Focused Near-Field Testing for Multiphase-Center Adaptive Radar Systems," *IEEE Trans. Antennas Propagat.*, Vol. 40, No. 8, 1992, pp. 878-887.

[11] Monzingo, R.A., and T.W. Miller, *Introduction to Adaptive Arrays*, New York: Wiley, 1980.

[12] Strang, G., *Linear Algebra and Its Applications*, New York: Academic, 1976, p. 213.

[13] Scharfman, W.E., and G. August, "Pattern Measurements of Phased-Arrayed Antennas by Focusing into the Near Zone," in *Phased Array Antennas (Proc. of the 1970 Phased Array Antenna Symposium)*, A.A. Oliner and G.H. Knittel, (eds.), Dedham, MA: Artech House, 1972, pp. 344-350.

[14] Walter, C.H., *Traveling Wave Antennas*, New York: Dover, 1970, pp. 36-50.

4

Moment Method Analysis of Focused Near-Field Adaptive Nulling

4.1 INTRODUCTION

As described in Chapter 3, phased array antennas having adaptive nulling capability are often desirable for radar or communications applications. The adaptive nulling performance of these antennas is principally tested using conventional far-field antenna ranges with far-field (or plane wave) interferers. For electrically large aperture antennas at microwave frequencies, significant far-field range distances can be involved which require that testing be made outdoors. Multiple, widely separated interferers make the far-field range design more difficult. Near-field testing, suitable for indoor measurements, is desirable (as has been demonstrated in the case of near-field scanning as discussed by Yaghjian [1]) for far-field antenna radiation pattern measurements and compact range reflector techniques as discussed by Johnson [2] for far-field radiation pattern and radar cross-section measurements. In both the near-field scanning and compact range methods, the goal is to evaluate the plane wave response of a test article.

Based on the results for arrays of isotropic elements presented in Chapter 3, if the requirement for plane wave test conditions is removed and spherical wave incidence is allowed (as will be shown theoretically) for a focused phased array adaptive antenna, near-field testing with interference sources at *one- to two-aperture-diameter range* is possible. This chapter further develops the focused near-field testing approach by including array mutual coupling effects.

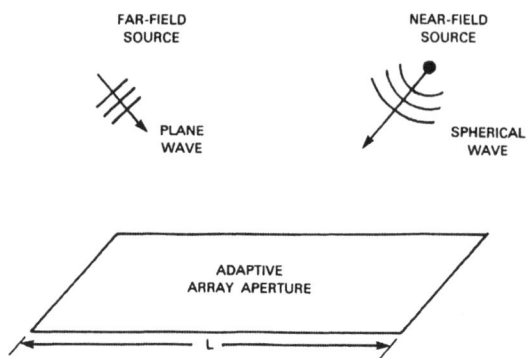

Figure 4.1 Contrast between plane wave incidence (far-field source) and spherical wave incidence (near-field source) for an adaptive array aperture. From [10] with permission, IEE.

The contrast between plane wave and spherical wave incidence is depicted in Figure 4.1. The amount of wavefront dispersion observed by the array is a function of the nulling bandwidth, array length L, and angle of incidence. Interference wavefront dispersion is an effect that can limit the depth of null (or cancellation) achieved by an adaptive antenna, as discussed by Mayhan [3]. An accurate dispersion model essentially quantifies the characteristics of the interference covariance matrix – namely its eigenvalues or degrees of freedom [3, 4] as discussed in Chapter 1. The covariance matrix contains all wavefront dispersion presented to the adaptive channels. The interference covariance matrix eigenvalues can be used to quantify and to compare the dispersion present for plane wave and spherical wave incidence. For near-field (NF) adaptive nulling to be equivalent to far-field (FF) adaptive nulling, it is assumed that the NF/FF interference covariance matrix eigenvalues must be equivalent. Additionally, it is assumed that the NF/FF adaptive array weights, cancellation of interference power, and radiation patterns must also be equivalent.

As discussed in Chapter 3 investigations of NF/FF adaptive nulling for focused linear arrays of isotropic receive elements have been performed [5-9]. Near-field adaptive nulling results for a single interferer at a range of $1.7L$ for sidelobe canceller [5] and fully adaptive [6, 7] arrays have been presented. The effects of nulling bandwidth were taken into account. Comparisons with far-field adaptive nulling indicated an excellent NF/FF equivalence. A detailed analysis of near-field nulling for a fully adaptive array with single and multiple interferers, in the range of one to two aperture diameters, has been made [7]. Application of this near-field technique to testing main beam clutter cancellation has also been examined [8, 9]. Having shown the usefulness of focused near-field nulling under the conditions of isotropic receive antennas and isotropic interference, it is now appropriate to consider the effects of polarization and mutual coupling [10]. References [5-8] assumed that all sources are constrained to be located at a constant range with respect to the phase center of the antenna under test. The present chapter seeks to further develop an adaptive phased array testing methodology [9] that has compatibility with planar near-field scanner hardware. Thus, all sources of interference are assumed to be located on a test plane.

A theory for analyzing and comparing both near-field and far-field adaptive nulling, including mutual coupling effects, is presented in Section 4.2. Near-field focusing and the near-field nulling concept are described first. General adaptive nulling concepts are then addressed. A method-of-moments formulation for the voltages received by the array elements is then given. Computation of the interference covariance matrix and array radiation patterns is described. Section 4.3 presents results that show that an adaptive array responds in the same manner to near-field sources as it does to far-field sources. A summary is given in Section 4.4.

4.2 THEORY

4.2.1 FOCUSED NEAR-FIELD NULLING CONCEPT

In the near-field nulling technique described here, it is assumed that the quiescent near-field radiation pattern of the array should have the same characteristics as the quiescent far-field radiation pattern of the array. This means typically that a main beam and sidelobes should be formed. To produce an array near-field pattern that is approximately equal to that of the far field, phase focusing can be used [11]. Consider Figures 4.2 and 4.3, which show a CW calibration source located at a desired focal point of the array. The array can maximize the signal received from the calibration source by adjusting its phase shifters such that the spherical wavefront phase variation is removed. One way to do this is to choose a reference path length as the distance from the focal point to the center of the array. This distance is denoted r_F, and the distance from the focal point to the nth array element is denoted r_n^F. The voltage received at the nth array element relative to its center element is computed here using the method of moments. To maximize the received voltage at the array output, it is necessary to apply weighting equal to the phase conjugate of the incident wavefront at the array elements, as depicted conceptually in Figure 4.2. The resulting radiation pattern on the test plane $z = z_F$ looks similar to a far-field pattern. A main beam will be pointed at the array focal point. Sidelobes will exist at angles away from the main beam.

Interferers can then be placed on near-field sidelobes in the test (or focal) plane, as depicted in Figure 4.3.

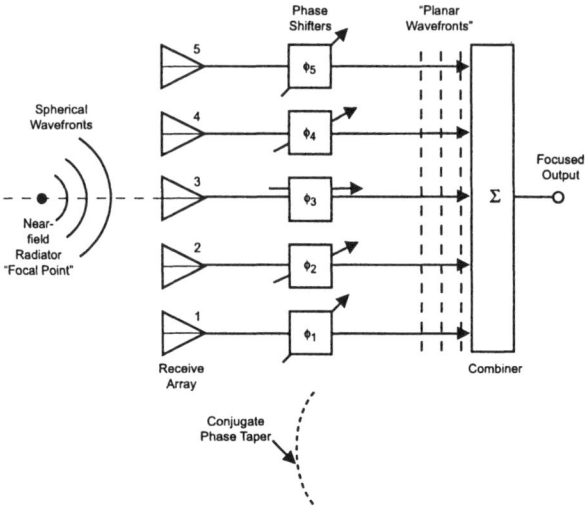

Figure 4.2 Conceptual diagram depicting a receive array and phase shifters that apply a conjugate phase taper to convert a spherical wavefront generated by a near-field radiator into a plane wavefront that produces a focused output.

4.2.2 ADAPTIVE ARRAY CONCEPTS

Consider the array and interferer geometry as shown in Figure 4.3. In general, the array will contain a total of N elements, with only N_r elements used to form the receive main channel. This is often the case with arrays having a guard band of passively terminated elements, which is used to provide impedance matching to the active elements and/or isolation from ground plane edges. The output from each of the N_r array elements is summed in the power combiner to form the main channel. Let a wavefront (either planar or spherical) due to the ith interference source be impressed across the array, which results in a set of array element received voltages denoted $v_1^i, v_2^i, \cdots, v_N^i$. The number of adaptive channels is denoted M. For a sidelobe canceller $M = 1 + N_{\text{aux}}$ where N_{aux} is the number of auxiliary channels. The main and auxiliary channel voltages are selected from among this set of array received voltages. In this chapter, ideal weights are assumed with $\boldsymbol{w} = (w_1, w_2, \cdots, w_M)^T$ denoting the adaptive channel weight vector and $\boldsymbol{W} = (W_1, W_2, \cdots, W_N)^T$ denoting the sidelobe canceller array element

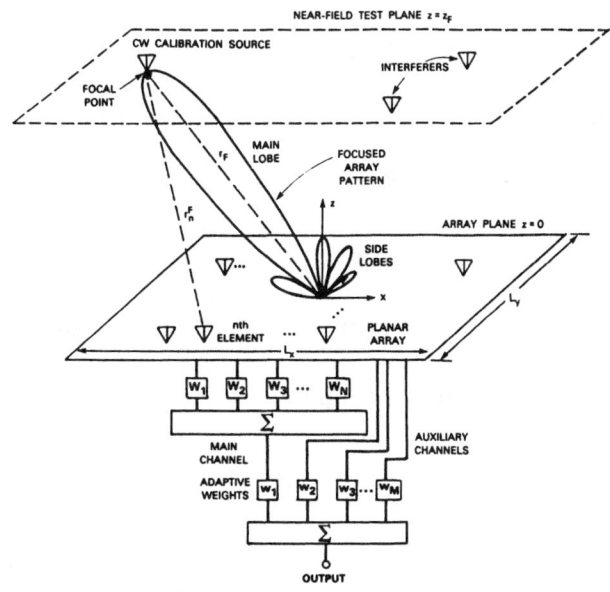

Figure 4.3 Adaptive phased array antenna near-field focusing concept. From [10] with permission, IEE.

weight vector, as shown in Figure 4.3. (Superscript T means transpose.) The fundamental quantities required to fully characterize the incident field for adaptive nulling purposes are the adaptive channel cross correlations.

The cross correlation R^i_{mn} of the received voltages in the mth and nth adaptive channels, due to the ith source, is given by

$$R^i_{mn} = E(v_m v_n^*) \qquad (4.1)$$

where * means complex conjugate and $E(\cdot)$ means mathematical expectation. Since v_m and v_n represent voltages of the same waveform, but at different times, R^i_{mn} is also referred to as an autocorrelation function. Note: For convenience, in (4.1) the superscript i in v_m and in v_n has been omitted.

In the frequency domain, assuming the interference has a band-limited white noise power spectral density, (4.1) can be expressed as the frequency average

$$R^i_{mn} = \frac{1}{B} \int_{f_1}^{f_2} v_m(f) v_n^*(f) \, df \qquad (4.2)$$

where $B = f_2 - f_1$ is the nulling bandwidth and f_c is the center frequency. It should be noted that $v_m(f)$ takes into account the wavefront shape which can be spherical or planar.

Let the channel or interference covariance matrix be denoted R. If there are J uncorrelated broadband interference sources, then the J-source covariance matrix is the sum of the covariance matrices for the individual sources, that is,

$$R = \sum_{i=1}^{J} R_i + I \qquad (4.3)$$

where R_i is the covariance matrix of the ith source and I is the identity matrix used to represent the thermal noise level of the receiver.

In this chapter, prior to generating an adaptive null, the adaptive channel weight vector, w, is chosen to synthesize a desired quiescent radiation pattern. When interference is present, the optimum set of weights, denoted w_a, to form an adaptive null is computed by [12]

$$w_a = R^{-1} w_q \qquad (4.4)$$

where $^{-1}$ means inverse and w_q is the quiescent weight vector (steering vector). For a sidelobe canceller, the quiescent weight vector is chosen to be $w_q = (1, 0, 0, \cdots, 0)^T$, that is, the main channel weight is unity and the auxiliary channel weights are zero.

The output power at the adaptive array summing junction is given by

$$p = w^\dagger R w \qquad (4.5)$$

where $\dagger$ means complex conjugate transpose. The interference-to-noise ratio, denoted INR, is computed as the ratio of the output power [defined in (4.5)] with the interferer present to the output power with only receiver noise present, that is,

$$\text{INR} = \frac{w^\dagger R w}{w^\dagger w} \qquad (4.6)$$

The adaptive array cancellation ratio, denoted C, is defined here as the ratio of interference output power after adaption to the interference output power before adaption, that is,

$$C = \frac{p_q}{p_a} \qquad (4.7)$$

Substituting (4.5) in (4.7) yields

$$C = \frac{w_q^\dagger R w_q}{w_a^\dagger R w_a} \qquad (4.8)$$

Next, the covariance matrix defined by the elements in (4.2) is Hermitian (that is, $R = R^\dagger$) which, by the spectral theorem, can be decomposed in eigenspace as [13]

$$R = \sum_{k=1}^{M} \lambda_k e_k e_k^\dagger \qquad (4.9)$$

where $\lambda_k, k = 1, 2, \cdots, M$ are the eigenvalues of R, and $e_k, k = 1, 2, \cdots, M$ are the associated eigenvectors of R. The interference covariance matrix eigenvalues $(\lambda_1, \lambda_2, \cdots, \lambda_M)$ are a convenient quantitative measure of the utilization of the adaptive array degrees of freedom as discussed in Chapter 1.

4.2.3 MOMENT METHOD FORMULATION

This section considers using the method of moments [14] to compute the array element received voltages in (4.2) due to near- or far-field sources. The array element received voltages are used as input signals to the adaptive nulling process. The far-field formulation given here is analogous to that which has been developed by Gupta [15]. Referring to Figure 4.4, assume that each element is terminated in the load impedance Z_L which is known. Let $v_{n,j}^{o.c.}$ represent the open-circuit voltage in the nth array element due to the jth source. Here, the jth source can denote either the CW calibrator or one of the broadband noise interferers. Next, let $Z^{o.c.}$ be the open-circuit mutual impedance matrix for the N-element array.

Consider Figure 4.4, which depicts the circuit model for a receive array and a source antenna. Let $v_{n,j}^{\text{rec}}$ be the voltage received in the nth array element due to the jth source. The array elements are assumed to be terminated in a load impedance denoted Z_L, which in general is complex. The open-circuit mutual impedance between the mth and nth array elements is denoted by $Z_{m,n}^{o.c.}$. Similarly, the open-circuit mutual impedance between the nth array element and the jth source is denoted $Z_{n,j}^{o.c.}$. Now, $i_{1,j}, i_{2,j}, \cdots, i_{n,j}, \cdots, i_{N,j}$ are the received terminal currents for the N array elements. The received voltages are related to the terminal currents and load impedances using

$$v_{n,j}^{\text{rec}} = -i_{n,j}^{\text{rec}} Z_L, \quad n = 1, 2, \cdots, N \qquad (4.10)$$

Let i_j be the terminal current of the jth source. The received voltages can be written as

$$v_{n,j}^{\text{rec}} = i_{1,j}^{\text{rec}} Z_{n,1}^{o.c} + \cdots + i_{n,j}^{\text{rec}} Z_{n,n}^{o.c.} + \cdots + i_{N,j}^{\text{rec}} Z_{n,N}^{o.c.} + i_j Z_{n,j}^{o.c.} \qquad (4.11)$$

In (4.11), the term $i_j Z_{n,j}^{o.c.}$ is the open-circuit voltage at the nth array element. Note: The index j for the jth source should not be confused with the indices used for the array elements.

Now, define

$$v_{n,j}^{o.c.} = i_j Z_{n,j}^{o.c.} \qquad (4.12)$$

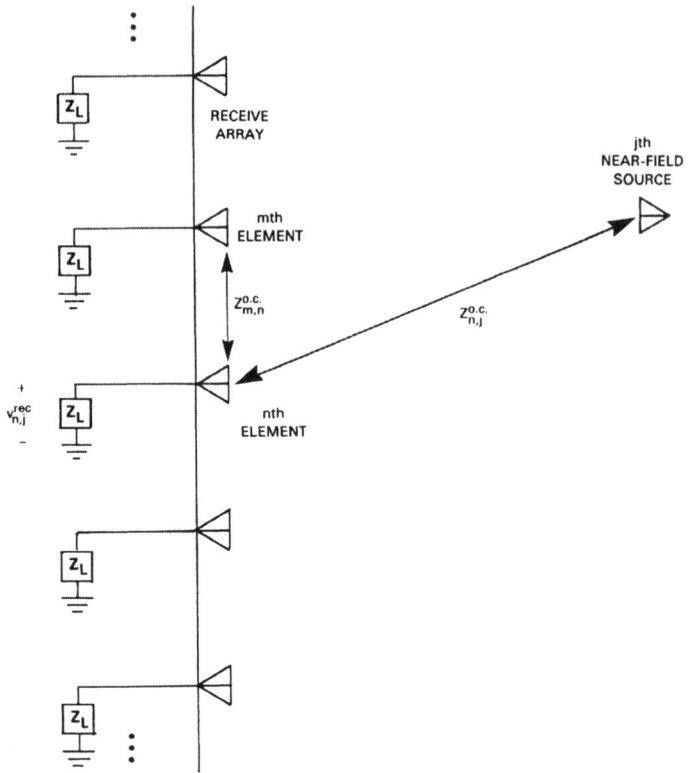

Figure 4.4 Circuit model for receive array with near-field source. From [10] with permission, IEE.

and using (4.10) and (4.11) in (4.12), rearrange the terms to yield

$$-v_{n,j}^{o.c.} = i_{1,j}^{rec} Z_{n,1}^{o.c.} + \cdots + i_{n,j}^{rec}(Z_{n,n}^{o.c.} + Z_L) + \cdots + i_{N,j}^{rec} Z_{n,N}^{o.c.} \quad (4.13)$$

where $n = 1, 2, \ldots, N$. Equation (4.13) can be written in a more compact form as

$$-v_j^{o.c.} = [\mathbf{Z}^{o.c.} + Z_L \mathbf{I}] \, i_j^{rec} \quad (4.14)$$

where $v_j^{o.c.}$ is the open-circuit voltage matrix, $\mathbf{Z}^{o.c.}$ is the open-circuit mutual impedance matrix, $\mathbf{I}$ denotes the identity matrix, and i_j^{rec} is the received terminal current matrix. From (4.10) it is clear that

$$i_j^{rec} = -\frac{v_j^{rec}}{Z_L} \quad (4.15)$$

Substituting (4.14) in (4.15) and solving for v_j^{rec} yields

$$v_j^{rec} = Z_L \left[\mathbf{Z}^{o.c.} + Z_L \mathbf{I} \right]^{-1} v_j^{o.c.} \quad (4.16)$$

which is the desired result.

The moment method expansion and testing functions are assumed to be piecewise sinusoidal, which is appropriate to thin cylindrical-wire monopole/dipole antennas. The earlier open-circuit mutual impedances are computed based on subroutines from a well-known moment method computer code [16]. In evaluating $Z_{n,j}^{o.c.}$ for the jth interferer, double precision computations were required. For far-field sources, $v_{n,j}^{o.c.}$ is evaluated by assuming plane wave incidence.

As mentioned earlier, the array is calibrated (phased focused) initially using a CW radiating dipole. To accomplish this numerically, having computed v_{CW}^{rec}, the receive array weight vector w will have its phase commands set equal to the conjugate of the corresponding phases in v_{CW}^{rec}. Receive antenna radiation patterns are obtained by scanning (moving) a dipole with half-length l in either the far or near field and computing the antenna response. Far-field received patterns are computed using a $\hat{\theta}$-polarized dipole source at infinity, which generates plane wave illumination of the array. Principal plane near-field radiation pattern cuts (versus angle) are obtained by computing the near field on the line $(x, y = 0, z = z_F)$ and using the relation $\theta(x) = \tan^{-1}(x/z_F)$. The near-field source is a one-half wavelength dipole, which is $\hat{x}$-polarized. Let the voltage received by the array, due to the x-directed near-field dipole, be denoted $v_x^{NF}(\theta)$. Let $p_\theta(\theta)$ denote the $\hat{\theta}$ component of the dipole probe pattern. Then the probe-compensated array near-field received pattern is expressed as

$$E_\theta(\theta)^{NF} = \frac{v_x^{NF}(\theta)}{p_\theta(\theta)} \tag{4.17}$$

where

$$p_\theta(\theta) = \frac{\cos(kl\sin\theta) - \cos(kl)}{\cos\theta} \tag{4.18}$$

where $k = 2\pi/\lambda$ is the angular wavenumber. Equation (4.17) is correct provided that the radial component of the electric field is zero. For the test distances involved in this chapter, the radial component E_r is typically -20 dB below the E_θ component, according to theoretical calculations. Thus, for all practical purposes the equality in (4.17) is valid. Further discussion of the near-field radial component is deferred to Chapter 10. Let $\theta_{\max}$ denote the maximum angle of interest for the antenna radiation pattern. As depicted in Figure 4.5, the required near-field scan length for pattern coverage of $\pm\theta_{\max}$ is given in terms of the F/L ratio as

$$D_x = 2L\left(\frac{F}{L}\right)\tan\theta_{\max} \tag{4.19}$$

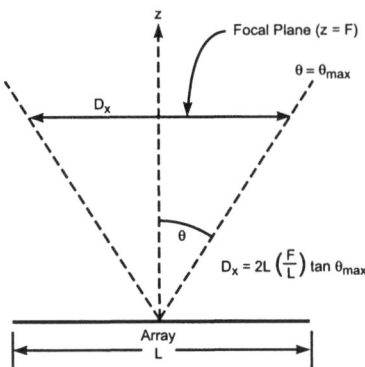

Figure 4.5 Near-field scan length as a function of F/L.

As an example, using F/L ratios of 1, 1.5, and 2, the required scan lengths (D_x) for 60° and 120° field-of-view coverage are $1.2L, 1.7L, 2.4L$, and $3.5L, 5.2L$, and $7.0L$, respectively. It is clear that to reduce the scan length (or source deployment length) it is desirable to keep the F/L ratio as small as possible. The array received voltage matrix for the jth interferer (denoted v_j^{rec}) is computed at M frequencies across the nulling bandwidth. Thus, the received voltages $v_j^{\text{rec}}(f_1), v_j^{\text{rec}}(f_2), \cdots, v_j^{\text{rec}}(f_M)$ are needed. In this analysis, the impedance matrix is computed at M frequencies and is inverted M times. The interference covariance matrix elements are computed by evaluating (4.2) numerically, using Simpson's rule of numerical integration. For multiple interferers, the covariance matrix is evaluated using (4.3). Adaptive array radiation patterns are computed by superimposing the quiescent radiation pattern with the weighted sum of auxiliary channel received voltages.

4.3 RESULTS

4.3.1 FOCUSED ARRAY QUIESCENT CONDITIONS

Consider a planar array of thin resonant monopoles having element length 0.275λ with wire radius 0.007λ and 0.473λ element spacing at center frequency 1.3 GHz. Such an array is useful for wide-angle scanning [17-19] as described in Chapter 9. In standard spherical coordinates the principal polarization of a monopole array (z-directed elements) is the E_θ component (or vertical polarization). The array is assumed to have 180 elements in a square lattice with 5 rows and 36 columns, and the array operates over an

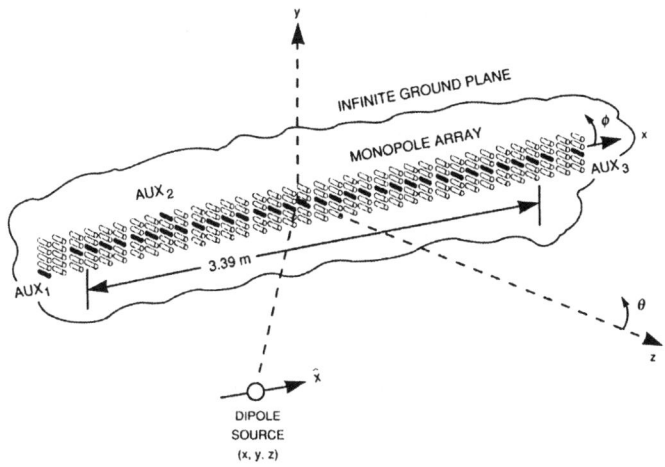

Figure 4.6 Planar array of monopoles with dipole source. From [10] with permission, IEE.

infinite ground plane as depicted in Figure 4.6. One current expansion function per array element is used, and so there are 180 unknowns in (4.16). Assume that 32 of the elements in the center row are used to form the main channel output for a sidelobe canceller, that is, $N_r = 32$. The surrounding elements are passively terminated (load impedance $Z_L = 50$ Ω resistive) to guard the center elements from finite array edge effects. (Note: Auxiliary elements for the sidelobe canceller auxiliary channels will be selected from this guard band.) The active receive array length for this case is 3.39m. Referring to Figure 4.3, let $z_F = F$ be the focal distance. Focal lengths of 1, 1.5, and 2 L will be examined. Referring to Figure 4.5, the near-field desired field of view is assumed to be 120°. A scan angle of 30° from broadside and a Chebyshev illumination which generates −40 dB uniform far-field sidelobes are assumed. A 12m near-field scan length was used at $F/L = 1$, which provided a ±60° field of view.

The monopole array near-field focused/near-field observation quiescent patterns are shown in Figure 4.7. Figure 4.7(a) shows the near-field result (dashed curve) obtained at $F/L = 1$. Included in this figure is the conventional far-field pattern (solid curve) observed at infinite range under the condition of focusing at infinite range. Figures 4.7(b) and (c) give the corresponding results at $F/L = 1.5$ and $F/L = 2$, respectively. The half-power beam width is observed to be 5.5°. For $F/L = 1$, considerable degradation of the near-field main beam shape occurs below the half-power points. It can be noted that the near-field sidelobe envelope behaves much like that of the far-field sidelobe envelope, except in the vicinity of the main beam.

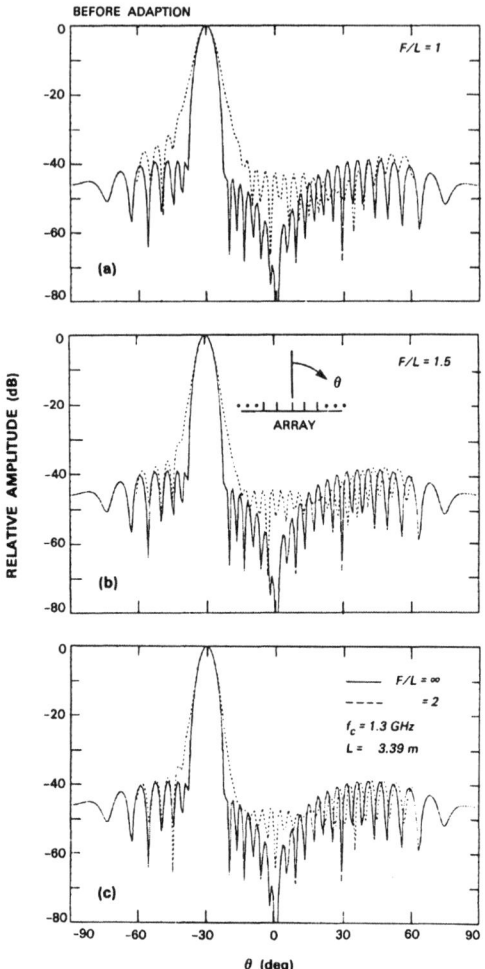

Figure 4.7 Monopole array NF/FF radiation patterns before nulling: (a) $F/L = 1$, (b) $F/L = 1.5$, and (c) $F/L = 2$. From [10] with permission, IEE.

Also, the near- and far-field nulls are not aligned. In particular, due to the finite range, the monopole element broadside null has filled in. As the near-field distance increases from L to $2L$, it is observed that the near-field pattern behaves more like that of the far-field pattern.

4.4 SIDELOBE CANCELLER ADAPTIVE ARRAY BEHAVIOR

In this section the adaptive nulling characteristics of a sidelobe canceller adaptive array are investigated. It is assumed that there are three auxiliary channels ($N_{\text{aux}} = 3$), so the covariance matrix size is 4×4 in this case, and there are four eigenvalues or degrees of freedom. The auxiliary element positions are designated by (row, column) and were chosen to be (1, 1), (5, 9), and (3, 36). Notice that the degrees of freedom have been allocated in two dimensions. The array quiescent conditions are the same as those described in the previous section. The quiescent radiation patterns were shown in Figure 4.7. Near-field ranges of $F/L = 1$, 1.5, and 2 are examined. In all near-field examples, the interference source range and focal range are equal. For the example array size, the actual near-field test distances are 3.39m ($F/L = 1$), 5.08m ($F/L = 1.5$), and 6.77m ($F/L = 2$). The far-field test distance is assumed to be at range $F/L = \infty$. A range of nulling bandwidths will be considered: $B = 1$ MHz (narrowband, $BL\cos\theta/c = 0.01$) to $B = 100$ MHz (wideband, $BL\cos\theta/c = 1$). Nonstressing and stressing interference scenarios will be examined. Nonstressing interference is defined as where the adaptive array degrees of freedom are sufficient to cancel the interference down to the noise level of the receiver. Stressing interference refers to the situation where the adaptive array degrees of freedom are insufficient for the adaptive output to reach the receiver noise floor.

4.4.1 NONSTRESSING INTERFERENCE: ONE SOURCE

Consider the case of one interference source ($J = 1$). Let an interferer, with power 50 dB above receiver noise at the array output, be located at $\theta = 42°$ both for finite and infinite range focused arrays. The covariance matrix elements defined in (4.2) are evaluated using a 15-point Simpson's rule numerical integration.

For $B = 1$ MHz, the adaptive array radiation patterns at center frequency 1.3 GHz are shown in Figure 4.8. Figure 4.8(a) is for a near-field distance of one aperture diameter ($F/L = 1$). Figures 4.8(b) and (c) are for near-field ranges of $1.5L$ and $2L$, respectively. For each case, the adaptive cancellation ratio was computed to be 50 dB. The consumption of adaptive array degrees of freedom is depicted in Figure 4.9. It is seen that two eigenvalues are significantly above the receiver noise level. The remaining eigenvalues are at the receiver noise level (0 dB); thus, only two degrees of freedom are engaged. The near-field eigenvalues are in good agreement with the corresponding far-field eigenvalues indicating that the degrees of freedom

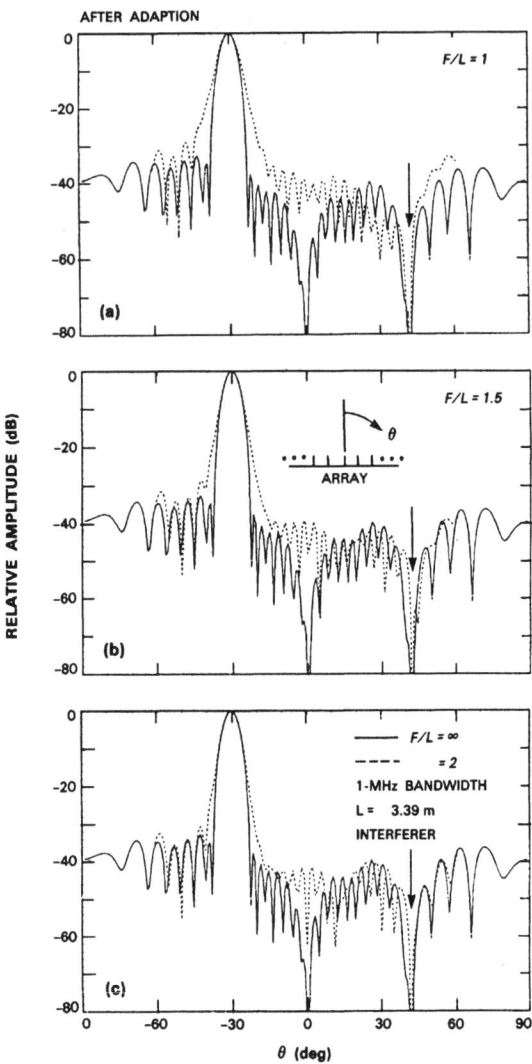

Figure 4.8 Monopole array NF/FF radiation patterns after nulling: (a) $F/L = 1$, (b) $F/L = 1.5$, and (c) $F/L = 2$. From [10] with permission, IEE.

are consumed the same. The amplitude of the adaptive array NF/FF weights (w_a) is given in Figure 4.10 and good agreement is evident. To show the behavior of a near-field null on the test plane, a two-dimensional contour radiation pattern has been computed for the $F/L = 2$ case ($z_F = 6.77$m) and is given in Figure 4.11.

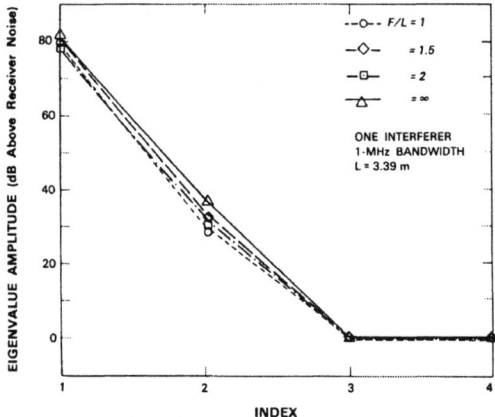

Figure 4.9 Covariance matrix eigenvalues for a monopole array with one NF/FF interferer. From [10] with permission, IEE.

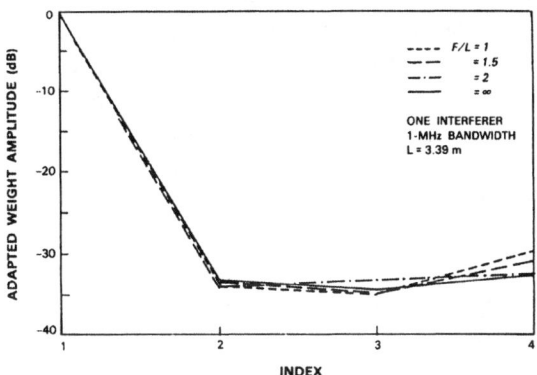

Figure 4.10 Amplitude of adaptive weights for for a monopole array with one NF/FF interferer. From [10] with permission, IEE.

4.4.2 STRESSING INTERFERENCE: TWO SOURCES

To demonstrate the validity of the NF/FF adaptive nulling equivalence for multiple sources, consider the previous array ($N_r = 32$) and the present case with two interferers ($J = 2$). Let the interferers be equal in power, uncorrelated, and located at $\theta = 42°, 47°$. (Note: The sources are separated by approximately one half-power beamwidth.) The quiescent radiation patterns were shown in Figure 4.7. The total amount of interference power before adaption is set to 50 dB above noise at the array output.

For $B = 1$ MHz, the NF/FF adaptive radiation patterns ($F/L = 1, 1.5, 2$), at center frequency 1.3 GHz, are shown in Figure 4.12. It is observed

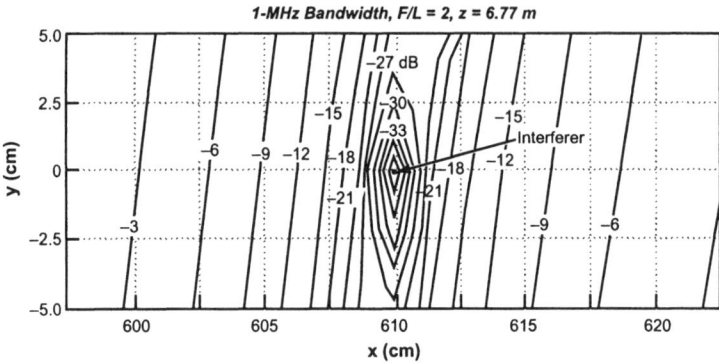

Figure 4.11 Two-dimensional near-field radiation pattern for a monopole array (Figure 4.6) with one interferer: after adaption.

that adaptive nulls are formed at the interferer positions. To compare near- and far-field consumption of the adaptive array degrees of freedom as a function of nulling bandwidth, the interference covariance matrix eigenvalues are presented in Figure 4.13 in the range $(1 \leq F/L \leq 2)$. Notice that each of the near- and far-field eigenvalues are in good agreement, that is, $\lambda_1^{NF} \approx \lambda_1^{FF}, \lambda_2^{NF} \approx \lambda_2^{FF}, \cdots, \lambda_4^{NF} \approx \lambda_4^{FF}$. Similarly, Figure 4.14 presents the adaptive cancellation as a function of nulling bandwidth. Complete interference cancellation, $C = 50$ dB, is not achieved here for bandwidths greater than CW, because the interference is significantly stressing the adaptive array degrees of freedom. This is observed in Figure 4.13, which shows that for bandwidths greater than CW, all the covariance matrix eigenvalues are above receiver noise. At two aperture diameters source distance, the NF cancellation is equal to the FF cancellation over the entire bandwidth shown. The interference cancellation degrades by only about 3 dB, compared to the FF result, when the sources are at one aperture diameter distance. In this situation, the equivalence between near- and far-field sources is approximate. Finally, the adaptive array weights are shown in Figure 4.15 for the wideband case (100 MHz). Although there are some differences (in particular w_3), the NF/FF weights are in good agreement.

From these results, which have taken account of array mutual coupling and polarization, it is concluded that two near-field (or spherical wave) interferers, arranged equivalently in terms of angle, are equivalent to two far-field (or plane wave) interferers. A generalization of this statement would be that J near-field interferers are equivalent to J far-field interferers, which is consistent with the results presented in Chapter 3.

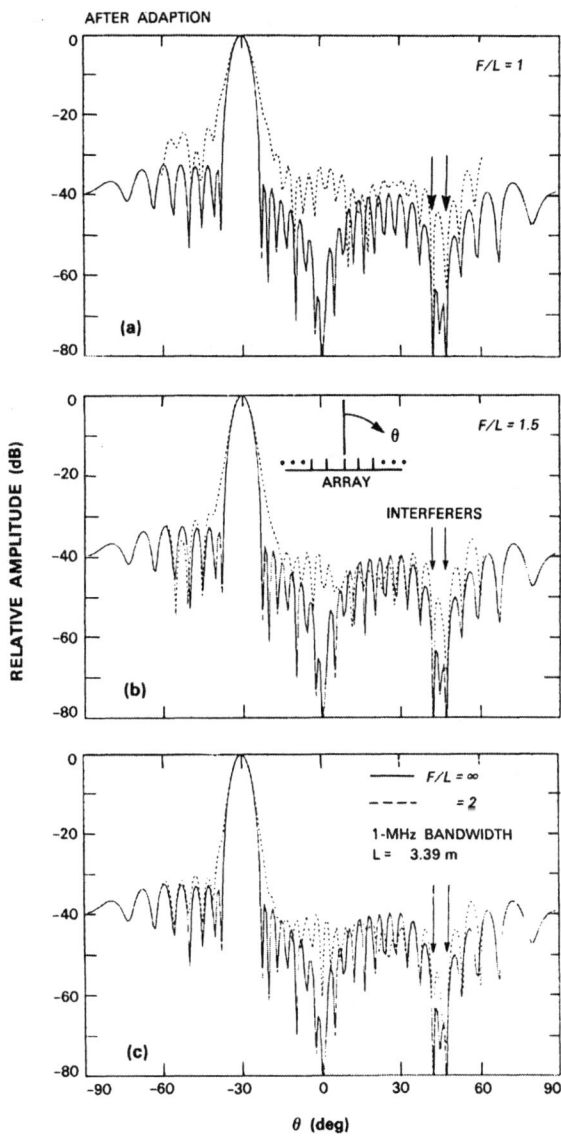

Figure 4.12 Monopole array NF/FF radiation patterns (two interferers) after nulling: (a) $F/L = 1$, (b) $F/L = 1.5$, and (c) $F/L = 2$. From [10] with permission, IEE.

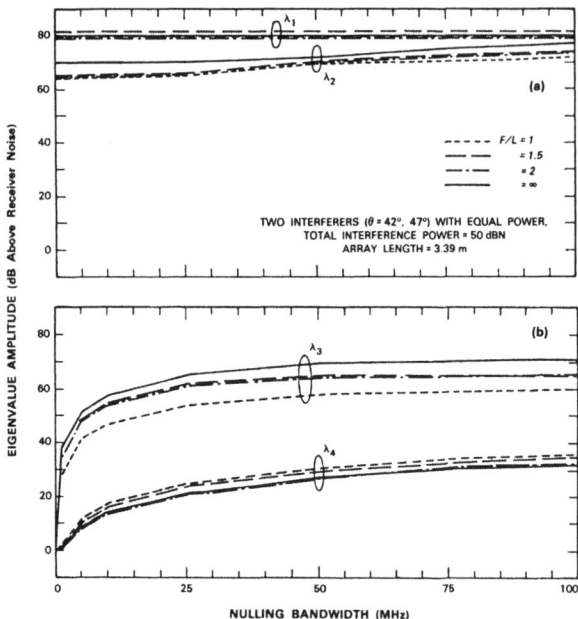

Figure 4.13 Covariance matrix eigenvalues for a monopole array with two NF/FF interferers versus nulling bandwidth: (a) λ_1, λ_2, and (b) λ_3, λ_4. From [10] with permission, IEE.

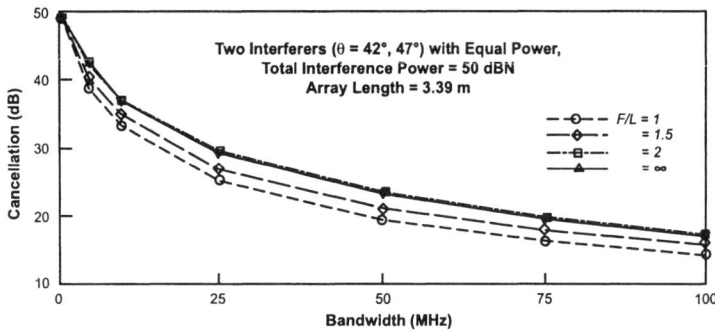

Figure 4.14 Adaptive cancellation for a monopole array with two NF/FF interferers versus nulling bandwidth. From [10] with permission, IEE.

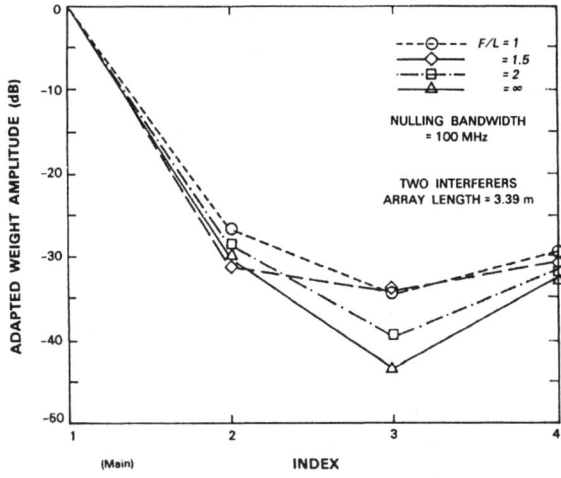

Figure 4.15 Amplitude of the adaptive weights for a monopole array with two NF/FF interferers. From [10] with permission, IEE.

4.5 SUMMARY

This chapter has further analyzed an approach (described in Chapter 3) to testing adaptive arrays in the near field. A theory for analyzing the behavior of an adaptive array, including bandwidth, polarization, and mutual coupling effects, in the presence of near-field interference was developed. Near-field focusing has been used to establish effectively a far-field pattern in the near zone. The near-field range of interest here has been taken to be one to two aperture diameters of the antenna under test. Equations for calculating the adaptive array covariance matrix and antenna radiation patterns, for near-field (spherical wave) and far-field (plane wave) interference, were given. Numerical simulations of a partially adaptive planar array indicate that the radiation patterns, adapted weights, cancellation, and covariance matrix eigenvalues (degrees of freedom) are effectively the same for near- and far-field interference. Both nonstressing (single source) and stressing (multiple source) interference conditions have been analyzed. From the results shown, it can be inferred that J near-field interferers in the presence of a near-field focused adaptive array are equivalent to J far-field interferers in the presence of a far-field focused adaptive array. That is, a one-to-one correspondence can be made between near- and far-field interferers. Thus, a phased array antenna adaptive nulling system designed for far-field conditions can potentially be evaluated more conveniently using near-field interference sources. The adaptive antenna under test and interference sources can all be positioned within an anechoic chamber with chamber dimensions on the order of one to two times the antenna length. Experimental verification of this technique is described in Chapter 6.

In this chapter, all sources (calibration and interference) were assumed to lie on a common test plane. This is not to restrict the technique but rather to have compatibility with planar near-field scanner hardware. The array had monopole elements and the interferer antenna was a one-half wavelength dipole. Not considered in this chapter are other, more conventional types of array antenna elements (broadside radiators) such as dipoles or waveguides. Broadside array elements (rectangular waveguides), in the application of focused near-field adaptive nulling, are addressed with measurements as presented in Chapter 6. Multiple phase centers, including array and source mutual coupling effects with main beam clutter and sidelobe interference, are investigated in the next chapter (Chapter 5).

References

[1] Yaghjian, A.D., "An Overview of Near-Field Antenna Measurements," *IEEE Trans. Antennas Propag.*, Vol. 34, No. 1, 1986, pp. 30-45.

[2] Johnson, R.C., H.A. Ecker, and R.A. Moore, "Compact Range Techniques and Measurements," *IEEE Trans. Antennas Propag.*, Vol. 17, No. 5, 1969, pp. 568-576.

[3] Mayhan, J.T., "Some Techniques for Evaluating the Bandwidth Characteristics of Adaptive Nulling Systems," *IEEE Trans. Antennas Propag.*, Vol. AP-27, 1979, pp. 363-373.

[4] Gabriel, W.F., "Adaptive Arrays – An Introduction," *Proc. IEEE*, Vol. 64, 1976, pp. 239-271.

[5] Fenn, A.J., "Theory and Analysis of Near Field Adaptive Nulling," *IEEE Antennas Propag. Soc. 1986 Symp. Digest Vol. 2*, pp. 579-582.

[6] Fenn, A.J., "Theory and Analysis of Near Field Adaptive Nulling," *1986 Asilomar Conf. on Signals, Systems, and Computers*, Washington, D.C.: Computer Society Press of the IEEE, 1986, pp. 105-109.

[7] Fenn, A.J., "Evaluation of Adaptive Phased Array Far-Field Nulling Performance in the Near-Field Region," *IEEE Trans. Antennas Propagat.*, Vol. 38, No. 2, 1990, pp. 173-185.

[8] Fenn, A.J., "Theoretical Near Field Clutter and Interference Cancellation for an Adaptive Phased Array Antenna," *IEEE Antennas Propagat. Soc. 1987 Symp. Digest, Vol. 1*, pp. 46-59.

[9] Fenn, A.J., "Analysis of Phase-Focused Near-Field Testing for Multiphase-Center Adaptive Radar Systems," *IEEE Trans. Antennas Propagat.*, Vol. 40, No. 8, 1992, pp. 878-887.

[10] Fenn, A.J., "Moment Method Analysis of Near Field Adaptive Nulling," *IEE Sixth International Conference on Antennas and Propagation, ICAP 89*, April 4-7, 1989, pp. 295-301.

[11] Scharfman, W.E., and G. August, "Pattern Measurements of Phased-Arrayed Antennas by Focusing into the Near Zone," in *Phased Array Antennas (Proc. 1970 Phased Array Antenna Symp.)*, A.A. Oliner and G.H. Knittel, (eds.), Dedham, MA: Artech House, 1972, pp. 344-350.

[12] Monzingo, R.A., and T. W. Miller, *Introduction to Adaptive Arrays*, New York: Wiley, 1980.

[13] Strang, G., *Linear Algebra and Its Applications*, New York: Academic Press, 1976, p. 213.

[14] Stutzman, W.L., and G.A. Thiele, *Antenna Theory and Design*, New York: Wiley, 1981, pp. 306-374.

[15] Gupta, I.J., and A.A. Ksienski, "Effect of Mutual Coupling on the Performance of Adaptive Arrays," *IEEE Trans. Antennas Propag.*, Vol. 31, No. 5, 1983, pp. 785-791.

[16] Richmond, J.H., "Radiation and Scattering by Thin-Wire Structures in a Homogeneous Conducting Medium (Computer Program Description)," *IEEE Trans. Antennas Propag.*, Vol. 23, 1975, pp. 412-414.

[17] Herper, J.C., and A. Hessel, "Performance of $\lambda/4$ Monopole in a Phased Array," *IEEE Antennas Propag. Soc. 1975 Symp. Digest*, pp. 301-304.

[18] Fenn, A.J., "Theoretical and Experimental Study of Monopole Phased Array Antennas," *IEEE Trans. Antennas Propag.*, Vol. 33, No. 10, 1985, pp. 1118-1126.

[19] Fenn, A.J., H.M. Aumann, and F.G. Willwerth, "Linear Array Characteristics with One-Dimensional Reactive-Region Near-Field Scanning: Simulations and Measurements," *IEEE Trans. Antennas Propag.*, Vol. 39, No. 9, 1991, pp. 1305-1311.

5

Focused Near-Field Testing of Multiphase-Center Adaptive Array Radar Systems

5.1 INTRODUCTION

As discussed in Chapter 3, in developing any radar (or communications) system it is necessary to perform tests of the associated hardware and software at various levels of the design. Prior to achieving a deployed or fielded radar system, design specifications must be verified at the subsystem development level, system prototype level, and final system level. For ground-based or airborne radar systems, the final system can be tested in the field and be modified or upgraded as necessary. However, in the case of a spaceborne radar, the deployed system hardware is not accessible, which requires that comprehensive prelaunch testing and modifications (as may be required) be performed on the ground. This chapter expands on Chapters 3 and 4 and further describes and develops a focused near-field technique that can be used to measure the performance characteristics of adaptive phased array antenna systems within a ground-test facility. Although this technique is especially suitable for space-based radar (SBR) phased array antenna systems [1-4], the technique is applicable to most adaptive airborne and ground-based radar systems and communications systems as well.

The important subsystems of an adaptive radar system typically consist of an antenna, a multichannel receiver, and a signal processor. In addition to receiving desired target signals, the radar must contend with interference which consists of noise, background clutter, and sidelobe jamming as depicted in Figure 5.1. The antenna collects signals that subsequently are filtered, downconverted, and digitized within the radar receiver. The digitized data are then processed, by the signal processor, in such a way that undesired signals are suppressed, leaving desired target reports.

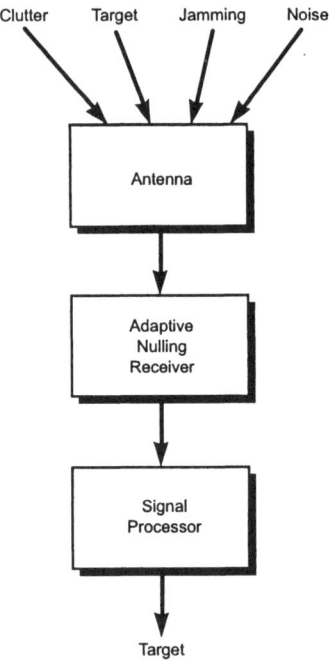

Figure 5.1 The important subsystems of an adaptive-nulling radar system. Noise, clutter, jamming, and desired target signals are received by the antenna. These signals are downconverted and digitized by the receiver and are then processed within the signal processor, that provides desired target information. © 1992 IEEE [4].

Typically, a spaceborne radar system could utilize a deployable planar phased array antenna structure having a largest dimension on the order of 10 to 100 meters and a nominal operating range of several thousand kilometers [1-3]. This means that the antenna would receive planar-shaped wavefronts from targets, clutter, and jamming. At microwave frequencies, to approximate plane-wavefront conditions a conventional far-field test distance can be many kilometers. The minimum far-field distance is computed according to $2D^2/\lambda$, where D is the antenna diameter and λ is the wavelength. For low-sidelobe antenna pattern measurements, to reduce the phase curvature of the incident spherical wave an even longer far-field test distance is often used. To place many radiating test sources on a far-field ground test range over a wide field of view and demonstrate low sidelobes together with jammer/clutter suppression is difficult if not impossible. Thus, an alternate ground test configuration is necessary. Near-field testing, where the radar system is positioned within a high-quality controlled-environment anechoic chamber is a desirable method for evaluating the radar's performance. Figure 5.2 shows the proposed method of testing, where the antenna and test sources (jammer, clutter, and target) are placed within the anechoic chamber. Subject to practicality, the adaptive nulling receiver and signal processor are placed either inside or outside the anechoic chamber. The details of how this technique naturally develops is described next.

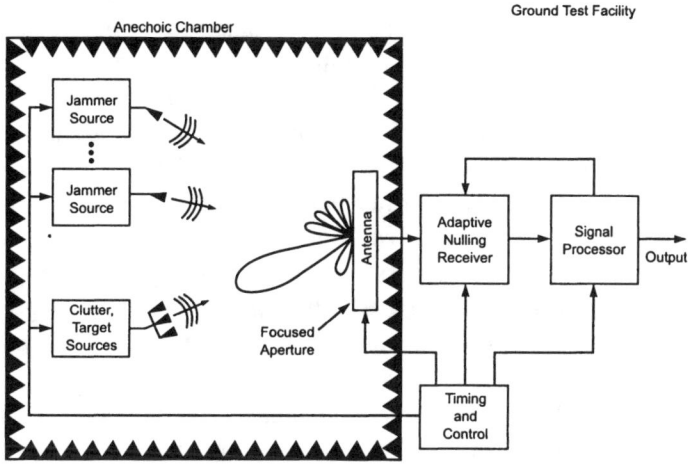

Figure 5.2 Focused near-field nulling concept for ground testing an adaptive radar system. The antenna under test and radiating sources (clutter, jammer, targets) are positioned within a high-quality anechoic chamber. Focusing the test antenna in the near field produces far-field equivalent conditions. Utilizing proper timing and control, a signal environment comparable to fielded radar system conditions can be achieved. © 1992 IEEE [4].

Consider Figure 5.3, which depicts the various regions in front of an antenna system under test. The horizontal axis is the test antenna length D/λ and the vertical axis is the normalized test distance z/D. The far-field test region is indicated in this figure, but will not be considered further in this chapter as a potential test technique for the reasons cited earlier. Planar scanning with a probe antenna at a distance of a few wavelengths (typically 3λ or less) from the test antenna is a conventional near-field technique for calibration and far-field radiation pattern measurement [5]. Planar near-field scanning is discussed in Chapters 11 and 12. This form of near-field scanning is a CW method and a nonreal-time measurement technique where near-field data are collected and far-field data are computed. However, an adaptive nulling test requires real-time signal wavefronts, which precludes using the planar near-field scanning technique. The Fresnel region [6] (or transition region between the near field and far field), which extends from about $0.6\sqrt{D^3/\lambda}$ to $2D^2/\lambda$, offers reduced range testing but still not short enough for testing large antennas in an anechoic chamber. The near-field region is defined here to be at ranges less than $0.6\sqrt{D^3/\lambda}$. A compact-range reflector [7] (see Chapter 15 for an example) can be used to reduce the test distance down to about two to four aperture diameters ($2D$ to $4D$), which is within the near-field region. This technique utilizes a parabolic reflector to convert the spherical wavefront from a feed horn into a planar wavefront. A compact-range reflector could be used for adaptive nulling tests provided the wavefront is sufficiently free of multipath. However, the required reflector diameter is large (about $2D$) for the typical radar bands from UHF (about 400 MHz) to X-band (about 10 GHz) and sources widely spaced in the test antenna field of view would be difficult to achieve. Thus, the compact range reflector is also not pursued further here. The basis of these measurement techniques is to create plane wave illumination for the antenna, which can result in impractical test geometries for adaptive nulling. To develop a suitable testing method for anechoic chamber testing of large-aperture adaptive phased arrays, it is necessary to drop the plane wave constraint and instead consider spherical waves.

The technique described here implements conventional near-field phase focusing to establish an instantaneous or real-time antenna radiation pattern that is equivalent to its far-field pattern [8]. This technique appears to be ideal for ground-based testing. The test distance can be from about one to two aperture diameters (D to $2D$) of the adaptive antenna under evaluation as shown in Figure 5.3. For large antenna diameters, this test distance is located well within the near-field boundary. The incident wavefront from radiating sources in the near field is spherical rather than planar. Thus, the radiating source antennas can be simple horns or dipoles. A number of papers describing focused near-field adaptive nulling have been published by the author [4, 8-11]. The equivalence between conventional far-field adaptive nulling and focused near-field adaptive nulling has been demonstrated for sidelobe canceller [8] and fully adaptive arrays [9]. Near-field clutter and jamming for a sidelobe canceller have also been addressed [4, 10]. In [8-10], the analysis assumed that the array elements and radiating sources were isotropic. The effects of array polarization and mutual coupling have also been studied and it has been demonstrated that the equivalence between near-field and far-field adaptive nulling still holds, for single and multiple jammers [11]. The present chapter expands the mutual coupling formulation presented in Chapter 4 to include simulated clutter and jamming [4].

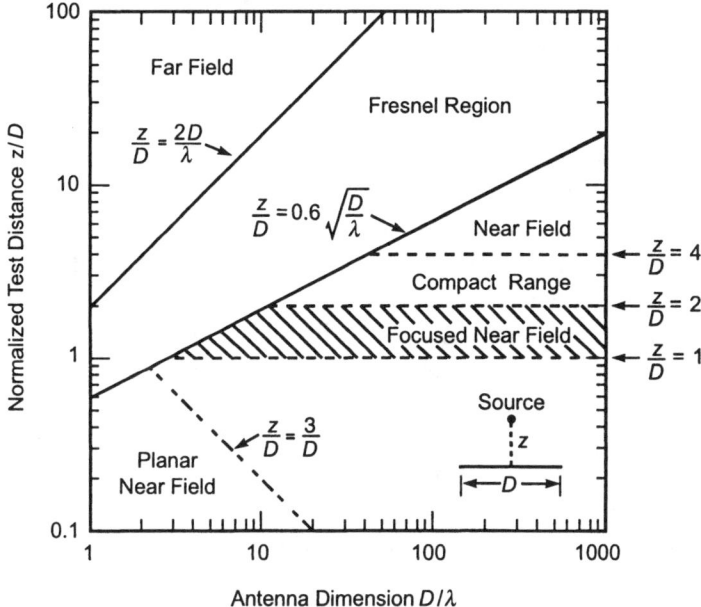

Figure 5.3 Summary of antenna test regions. The hemispherical volume in front of an antenna can be divided into a number of test regions. The three basic regions are far field, Fresnel, and near field. The near-field region can be further divided into compact range (a reflector-based method), planar near field, and focused near field, which is addressed in this chapter. The focused near-field region is taken here to be located from between one and two aperture diameters from the antenna under test. © 1992 IEEE [4].

In the next section, the characteristics of an incident signal wavefront are investigated as a function of source distance and angle of arrival. This is followed by a description of how focused near-field adaptive nulling is used to establish appropriate quiescent conditions. An application of the technique to a displaced phased center antenna (DPCA) is made, for which details of the theoretical formulation are given. Antenna modeling is accomplished using the method of moments, which inherently includes array mutual coupling effects. The theory is applied to a linear array of dipole elements with dipole near-field sources (main beam clutter and sidelobe jamming). The results will show that focused near-field adaptive nulling is a viable approach to testing full-scale adaptive radar systems.

5.2 NEAR-FIELD/FAR-FIELD SOURCE WAVEFRONT DISPERSION

An explanation of why near-field nulling can be equal to far-field nulling is as follows: Signal wavefront dispersion (time-bandwidth product) is an effect that can limit the depth of null (or cancellation) achieved by an adaptive antenna [12, 13]. The amount of wavefront dispersion, denoted γ, observed by a linear array is a function of the nulling bandwidth, array length, source range, and angle of incidence.

A basic dispersion model for spherical wave incidence and plane wave incidence can be made by considering the wavefront dispersion observed across the end points of an adaptive array. This calculation is useful in gaining some initial insight into how near-field (NF) nulling will relate to far-field (FF) nulling. Consider first, a plane wave arriving from infinity and an array of length L.

The dispersion for this case is denoted γ_{FF} and is computed according to the product of bandwidth and time delay as

$$\gamma_{FF} = \frac{BL}{c} \sin\theta_i \quad (5.1)$$

where B is the nulling bandwidth, c is the speed of light, and θ_i is the angle of incidence with respect to broadside. Note that the dispersion is maximum for endfire incidence ($\theta_i = 90°$) and is zero for broadside incidence ($\theta_i = 0°$). Next, consider a point source at a constant distance $z = z_i$ and variable angle $\theta = \theta_i$, which produces an incident spherical wavefront. The distances between the source and the two end points are denoted r_1 and r_2. The near-field dispersion, denoted γ_{NF}, is given by

$$\gamma_{NF} = \frac{BL}{c}\frac{(r_1 - r_2)}{L} \quad (5.2)$$

where the quantity $(r_1 - r_2)/L$ is the normalized range difference. In comparing (5.1) and (5.2) it is seen that the far-field and near-field dispersions have a common factor BL/c. If near-field nulling is to have any possibility of being equivalent to far-field nulling then γ_{NF} must be equivalent (or nearly equivalent) to γ_{FF}. This is clearly satisfied when $(r_1 - r_2)/L = \sin\theta_i$. Figure 5.4 shows a plot of the normalized dispersion $\gamma/(BL/c)$ as a function of the angle of incidence for values of normalized source distance from $z/L = 0.25$ to 2 (that is, aperture lengths $0.25L$ to $2L$). From this figure it is seen that the near-field dispersion approaches the value of the far-field dispersion for source distances greater than approximately one aperture diameter ($z/D \geq 1$). At one diameter, the percent difference between near-field and far-field dispersion is less than 10%. At two diameters, the percent difference is less than 3%. Clearly, at source distances such that $z/D \leq 0.5$ (one-half aperture diameter), the near-field dispersion is significantly different (by as large as 30%) compared to the far-field dispersion. Thus, for this simple dispersion model it is expected that near-field adaptive nulling can be similar to far-field nulling at source distances greater than one aperture diameter.

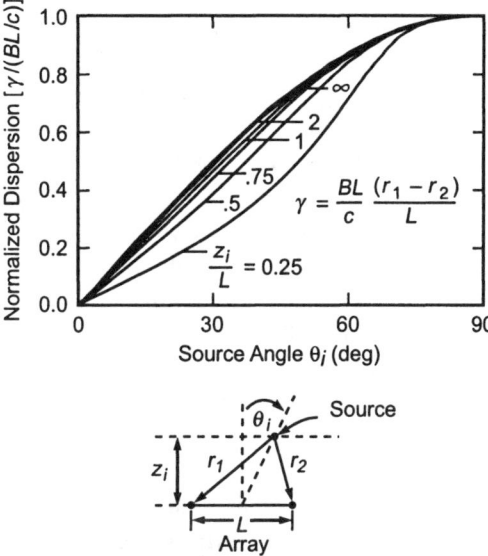

Figure 5.4 Normalized wavefront dispersion as a function of source angle of arrival. The wavefront dispersion (time-bandwidth product) observed relative to the end points of an antenna is a simple model for comparing the characteristics of near-field and far-field sources.

At source distances of one to two aperture diameters, the incident near-field wavefront dispersion appears to be comparable to that of a far-field wavefront. However, before attempting to perform near-field adaptive nulling, it is necessary to make the receive antenna quiescent conditions appear the same as far-field quiescent conditions. This is achieved by phase focusing the antenna under test as described in the next section.

5.3 FOCUSED NEAR-FIELD TESTING CONCEPT

In the near-field testing technique described here, it is assumed that the quiescent near-field radiation pattern of the array should have the same characteristics as the quiescent far-field radiation pattern of the array. The reason for this is that the dynamic range of received signals from sources distributed across the radar field of view is dependent upon the antenna quiescent conditions. This means typically that a main beam and sidelobes should be formed. To produce an array near-field pattern that is approximately equal to that of the far-field pattern, phase focusing can be used as demonstrated by Scharfman and August [14]. Consider Figure 5.5, which shows a CW calibration source located at a desired focal point of the array. The array can maximize the signal received from the calibration source by adjusting its phase shifters so that the spherical wavefront phase variation is removed (refer to Figure 4.2). One way to do this is to choose a reference path length as the distance from the focal point to the center of the array. This distance is denoted r_F (refer to Figure 3.4), and the distance from the focal point to the nth array element is denoted r_n^F. The voltage received at the nth array element relative to its center element is computed in this chapter by using the method of moments [15]. To maximize the received voltage at the array output, it is necessary to apply the phase conjugate of the incident wavefront at the array elements. The resulting radiation pattern on the test plane $z = z_F$ looks similar to a far-field pattern [11]. A main beam will be pointed at the array focal point. Sidelobes will exist at angles away from the main beam. Interferers can then be placed on near-field sidelobes in the test (or focal) plane, as depicted in Figure 5.5. Similarly, clutter sources are positioned on the near-field main beam.

It is known that near-field focusing using only phase control will result in some distortion of the main beam and first sidelobes (examples are shown in Chapter 3). Amplitude control can be used to make the near-field pattern main beam and first sidelobes more closely resemble a far-field pattern as demonstrated by Aumann and Willwerth in [16]. For convenience, this chapter will investigate phase focusing only. The minimum size of the required ground test facility can be inferred by considering the near-field geometry. A flat test plane is assumed in order to have compatibility with conventional planar near-field scanning equipment. Let θ_{max} denote the maximum angle of interest for the antenna radiation pattern. The required near-field scan length for pattern coverage of $\pm\theta_{max}$ is given in terms of the F/L ratio as

$$D_x = 2L(\frac{F}{L}) \tan \theta_{max} \tag{5.3}$$

As an example, Figure 5.6 depicts the required scan lengths for 60° and 120° field-of-view coverage using F/L ratios of 1 and 2. To reduce the scan length (or source deployment length), it is desirable to keep the F/L ratio as small as possible. Clearly, the ground test facility must be large enough to encompass both the length of the desired scan (D_x) and the focal length (F).

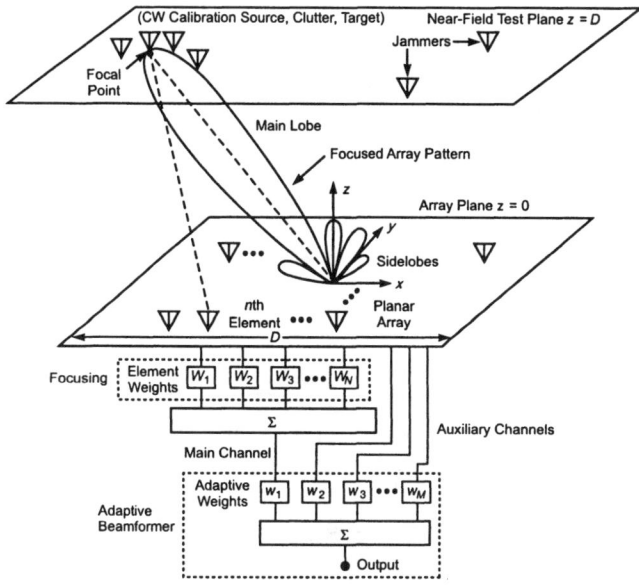

Figure 5.5 Focused near-field adaptive nulling test concept. A CW radiating source is used as a calibration signal for a phased array antenna. The antenna element phase shifters are used to focus the receive antenna radiation pattern in the direction of the calibration source. This creates an antenna radiation pattern that has a main beam and sidelobes that are very similar to a far-field radiation pattern. Clutter and target sources can be placed across the main beam as desired. Similarly, jammers can be positioned at the location of the sidelobes. © 1992 IEEE [4].

Before applying the focused near-field testing concept to an example radar system, it is worthwhile summarizing the important assumptions made. These are that: (1) the incident near-field wavefront must be reasonably well-matched to a far-field wavefront (refer to Figure 5.4), and (2) the adaptive antenna under test must be focused at the range of the test sources. An important assumption not yet mentioned is that (3) the antenna and beamformer characteristics and receiver characteristics, such as channel mismatch, are independent of the type of wavefront (near field or far field). Another assumption is that (4) the technique is applicable to both analog and digital adaptive nulling systems.

5.4 ADAPTIVE DPCA RADAR CONCEPT

The technique referred to as displaced phase center antenna (DPCA) has application in airborne or spaceborne radar systems requiring adaptive

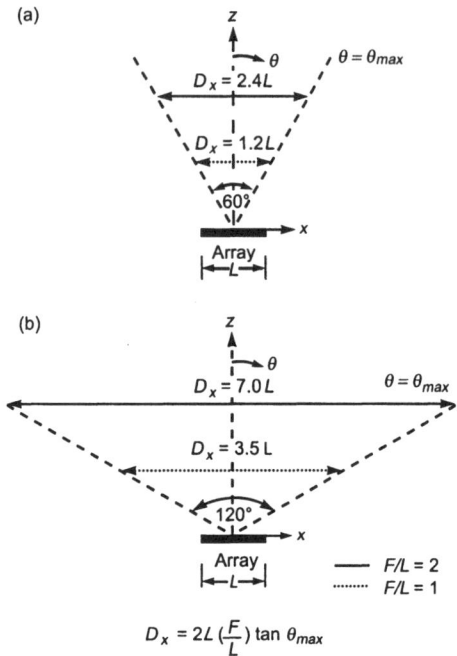

Figure 5.6 Example near-field scan lengths: (a) 60° and (b) 120° fields of view. © 1992 IEEE [4].

suppression of jamming and clutter as described by Kelly and Tsandoulas [2]. Chapter 2 described DPCA for the case of clutter suppression, and this chapter will include both jamming and clutter. With a displaced phase center antenna, stationary ground clutter is cancelled from a moving platform by employing two or more independent receive phase centers having well-matched main beams. A two-phase center adaptive DPCA system is depicted in Figure 5.7, where a moving target and a moving radar are shown. Here, the full aperture is used for two successive pulse transmissions (positions A_T and B_T) and, on receive, two displaced portions of the aperture are used. The radar illuminates moving targets and fixed ground terrain which, due to the radar cross section of the targets and ground terrain, produce the desired signals and undesired clutter. Ground-based emitters represent a source of interference or jamming. On reception, the antenna phase center displacement between receive apertures is adjusted to compensate for the platform velocity. For two incident pulses separated in time by one pulse repetition interval (PRI) the first reception occurs at the forward phase center, denoted A. A second reception occurs at the trailing phase center, which is denoted B. This slightly bistatic radar system is equivalent to a monostatic radar system having two

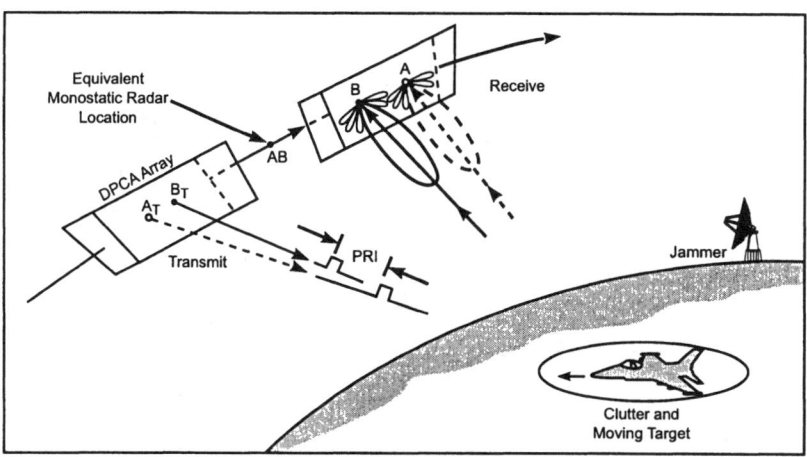

Figure 5.7 Two-phase center DPCA radar platform showing transmit and receive phase centers for consecutive pulses. Receive phase centers (or subarrays) A and B compensate for the radar motion by creating a phase-center displacement.

measurements of the signal environment at a single point in space. During a PRI, the clutter is effectively stationary; however, during this interval the target moves. As a result of this movement, the target has a relative phase shift. There is no such phase shift from the clutter during this time. The clutter is assumed to be correlated between the two phase centers. In contrast, jamming is assumed to be uncorrelated between the two phase centers due to the one PRI delay imposed in the signal processing. When the signals received by the two phase centers are adaptively combined in two stages as shown in Figure 5.8, the clutter and jammer are significantly cancelled. The target signal strength received, after adaptive signal processing, depends on the amount of target phase shift in one PRI. The DPCA quiescent main beam pattern match is affected by array geometry and scan conditions (due to array element mutual coupling), and hardware tolerances (such as the quantization and random errors of the transmit/receive (T/R) modules). In this chapter, both the effects of receive module errors and array mutual coupling are considered. Next, the details of the adaptive DPCA formulation are given.

5.5 ADAPTIVE DPCA ARRAY FORMALISM

Consider the DPCA array and beamformer as shown in Figure 5.8. In general, the array will contain a total of N elements that are used to form the receive main channel. In addition to these N elements, a guard band of

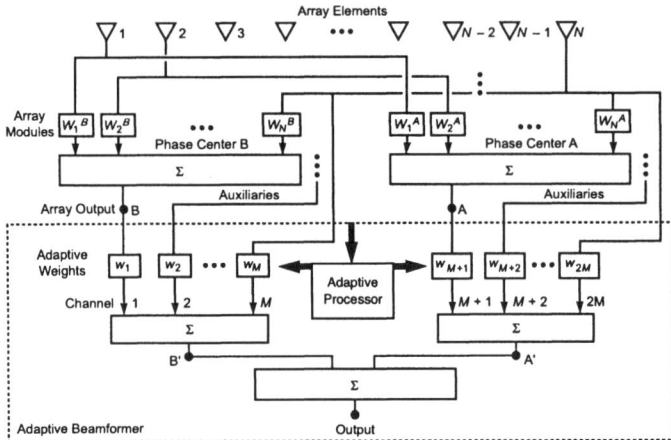

Figure 5.8 Adaptive beamformer arrangement for DPCA operation. The vertical arrow entering the adaptive processor refers to the input data vector consisting of samples of the main and auxiliary channels. The horizontal arrows exiting the adaptive processor refer to the adaptive weight commands. The jamming signals in phase centers A and B are cancelled at points A' and B', while the clutter is cancelled at the final output of the adaptive beamformer.

passively terminated elements can be used to provide impedance matching to the active elements and/or isolation from ground plane edges. As shown in Chapter 2, the guard band elements serve to maintain a good pattern match between the two displaced phase centers. The output from each of the array elements is first split into two paths that are weighted and then summed in separate power combiners to form two independent subarray main channels (or moveable phase centers). In each element channel is a transmit/receive (T/R) module, which has amplitude and phase control. The amplitude control is used to provide the desired low-sidelobe array illumination function and to achieve phase-center displacement. The modules utilize phase shifters, which are used to steer the main beam to a desired angle. Let $\boldsymbol{W}_A = (W_1^A, W_2^A, \cdots, W_N^A)^T$ and $\boldsymbol{W}_B = (W_1^B, W_2^B, \cdots, W_N^B)^T$ denote the array element weight vectors (including quantization and random errors) of phase centers A and B, respectively (superscript T means transpose). To effect phase center displacement, a portion of each subarray is turned off by applying a large value of attenuation for a group of antenna elements. This amplitude weighting will essentially move the electrical phase center to the center of gravity for the remaining elements. Thus, there is an effective number of elements actually used in receiving signals in phase centers A and B and are denoted as N_A and N_B, respectively. Let a wavefront (either planar or spherical) due to the jth source (either clutter or jammer), be impressed across the array, which results in a set of array element received voltages denoted $v_1^j, v_2^j, \cdots, v_N^j$. The number of adaptive channels per phase center is denoted M. For a sidelobe canceller $M = 1 + N_{aux}$ where N_{aux} is the number of auxiliary channels in each phase center. This adaptive system has M degrees of freedom in each phase center and, thus, a total of $2M$ degrees of freedom. For each phase center, the main and auxiliary channel voltages are derived from among the earlier set of array received voltages. In this chapter, ideal adaptive weights (no quantization or random errors) are assumed with $w = (w_1, w_2, \cdots, w_{2M})^T$ denoting the adaptive channel weight vector. The fundamental quantities required to fully characterize the incident field for adaptive nulling purposes are the adaptive channel cross correlations.

The cross correlation R_{mn}^j of the received voltages in the mth and nth adaptive channels, due to the jth source, is given by

$$R_{mn}^j = E(v_m v_n^*) \qquad (5.4)$$

where $*$ means complex conjugate and $E(\cdot)$ means mathematical expectation. Since v_m and v_n represent voltages of the same waveform, but at different times, R_{mn}^j is also referred to as an autocorrelation function. Note: For convenience, in (5.4) the superscript j in v_m and in v_n has been omitted.

In the frequency domain, assuming the source has a band-limited white noise power spectral density, (5.4) can be expressed as the frequency average

$$R_{mn}^j = \frac{1}{B} \int_{f_1}^{f_2} v_m(f) v_n^*(f) \, df \qquad (5.5)$$

where $B = f_2 - f_1$ is the nulling bandwidth and f_c is the center frequency. It should be noted that $v_m(f)$ takes into account the wavefront shape, which can be spherical or planar.

Let the channel or source covariance matrix be denoted R. If there are J uncorrelated narrowband or broadband interference sources, then the J-source covariance matrix is the sum of the covariance matrices for the individual sources, that is,

$$R = \sum_{j=1}^{J} R_j + I \qquad (5.6)$$

where R_j is the covariance matrix of the jth source, I is the identity matrix which is used to represent the thermal noise level of the receiver.

Prior to generating an adaptive null, the adaptive channel weight vector, w, is chosen to maintain a desired quiescent radiation pattern. When undesired signals are present, the optimum set of weights, denoted w_a, to form one or more adaptive nulls is computed by [12]

$$w_a = R^{-1} w_q \tag{5.7}$$

where $^{-1}$ means inverse and w_q is the quiescent weight vector. For a side-lobe canceller, the quiescent weight vector is chosen to be $w_q = (1, 0, 0, \cdots, 1, 0, 0, \cdots, 0)^T$, that is, each main channel weight is unity and the auxiliary channel weights are zero.

The output power at the adaptive array summing junction is given by

$$p = w^\dagger R w \tag{5.8}$$

where † means complex conjugate transpose. The interference-plus-noise-to-noise ratio, denoted INR, is computed as the ratio of the output power [defined in (5.8)] with the interferer present to the output power with only receiver noise present, that is,

$$\text{INR} = \frac{w^\dagger R w}{w^\dagger w} \tag{5.9}$$

The adaptive array cancellation ratio, denoted C, is defined here as the ratio of interference output power before adaption to the interference output power after adaption, that is,

$$C = \frac{p_q}{p_a} \tag{5.10}$$

Substituting (5.8) in (5.10) yields

$$C = \frac{w_q{}^\dagger R w_q}{w_a{}^\dagger R w_a} \tag{5.11}$$

Next, the covariance matrix defined by the elements in (5.4) is Hermitian (that is, $R = R^\dagger$) which, by the spectral theorem, can be decomposed in eigenspace as [17]

$$R = \sum_{k=1}^{2M} \lambda_k e_k e_k^\dagger \tag{5.12}$$

where $\lambda_k, k = 1, 2, \cdots, 2M$ are eigenvalues of R, and $e_k, k = 1, 2, \cdots, 2M$ are the associated eigenvectors of R. The multiple-source covariance matrix eigenvalues $(\lambda_1, \lambda_2, \cdots, \lambda_{2M})$ are a convenient quantitative measure of the utilization of the adaptive array degrees of freedom as discussed in Chapter 1. Expressed in decibels, the minimum amplitude that an eigenvalue can have is 0 dB (the receiver noise level) – this is because the identity matrix has been added to the covariance matrix. Counting the number of eigenvalues above the receiver noise level is a direct indication of how many degrees of freedom are being used to suppress the undesired signals [13, 18].

5.6 ARRAY ANTENNA/SOURCE MODELING

This section applies the method of moments [15] in computing the array element received voltages, in (5.4), due to near- or far-field sources. The far-field formulation given here is analogous to that which has been described by Gupta and Ksienski in [19]. Referring to Figure 5.9, assume that each element is terminated in a load impedance Z_L, which is known.

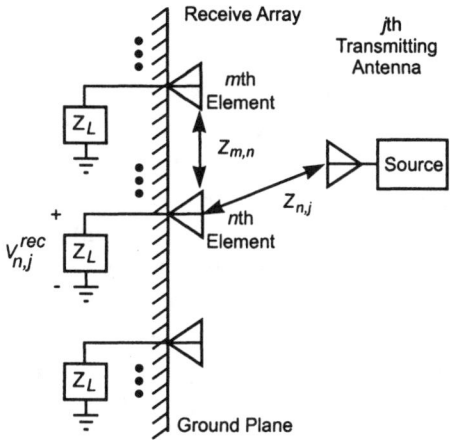

Figure 5.9 Receive array and near-field source antenna model. The quantity $Z_{m,n}$ represents the mutual impedance between array elements, while $Z_{n,j}$ represents the mutual impedance between the nth element and the jth transmitting antenna. © 1992 IEEE [4].

Let $v_{n,j}^{o.c.}$ represent the open-circuit voltage in the nth array element due to the jth source. Here, the jth source can denote either the CW calibrator or one of the jammer or clutter sources. Next, let $\mathbf{Z}^{o.c.}$ be the open-circuit mutual impedance matrix for the N-element array. The array elements are assumed to be dipoles over an infinite ground plane. As shown in Chapter 4, the array received voltage matrix, denoted $\mathbf{v}_j^{rec}$, due to the jth source, can be expressed as [20]

$$\mathbf{v}_j^{rec} = Z_L \left[\mathbf{Z}^{o.c.} + Z_L \mathbf{I}\right]^{-1} \mathbf{v}_j^{o.c.} \tag{5.13}$$

In (5.13), the nth element of $\mathbf{v}_j^{o.c.}$ is computed, for near-field sources, using the relation

$$v_{n,j}^{o.c.} = i_j Z_{n,j}^{o.c.} \tag{5.14}$$

where i_j is the terminal current for the jth source and $Z_{n,j}^{o.c.}$ is the open-circuit mutual impedance between the jth source and the nth array element. For thin-wire array antenna elements, the moment method expansion and testing functions are assumed to be sinusoidal. The open-circuit mutual impedances are computed based on subroutines from a well-known moment method computer code developed by Richmond [21]. In evaluating $Z^{o.c.}_{n,j}$ for the jth interferer, double precision computations were required. For far-field sources, $v^{o.c.}_{n,j}$ is evaluated by assuming plane wave incidence. The main channel output is computed using $\boldsymbol{W}^\dagger_A \cdot \boldsymbol{v}^{rec.A}_j$ for phase center A and $\boldsymbol{W}^\dagger_B \cdot \boldsymbol{v}^{rec.B}_j$ for phase center B.

As mentioned earlier, each phase center of the array is calibrated (phased focused) initially using a CW radiating dipole. To accomplish this numerically, having computed $\boldsymbol{v}^{rec}_{CW}$, the receive array weight vector $\boldsymbol{W}_A$ (or $\boldsymbol{W}_B$) will have its phase commands set equal to the conjugate of the corresponding phases in $\boldsymbol{v}^{rec}_{CW}$. Receive antenna radiation patterns are obtained by scanning (moving) a dipole with half-length l in either the far- or near-field region and computing the antenna response. Far-field received patterns are computed by using a θ-polarized dipole source at infinity that generates plane wave illumination of the array. A point in front of the test antenna is specified by the coordinates (x, y, z). Principal plane near-field radiation pattern cuts (versus angle) are obtained by computing the near field on the line $(x, 0, z_F)$ and using the relation $\theta(x) = \tan^{-1}(x/z_F)$. The near-field source is a dipole with half-length l and is $\hat{x}$-polarized. Let the voltage received by the array, due to the x-directed near-field dipole, be denoted $v^{NF}_x(\theta)$. Let $p_\theta(\theta)$ denote the $\hat{\theta}$ component of the dipole probe pattern. Then the probe-compensated array near-field received pattern is expressed as

$$E_\theta(\theta)^{NF} = \frac{v^{NF}_x(\theta)}{p_\theta(\theta)} \qquad (5.15)$$

where

$$p_\theta(\theta) = \frac{\cos(kl \sin \theta) - \cos(kl)}{\cos \theta} \qquad (5.16)$$

where $k = 2\pi/\lambda$ is the angular wavenumber. Equation (5.15) is correct provided that the radial component of the electric field is zero. For the simulated test distances involved in this chapter, the radial component is typically at least 20 dB below the E_θ component (refer to Chapter 10). Thus, for all practical purposes, the equality in (5.15) is valid.

The array received voltage matrix for the jth source (denoted $\boldsymbol{v}^{rec}_j$) is computed at K frequencies across the nulling bandwidth. To obtain the received voltages $\boldsymbol{v}^{rec}_j(f_1), \boldsymbol{v}^{rec}_j(f_2), \cdots, \boldsymbol{v}^{rec}_j(f_K)$, the impedance matrix $\boldsymbol{Z}^{o.c.}$ is computed at K frequencies and the system of equations given by (5.13) is solved at each frequency. The interference covariance matrix elements are computed by evaluating (5.4) numerically, using Simpson's rule numerical integration. For multiple sources, the covariance matrix is evaluated by using (5.6). Adaptive array radiation patterns are computed by superimposing the quiescent radiation pattern with the weighted sum of auxiliary channel received voltages.

5.7 DPCA NEAR-FIELD SOURCE DISTRIBUTION

The positioning of near-field sources for two-phase center DPCA operation is shown in Figure 5.10. There are two sets of sources, one for phase center A and one for phase center B. Multiple clutter sources are distributed across the main beam of both phase centers. A desired target signal can be embedded within the clutter signals as shown. Jammer signals are assumed to radiate from antennas located within the sidelobe region. It has already been mentioned that the clutter is correlated between phase centers A and B. For example, denote the first clutter source in phase center A as $C_1^A(\theta_1)$ and the first clutter source in phase center B as $C_1^B(\theta_1)$. Thus, it is clear in theory that $C_1^A(\theta_1) = C_1^B(\theta_1)$. Equalities exist similarly for the remaining clutter sources. To achieve this in an experimental configuration will require digitally controlled arbitrary waveform generators. This form of signal generator will also be useful for the jammers and desired target signals. It is necessary that the two sets of sources be operated during different time intervals separated by the radar PRI delay. This means, for example, that a measurement of the phase center A signals is performed with just the "A" group of sources radiating. The next measurement (one PRI later) would be for the phase center B signals with just the "B" group of sources radiating. This would be implemented using timing and control, as was suggested earlier in Figure 5.2.

It is assumed that the near-field source antennas are spaced sufficiently far apart so that they do not interact (via mutual coupling) in such a way as to affect adaptive nulling performance. For example, with one source radiating, the surrounding source antennas represent possible multipath scattering. It is known that the near-field sources will couple, but this will not influence the adaptive weights provided that the coupled signal is reradiated and received at below the receiver noise level. The interaction between two antennas can be computed as follows [22]: Let i_1 be the terminal current generated on the active source antenna. Similarly, let i_2 denote the parasitic current generated on a second, nearby, antenna. From circuit theory it is straightforward to show that the ratio of the parasitic (induced) current to active current is given by

$$\frac{i_2}{i_1} = \frac{-Z_{21}}{Z_{22} + Z_L} \tag{5.17}$$

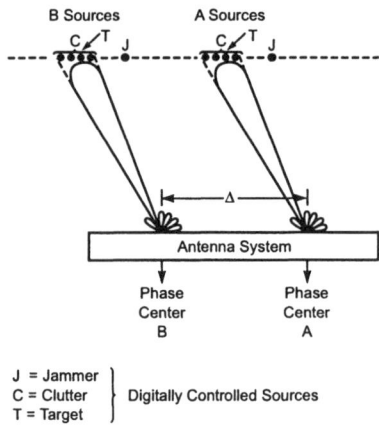

Figure 5.10 Near-field source positioning for a displaced phase center antenna. Two sets of sources, A and B, are used one set at a time to illuminate the test antenna. © 1992 IEEE [4].

where Z_{21} is the open-circuit mutual impedance between the two antennas, Z_{22} is the self-impedance of antenna number 2, and Z_L is the assumed load impedance. Clearly, a small value of mutual impedance is desired. Equation (5.17) will be used later to verify that the parasitic current induced by the source antennas is sufficiently small for a particular near-field source configuration. An important point to be stressed here is that interaction between source antennas in a particular near-field test configuration needs to be carefully evaluated. However, with proper source spacing and placement of anechoic material between source antennas, source interaction can be reduced to acceptable levels.

Having developed the necessary theory, an example adaptive array will be analyzed in the next section. The equivalence between near-field and far-field clutter and jammer suppression will be demonstrated.

5.8 NEAR-FIELD/FAR-FIELD SIMULATIONS

Consider a corporate-fed phased array antenna that consists of a single row of receive dipole elements spanning 16.0 meters. There are a total of $N = 148$ elements in the array with two elements at each end used as passive terminations. In this case, there are 144 active receive antenna elements. As depicted in Figure 5.11, the antenna elements are approximately one-half wavelength long, electrically thin dipoles that are center fed and spaced approximately one-quarter wavelength above an infinite ground plane. The dipoles array elements have a full length of 10.668 cm and a wire diameter of

0.3175 cm, and the dipole spacing to the ground plane is 5.334 cm. The center frequency is chosen to be 1.3 GHz (L-band) and the element spacing is 10.922 cm, which corresponds to 0.473λ. Thus, the active portion of the array spans 15.61 meters. It is assumed that the output from each active receive antenna element is divided into two paths, which are used to form two independent phase centers. The transmit/receive modules are chosen to have 5 bits of amplitude and phase control with rms errors of 0.3 dB and 3.0 degrees. The load impedance, Z_L, is assumed to be 50-ohms resistive at each array element.

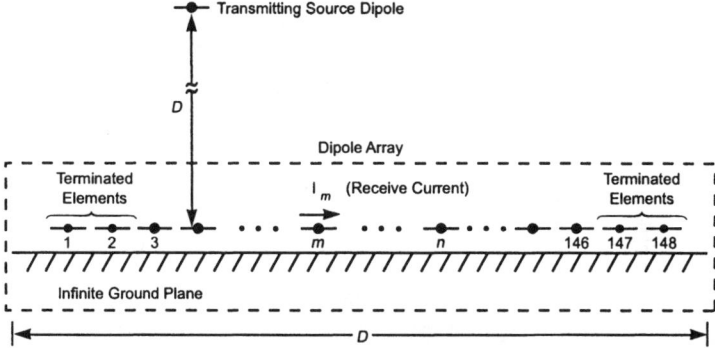

Figure 5.11 Geometry for dipole receive array and dipole source antenna. The transmitting source antenna can represent clutter, jammer, the target, and noise. © 1992 IEEE [4].

The near-field test distance is chosen to be $z = 15.61$ meters, corresponding to one active receive aperture diameter. Seven auxiliary channels selected from the element outputs of one of the phase centers are used to form a multiple sidelobe canceller configuration. To maintain the same dispersion relation for the adaptive array degrees of freedom, this random pattern of auxiliaries is assumed to repeat in the second phase center. The auxiliary elements chosen in phase center A were 3, 4, 10, 28, 29, 72, 74, and in phase center B they were 75, 76, 82, 100, 101, 144, 146. Thus, $N_{aux} = 7$ in each phase center, so the total number of degrees of freedom is sixteen. The channel covariance matrix has dimensions 16×16 and there are sixteen eigenvalues. The receiver bandwidth (also referred to as nulling bandwidth) is assumed to be 1 MHz. It should also be noted that the auxiliary channels are attenuated by 20 dB to have a signal output power comparable to the main channels. Auxiliary channel attenuation would likely be implemented in a practical radar in order to keep the dynamic range of signals within the limits of the adaptive nulling receiver.

Let the array illumination be chosen to synthesize a -40 dB Chebyshev radiation pattern (in the absence of T/R module errors) with a scan angle

$\theta_s = -30°$. Assume that the phase centers are fully split apart, that is, the effective number of receive elements per phase center is one half of 144 or $N_A = N_B = 72$, and this provides a phase center separation of 7.86 meters. Figure 5.12 shows the subarray amplitude illumination function for phase centers A and B. The random amplitude error of the T/R modules makes the illumination functions look slightly different from each other as expected. Notice that each illumination function is equal to zero over 7.86 meters (one-half of the overall aperture). This illumination (weighting) of the array shifts the phase center by 3.93 meters to the left of the antenna center for phase center B and 3.93 meters to the right of the antenna center for phase center A.

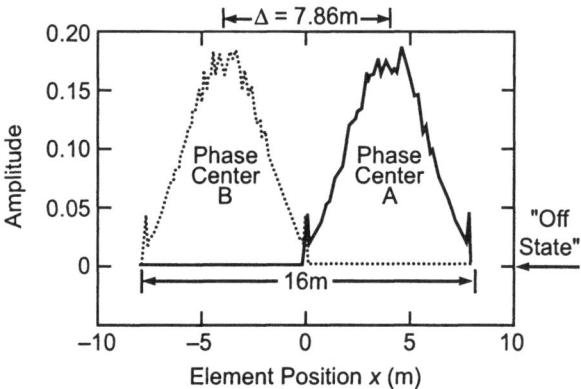

Figure 5.12 Simulated DPCA illumination functions, with T/R module errors, for a 16m linear array of dipoles. The phase center displacement Δ is created by turning off the left half of the array for phase center A and the right half of the array for phase center B. © 1992 IEEE [4].

To phase focus the DPCA array in the near field to the distance $z_s = 15.61$ meters, and angle $\theta_s = -30°$, a CW radiating dipole source is positioned at $x = -12.95$ meters for phase center B and at $x = -5.09$ meters for phase center A. The conjugate of the moment method calculated element phases is applied to the receive modules, which focuses the subarrays. The CW source is then scanned across 26.29 meters and the array output is computed at uniformly spaced probe positions. The resulting near-field received amplitude distribution for both phase centers is shown in Figure 5.13. Notice that the peak amplitudes occur at the desired locations. The main beams are fully separated so that when one phase center has a peak, the other phase center has a sidelobe. It should also be noted that while the test distance has been specified to be one aperture diameter for the full length of the array, the test distance appears to be effectively two subarray diameters for the displaced phase centers. In Figure 5.14, the near-field data are replotted as a function of angle with respect to each phase center. Here it is seen that the near-field main beams of phase centers A and B are well-matched as desired in a DPCA system. The corresponding far-field radiation patterns of phase centers A and B are shown in Figure 5.15 and a good main beam match is also apparent.

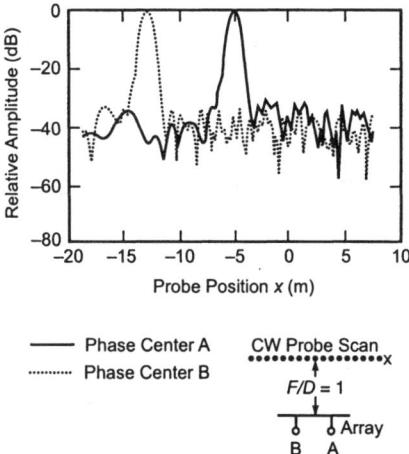

Figure 5.13 Simulated near-field probe scan at one active receive aperture diameter test distance for the DPCA 16-m linear test array. The CW source frequency is 1.3 GHz and the receive-array scan angle is $-30°$. © 1992 IEEE [4].

Next, a comparison of the near-field and far-field radiation patterns is made in Figure 5.16. Figure 5.16(a) shows the radiation patterns for phase center A and Figure 5.16(b) shows the radiation patterns for phase center B. Notice that the near-field patterns cover somewhat different angular sectors. The near-field main beam is in good agreement with the far-field main beam down to about the -20 dB level. There is a defocusing of the near-field first sidelobes for which, as mentioned earlier, amplitude calibration could be used to compensate. Although the near-field sidelobes do not match the far-field sidelobes on a point-by-point basis, the average sidelobe levels are the same.

Having established proper quiescent conditions, let seven clutter sources be uniformly distributed across the main beam of both near-field and far-field patterns. For this two-phase center example, there are seven sources per phase center. In each phase center, all sources are assumed to have equal power and all sources are uncorrelated. Note: Equal-power clutter sources are chosen here for convenience. In terms of angle, these seven clutter sources are distributed over an approximate 5-degree sector centered at the beam peak and they cover the main beam down to the approximate -20 dB level. The first

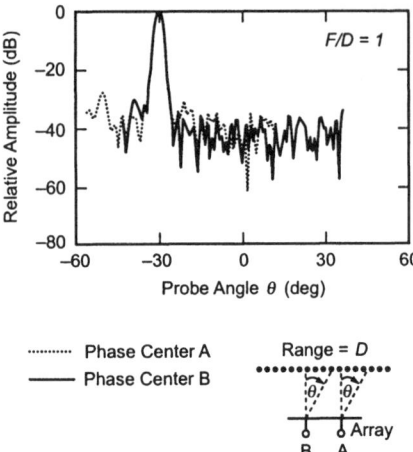

Figure 5.14 Near-field radiation patterns at one active receive aperture diameter test distance for phase centers A and B of the DPCA 16m linear array as a function of observation angle. The radiation patterns are computed from (5.15) and (5.16) using the simulated near-field data presented in Figure 5.13. © 1992 IEEE [4].

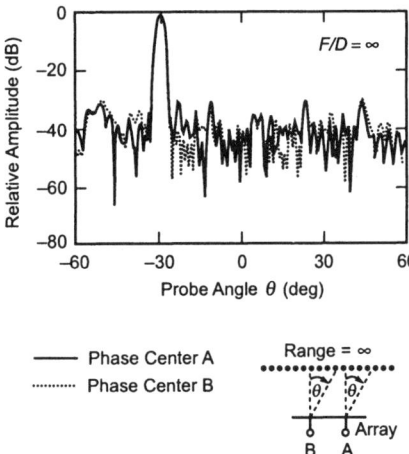

Figure 5.15 Simulated far-field radiation pattern for the DPCA 16m linear array. The focal distance and observation distance are both set to infinity. © 1992 IEEE [4].

clutter source in the near field of phase center B is located at $x = -13.843$m and the seven sources are distributed uniformly over a distance of 182.88 cm. Thus, the spacing between the clutter sources is 30.48 cm, and the rightmost

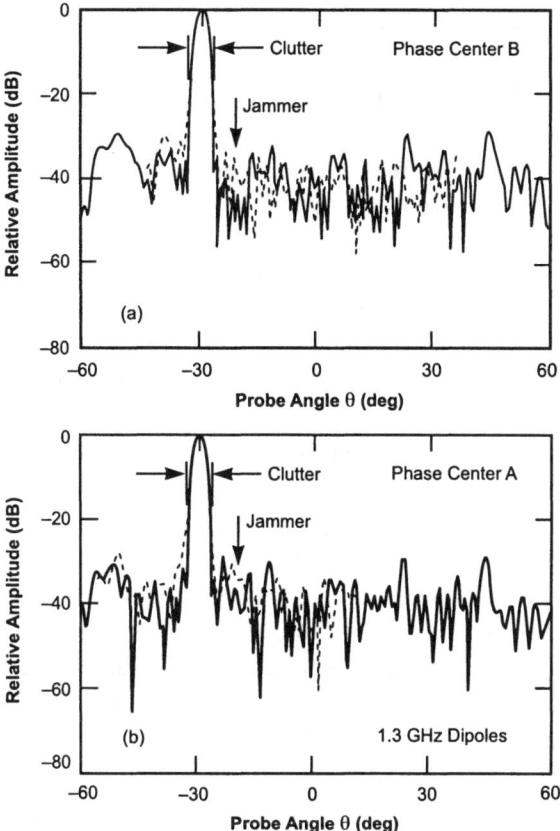

Figure 5.16 Comparison of simulated near-field and far field DPCA 16-m linear array radiation patterns (before adaption) previously shown in Figures 5.14 and 5.15, respectively. The near-field radiation patterns are at a simulated test distance of one active receive aperture diameter. (a) Phase center B radiation patterns and (b) phase center A radiation patterns. © 1992 IEEE [4].

clutter source (source number 7) is located at $x = -12.014$m. For phase center A, the first clutter source is located at $x = -5.983$m. The total power that these sources produce in one phase center is equal to 40 dB relative to receiver noise. Note: Although not shown here, increasing the number of clutter sources beyond seven does not significantly influence the adaptive nulling results that follow. Next, let one jammer be positioned at $\theta = -20°$ (corresponding to $x = -9.69$m) which produces an output power for the combined phase centers of 50 dB above noise. Thus, the jammer is located 2.32m from the nearest clutter source. It can be noted that with seven auxiliary channels per phase center, more than one jammer could be accommodated but will not be considered here.

As mentioned earlier, interaction between near-field signal sources can be an important consideration. For the current example, the interaction between the radiating jammer antenna and a clutter antenna is the most important case to consider. The reason for this is that the sidelobe jammer power is large. The sidelobe level at the jammer position is approximately -35 dB down from the main beam peak. The main beam amplitude at the nearest clutter antenna position is -20 dB. Thus, a parasitically generated jammer signal at the clutter antenna will effectively be increased by 15 dB at the test antenna due to the pattern directivity increase. Without yet taking into account mutual coupling, the parasitic jammer power in the main beam would be 65 dB above noise. The parasitic jammer current at the clutter antenna is computed using (5.17). The self-impedance of a one-half wavelength clutter dipole is $73 + j42$ ohms. The separation between the jammer and the nearest clutter antenna is 2.32m or approximately 10λ at 1.3 GHz. For this spacing, the mutual impedance is computed to be $-0.03748 - j0.000618$ ohms. Substituting these values and the load impedance ($Z_L = 50$ ohms) into (5.17) yields $|I_2|/|I_1| = 0.000289$. In decibels, the parasitic jammer current is down by -71 dB. The parasitic jammer signal is then 65 dB -71 dB$= -6$ dB, or 6 dB below receiver noise. This signal level will not affect the computation of the adaptive weights according to theoretical simulations.

For this signal environment, the covariance matrix was computed from which the ideal adapted weights were derived and then applied in order to cancel the interference. The near-field and far-field adapted radiation patterns are shown in Figure 5.17. In Figure 5.17(a), both the clutter and jammer are clearly suppressed by the pattern nulls, and it is noted that the near-field and far-field patterns are very similar. The total adaptive array cancellation of jamming and clutter power is 48.0 dB in the near field and 49.3 dB in the far field. As the amount of cancellation is large, this small difference (1.3 dB) is insignificant. The radiation patterns shown in Figure 5.17(a) are for the case of a stationary target, whose signal would effectively be cancelled. In contrast, a moving target which produces a 180° phase shift in one PRI would "see" the antenna radiation pattern shown in Figure 5.17(b). Here, full antenna gain is available in the main beam direction. Finally, the covariance matrix eigenvalues are plotted in Figure 5.18. There are a total of eight eigenvalues above receiver noise indicating that eight degrees of freedom are consumed. The near-field and far-field eigenvalues are in good agreement over a large dynamic range. Thus, the degrees of freedom are consumed in a similar manner for near-field and far-field sources.

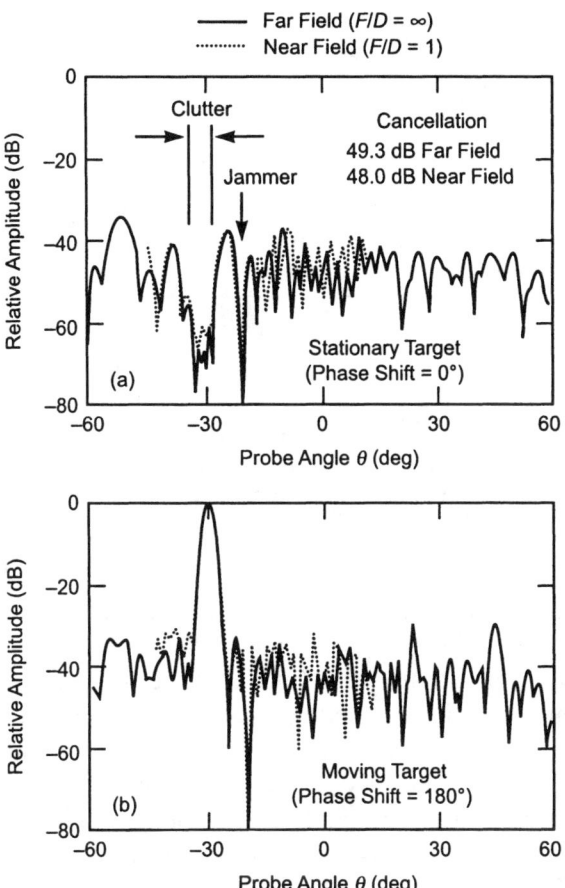

Figure 5.17 Adapted radiation patterns for the combined phase centers of the DPCA 16-m linear array. The near-field simulation is at a one aperture diameter test distance and the far-field simulation is at a test distance of infinity. Seven white-noise clutter sources are uniformly distributed across the main beam and one white-noise jammer is in a sidelobe. The receiver nulling bandwidth is 1 MHz. (a) Stationary target and (b) moving target. © 1992 IEEE [4].

5.9 SUMMARY

In this chapter, a theory for analyzing sources radiating in the near field of an adaptive radar system has been developed. Conventional phase focusing of an array has been used to create antenna far-field pattern conditions in the near-field region. Clutter sources are distributed across the main beam of the focused antenna pattern, and jammers are positioned within the sidelobes.

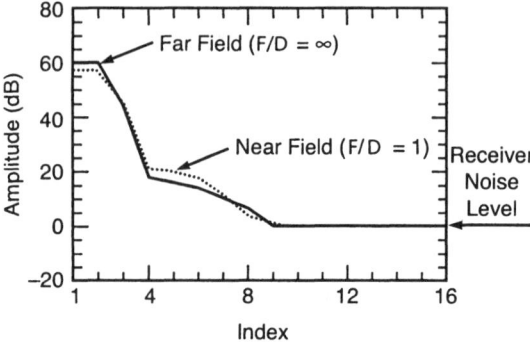

Figure 5.18 Covariance matrix eigenvalues (degrees of freedom) for near-field and far-field source distributions. Eigenvalues 1 to 8 are above the receiver noise level and represent the consumption of eight degrees of freedom. © 1992 IEEE [4].

The method of moments has been used to analyze a displaced phase center antenna linear array with near-field and far-field clutter and jamming. It has been shown that focused near-field adaptive nulling is equivalent to conventional far-field adaptive nulling. The near-field range distance can be one aperture diameter, which opens the possibility for indoor anechoic chamber testing. Thus, a radar system designed for far-field conditions can potentially be evaluated more conveniently using near-field sources. This technique is particularly attractive for space-based radar systems where prelaunch ground testing is desirable. With this method, it is possible to perform integrated testing of a phased array antenna, receiver, and adaptive signal processor. Array calibration, antenna radiation patterns, adaptive cancellation, and target detection could be verified with this technique. In practice, the source antennas can be implemented using dipoles or horns. Experimental verification of the jammer nulling portion of this technique for a single phase center is investigated in the next chapter. The interested reader may want to investigate other applications of focused near-field adaptive antenna systems as described by Fenn [23].

5.10 PROBLEM SET

5.1 Derive (5.17).

References

[1] Cantafio L.J., (ed.), *Space-Based Radar Handbook,* Norwood, MA: Artech House, 1989, pp. 481-528.

[2] Kelly, E.J., and G.N. Tsandoulas, "A Displaced Phase Center Antenna Concept for Space Based Radar Applications," *IEEE Eascon,* September 1983, pp. 141-148.

[3] Tsandoulas, G.N., "Space-Based Radar," *Science,* Vol. 237, July 17, 1987, pp. 257-262.

[4] Fenn, A.J., "Analysis of Phase-Focused Near-Field Testing for Multiphase-Center Adaptive Radar Sysems," *IEEE Trans. Antennas Propagat.,* Vol. 40, No. 8, 1992, pp. 878-887.

[5] Yaghjian, A.D., "An Overview of Near-Field Antenna Measurements," *IEEE Trans. Antennas Propagat.,* Vol. 34, No. 1, 1986, pp. 30-45.

[6] Walter, C.H., *Traveling Wave Antennas,* New York: Dover, 1970, pp. 36-50.

[7] Johnson, R.C., H.A. Ecker, and R.A. Moore, "Compact Range Techniques and Measurements," *IEEE Trans. Antennas Propagat.,* Vol. 17, No. 5, 1969, pp. 568-576.

[8] Fenn, A.J., "Theory and Analysis of Near Field Adaptive Nulling," *1986 IEEE AP-S Symposium Digest,* Vol. 2, IEEE, New York, 1986, pp. 579-582.

[9] Fenn, A.J., "Evaluation of Adaptive Phased Array Far-Field Nulling Performance in the Near-Field Region," *IEEE Trans. Antennas Propag.* Vol. 38, No. 2, 1990, pp. 173-185.

[10] Fenn, A.J., "Theoretical Near Field Clutter and Interference Cancellation for an Adaptive Phased Array Antenna," *1987 IEEE AP-S Symposium Digest, Vol. 1,* IEEE, New York, 1987, pp. 46-49.

[11] Fenn, A.J., "Moment Method Analysis of Near Field Adaptive Nulling," *IEE Sixth International Conference on Antennas and Propagation, ICAP 89,* April 4-7, 1989, pp. 295-301.

[12] Monzingo, R.A., and T.W. Miller, *Introduction to Adaptive Arrays,* New York: Wiley, 1980.

[13] Mayhan, J.T., "Some Techniques for Evaluating the Bandwidth Characteristics of Adaptive Nulling Systems," *IEEE Trans. Antennas Propagat.,* Vol. AP-27, No. 3, May 1979, pp. 363-373.

[14] Scharfman, W.E., and G. August, "Pattern Measurements of Phased-Arrayed Antennas by Focusing into the Near Zone," in *Phased Array Antennas (Proc. of the 1970 Phased Array Antenna Symposium),* A.A. Oliner and G.H. Knittel, (eds.), Dedham, MA: Artech House, 1972, pp. 344-350.

[15] Stutzman, W.L., and G.A. Thiele, *Antenna Theory and Design,* New York: Wiley, 1981, pp. 306-374.

[16] Aumann, H.M., and F.G. Willwerth, "Synthesis of Phased Array Far-Field Patterns by Focusing in the Near-Field," *Proceedings of the 1989 IEEE National Radar Conference,* March 29-30, 1989, pp. 101-106.

[17] Strang, G., *Linear Algebra and Its Applications,* New York: Academic, 1976, p. 213.

[18] Fenn, A.J., "Maximizing Jammer Effectiveness for Evaluating the Performance of Adaptive Nulling Array Antennas," *IEEE Trans. Antennas Propagat.*, Vol. 33, No. 10, October 1985, pp. 1131-1142.

[19] Gupta, I.J., and A.A. Ksienski, "Effect of Mutual Coupling on the Performance of Adaptive Arrays," *IEEE Trans. Antennas Propagat.*, Vol. 31, No. 5, 1983, pp. 785-791.

[20] Fenn, A.J., "Moment Method Analysis of Near-Field Adaptive Nulling," MIT Lincoln Laboratory, Technical Report 842, April 7, 1989.

[21] Richmond, J.H., "Radiation and Scattering by Thin-Wire Structures in a Homogeneous Conducting Medium (Computer Program Description)," *IEEE Trans. Antennas Propagat.*, Vol. 22, No. 2, 1974, p. 365.

[22] Stutzman, W.L., and G.A. Thiele, *Antenna Theory and Design*, New York: Wiley, 1981, pp. 155-159.

[23] Fenn, A.J., *Breast Cancer Treatment by Focused Microwave Thermotherapy*, Sudbury, MA: Jones and Bartlett, 2007, pp. 54-56.

6

Experimental Testing of Focused Near-Field Adaptive Nulling

6.1 INTRODUCTION

As discussed in Chapters 3 through 5, it is desirable to perform near-field testing and evaluation of integrated adaptive antenna systems, for radar or communications applications, prior to their deployment in the field [1-3]. Refering to Figure 5.1, typical adaptive radar or communications systems will consist of an antenna (for example, a corporate-fed phased array), an adaptive nulling receiver that can downconvert and digitize signals from the array main channel and from any auxiliary channels, and a signal processor. The amount of adaptive cancellation of one or more, possibly widely separated, interference sources can be used as a quantitative measure of adaptive system performance. In the situation where the radar or communications antenna has a large aperture and is required to adaptively suppress multiple sources of interference, the conventional far-field range distance $(2D^2/\lambda)$, where D is the maximum antenna aperture dimension and λ is the wavelength, requires large test distances (see Figure 5.3) leading to a requirement for outdoor measurements, and that measurement approach can make the testing of adaptive nulling difficult or impractical. Near-field testing on a short range either outdoors or, preferably, within an anechoic chamber is desirable in an adaptive nulling application, just as it is in antenna pattern and radar cross section measurements.

Chapters 3 to 5 have described the theoretical basis for a near-field adaptive nulling testing technique, which has been theoretically studied. The technique is referred to as *focused near-field adaptive nulling*, which utilizes the ability of an antenna to focus its radiation pattern at a distance of one to two aperture diameters from the adaptive antenna under test. For a phased array antenna, far-field or near-field focusing (see Figure 4.2) can be achieved using the existing phase shifters that are used normally to steer the main beam at a far-field range and angle [4]. Chapter 3 analyzed focused near-field adaptive nulling for the case of isotropic receive antenna elements and for single and multiple near-field interferers. In Chapter 3, both fully adaptive and sidelobe canceller arrays were analyzed. Chapter 4 described an analysis of near-field and far-field adaptive nulling that included array mutual coupling effects for the case of jamming. Chapter 5 analyzed the case of an adaptive displaced phase center antenna (DPCA) for a radar application involving focused near-field suppression of jamming and clutter. This chapter describes experiments that were conducted for focused near-field adaptive nulling for the case of a single jammer [3]. The adaptive nulling algorithms and receiver used in these experiments have been described previously by Johnson et al. [5]. The next section briefly discusses the adaptive nulling receiver and signal processing used in the focused near-field nulling experiments. Section 6.3 describes the prototype phased array antenna and antenna simulation model, and Section 6.4 describes the results.

6.2 ADAPTIVE NULLING SYSTEM DESCRIPTION

6.2.1 INTRODUCTION

With ongoing advances in high-speed digital signal processing technology, the sample matrix inversion (SMI) method [5-9] is a practical approach for adaptive nulling. Although the SMI algorithm has well-known theoretical advantages with regard to convergence, actual performance has been found to be sensitive to receiver channel tracking (frequency mismatch) errors [7]. This section describes an adaptive system used in investigating the channel equalization problem within the context of a digital sidelobe cancelling system. This approach relies heavily on digital filtering techniques to equalize the channels in the nulling system. The prototype adaptive nulling system described here was designed and built with the capability to test multimode adaptive nulling architectures consisting of feedforward (all digital), feedback (hybrid analog), and tandem modes (combined analog and digital). Its modes of operation are depicted in simplified block diagram form in Figure 6.1. In the feedforward mode (Figure 6.1(a)), the channels are equalized digitally and SMI-controlled adaptive nulling occurs in the digital domain.

In the feedback mode (Figure 6.1(b)), analog IF weighting and nulling occurs before narrowband filtering, which helps to avoid channel mismatch – equalization is accomplished digitally and is used in accurately determining the analog nulling weights. In the tandem mode (Figure 6.1(c)), the same SMI algorithm is used to determine analog and digital nulling weights that are applied to the input signals. Only the digital portion of the nulling receiver system is described in this chapter and the reader is encouraged to refer to [5] for further details of the analog feedback and tandem nulling system architectures.

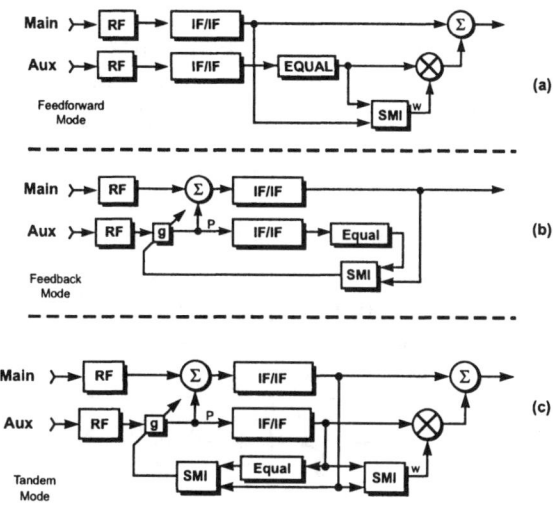

Figure 6.1 Adaptive nulling receiver modes of operation. (a) Feedforward mode in which channel equalization and SMI nulling are accomplished digitally. (b) Feedback mode where nulling is implemented at an intermediate frequency before narrowband filtering. (c) Tandem mode where the same SMI algorithm controls analog and digital nulling weights. © 1991 IEEE [5].

6.2.2 ADAPTIVE NULLING RECEIVER

A four-channel prototype adaptive nulling receiver system that utilizes the sample matrix inversion algorithm with channel equalization has been investigated by Johnson et al. [5]. Figure 6.2 shows a simplified block diagram of the four-channel adaptive nulling system, and Table 6.1 lists the system characteristics. The system has a main channel and three auxiliary channels. Each channel has an initial radio frequency (RF) to intermediate frequency (IF) conversion section. Analog nulling weights are applied in the three auxiliary channels in a mixing process that downconverts the signals to a second IF. After the second IF stage, the signals can pass directly to the final IF-to-IF downconversion section (third IF) and analog-to-digital conversion,

Table 6.1
Adaptive Nulling Receiver Test Bed Characteristics

Number of channels	4
Frequency range	1.25 to 1.35 GHz
Instantaneous bandwidth	1 MHz
Dynamic range	>65 dB
A/D converters	12-bit, 5 MHz
Nulling weights (IF)	16-bits Amp/Phase
First IF	200 MHz
Second IF	30 MHz
Third IF	1.5 MHz

or the signals can be combined in an analog nulling combiner. The switch shown in the main channel allows the main channel signal to pass either directly to the IF/IF section or to be passed to the IF/IF section after passing through the analog nulling combiner.

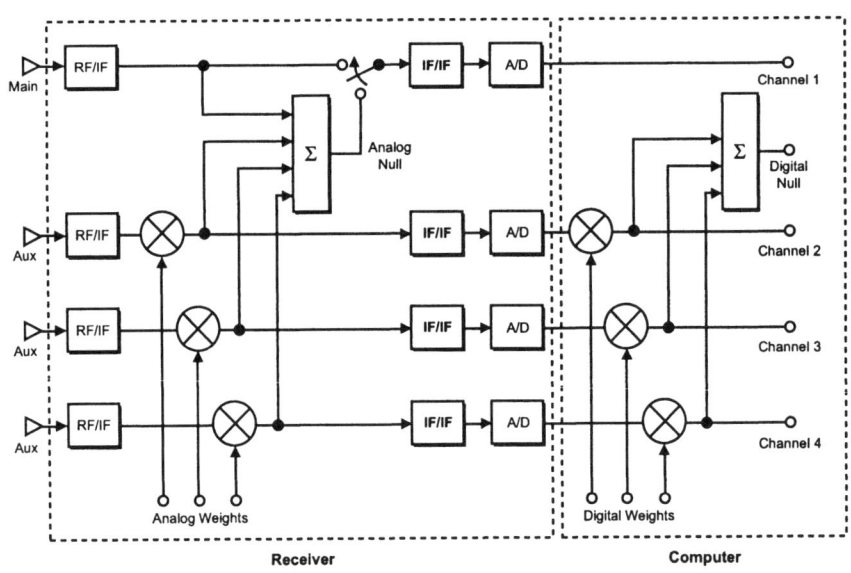

Figure 6.2 Four-channel adaptive nulling receiver block diagram. This receiver allows for both analog and digital nulling. © 1991 IEEE [5].

Consider Figure 6.3, which shows a detailed block diagram for one channel of the four-channel adaptive nulling receiver system. The adaptive nulling receiver described here has a 1-MHz instantaneous nulling bandwidth, and 50 dB or more of adaptive nulling was a design and demonstration goal of the overall system. The received signal in each channel of the receiver passes initially through the front-end LNA and then through an RF bandpass filter that covers the tunable bandwidth (1.25 GHz to 1.35 GHz) of the receiver. An adjustable (tunable) local oscillator (LO) mixes the received RF signal down to a 200-MHz intermediate frequency (first IF), which is then filtered by a 10-MHz bandpass filter. For analog nulling, amplitude weighting is achieved in this receiver using a voltage-controlled attenuator that can be varied over a 0-dB to 40-dB range of attenuation, but for the digital nulling mode it can be assumed that the attenuator is set to 0 dB. Phase weighting is achieved in the analog nulling portion of the receiver in the 170-MHz local oscillator path. For the all-digital nulling mode, the phase shifter value is ignored and can be considered fixed at 0°. The signal in the 200-MHz IF channel is mixed with the 170-MHz LO and then is filtered by a narrowband 1-MHz bandpass filter centered at 30 MHz (second IF). Prior to the narrowband filtering, the signal can be coupled to an analog nulling junction where analog nulling can take place. For digital nulling, once the signal has been been narrowband filtered at 30 MHz, it is then mixed with a 28.5-MHz LO to a third IF using a 2-MHz lowpass filter with the signal then covering the 1-MHz to 2-MHz band. Finally, the signal is digitized by a 12-bit analog-to-digital converter (A/D) sampling the signal at a 4.5 MHz rate.

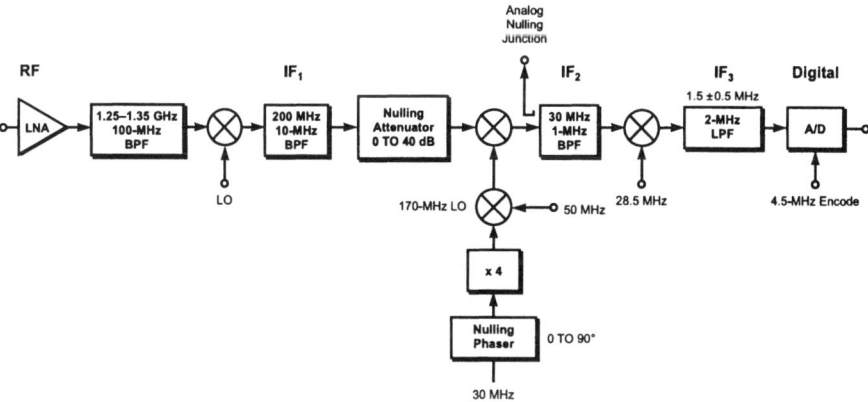

Figure 6.3 Block diagram for one channel of the four-channel adaptive nulling receiver. © 1991 IEEE [5].

6.2.3 RECEIVER CHANNEL EQUALIZATION

In any adaptive nulling receiver design, variations in the frequency response between the different channels will limit the amount of adaptive cancellation that can be achieved unless channel equalization is employed. A digital equalizer (also referred to as a matched filter or finite impulse response filter) can be used to weight and improve the channel match over a given nulling bandwidth [5].

In the all-digital feedforward nulling method described here, a digital equalization filter is implemented in each receiver channel. The main channel serves as a reference channel and the remaining auxiliary channels are equalized with respect to this main reference channel. Tapped delay lines can be utilized as transversal equalizers in wideband analog nulling systems, and the digital equivalent is referred to here as a finite impulse response (FIR) equalizer. Filter coefficients for a FIR equalizer are readily determined using least squares methods. The mathematical solution for FIR equalizer coefficients is similar to the SMI algorithm.

The channel equalization objective is to compute the best possible digital filter with a finite impulse response described by N complex coefficients. Assume that a wideband noise source can be applied to the receiver RF front end (input section) by either an external source or by injection into the receiver channels directly. After passing through the complete nulling circuit of the main and auxilliary channels, the baseband signal of each channel is digitized (sampled) by an analog-to-digital (A/D) converter. Assuming that N equalizer coefficients are sufficient for the desired null depth, digitally produced in-phase and quadrature (I/Q) data are arranged in rows made up of N contiguous samples. For an N-tap equalizer, theoretical results [8] indicate that $M = 5N$ statistically independent rows should yield nearly optimum performance.

The main channel of a sidelobe canceller provides a natural reference for equalizing the auxiliary nulling channels. Since the auxiliary channels are treated the same mathematically, it is sufficient here to consider only a typical auxiliary nulling channel as follows. As described in Johnson et al. [5], let the vectors X and Y represent the $M \times N$ signal matrices for the auxiliary and main channels, respectively. Let the equalizer weighting be represented by the matrix w, which is a column vector of reversed finite impulse response (FIR) coefficients

$$w_n = h_{N-n}, \quad n = 1, 2, \ldots, N \qquad (6.1)$$

In matrix form, the equalizer output can be expressed as Xw. The equalizer weight vector w is chosen to make the equalizer filter output match as closely as possible the reference signal which is assumed to be the signal in the main channel. In this formulation, the auxiliary channels are equalized with respect to the main channel by taking the rightmost column of Y as the equalization target signal. The remaining columns of the main-channel signal matrix Y taken in reverse order (that is, from right to left) represent a sequence of cases where the equalization delay increases from one to $N-1$ sampling intervals. Equalizer filter coefficients are given by the extended weight matrix W that minimizes the difference (denoted as E) between Y and XW, that is,

$$E = Y - XW \qquad (6.2)$$

in the least squares sense, or equivalently,

$$\min\|E\|^2 = \min\|Y - XW\|^2 \qquad (6.3)$$

The residual channel tracking error is given, as a function of equalization delay, by the main diagonal elements of the residual covariance matrix $E^\dagger E$, where † represents the complex conjugate transpose operator. The residual covariance matrix is recognized as a Hermitian matrix. The best equalization performance is obtained by identifying the filter coefficients with the column of W corresponding to the smallest diagonal element of $E^\dagger E$. The least squares solution is derived from the extended signal matrix, denoted Z, as

$$Z = [X \ Y] \qquad (6.4)$$

The QR decomposition of Z, where Q is a unitary matrix and R is an upper (right) triangular matrix, or, equivalently, the Cholesky factorization

$$Z^\dagger Z = \begin{bmatrix} X^\dagger X & X^\dagger Y \\ Y^\dagger X & Y^\dagger Y \end{bmatrix} = R^\dagger R$$

then leads directly to the desired solution. The upper right triangular matrix R can be partitioned similarly such that

$$R = \begin{bmatrix} U & V \\ 0 & T \end{bmatrix}$$

where U and T are upper triangular matrices. From Johnson et al. [5], the least squares solution for the equalizer weights is given by

$$W = U^{-1}V \qquad (6.5)$$

The matrix T is the Cholesky triangle of the residual covariance matrix, that is,

$$E^\dagger E = T^\dagger T \qquad (6.6)$$

Consequently, the residual channel tracking errors can be expressed as

$$|E_{nn}|^2 = \sum_{m=1}^{n} |T_{mn}|^2 \qquad (6.7)$$

and can be computed without having to first solve $UW = V$ for the filter coefficients.

It is of interest to determine the channel frequency response of the main channel and auxiliary channels, which is accomplished as follows. To apply the least squares procedure in the frequency domain at M frequencies, that is, $f_m, m = 1, 2, \ldots, M$, the signal matrix of the auxiliary channel is expressed as

$$\boldsymbol{X}_{mn} = \boldsymbol{X}(f_m) D^{-n}(f_m) \qquad (6.8)$$

where $\boldsymbol{X}(f)$ denotes the auxiliary channel frequency response and

$$D(f) = e^{-j2\pi f \Delta t} \qquad (6.9)$$

is referred to here as the unitary delay function and Δt is the sampling interval. The calibration frequencies and the sampling interval must be known explicitly in order to apply (6.9) in (6.8). From Johnson et al. [5], the unitary delay $D(f_m)$ can be determined directly from calibration data using a correlation technique. Thus, for the correlation technique let $x_m(k)$ represent the kth complex (I/Q data) sample data obtained from the mth calibration record. Each record represents the noise-corrupted response of a particular receiver channel to a CW calibration signal with some fixed but possibly unknown frequency. The correlation is expressed as

$$C_m = \sum_{k=1}^{K} x_m(k) x_m^*(k-1) \qquad (6.10)$$

where * denotes complex conjugate. Setting

$$D(f_m) = \frac{|C_m|}{C_m} \qquad (6.11)$$

effectively estimates the product $f_m \Delta t$ and the channel frequency response is then obtained by correlating the complex (I/Q) data for x with D^k, that is,

$$\boldsymbol{X}(f_m) = \sum_{k=1}^{K} x_m(k) D^k(f_m) \qquad (6.12)$$

6.3 PROTOTYPE PHASED ARRAY AND SIMULATION MODEL

The prototype phased array antenna used in the experiments is an L-band (1.25 to 1.35 GHz) horizontally polarized linear array containing 36 half-height open-ended rectangular waveguide elements with inside dimensions 4.07 cm by 16.51 cm. Two guard-elements at both ends of the array were terminated in 50-ohm resistive loads. The remaining 32 elements were connected to transmit/receive (T/R) modules. The element spacing (center-to-center) is 10.92 cm and so the length of the receive portion of the linear array is 3.4m. Three auxiliary channels at element positions 1, 16, 32 (left edge, middle, right edge) were used to collect the multiple sidelobe canceller signals. The aluminum ground plane has dimensions of 0.6m by 4.36m. Referring to Chapter 12, Figure 12.7, the T/R modules have 12 bits of amplitude and phase control [10]. Each module downconverts the L-band signal to a 30-MHz intermediate frequency (IF). Beamforming is implemented at the 30-MHz IF. The four-channel adaptive nulling receiver described in Section 6.2 was used in the measurements. Each receiver channel contains 1-MHz bandwidth narrowband filters and 12 bit analog-to-digital converters operating at a 4.5 MHz sampling rate. The sample matrix inversion algorithm together with channel equalization are used to digitally null the interference signal. A block diagram for the integrated phased array antenna/adaptive nulling receiver is shown in Figure 6.4. In Figure 6.4, the phase shifters in the T/R modules are denoted as PHS. Similarly, the attenuators in the array modules are denoted as ATT, and the pair of attenuators provide A and B channel outputs. The array has two independent beamformers that could be used to generate two displaced phase centers as would be used in a displaced phase center antenna (DPCA), as discussed in Chapters 2, 5, and 11, although only a single phase center was used in these tests. One of the beamformers was used to generate the main beam signal from the 32 driven waveguide array elements. The B-channel output of three modules connected to the auxiliary channels were fed directly into the four-channel nulling receiver. A source horn was located at a fixed range distance of 4.6m, which corresponds to a focal distance of 1.35 active-array diameters, and transmits either a CW tone for array calibration or wideband noise for interference. A photograph of the outdoor test configuration is shown in Figure 6.5.

The computer simulation model of the array and jammer is described as follows. Dipoles can behave similar to waveguide elements in an array environment as shown by Diamond and Bernella [11]. They showed that the theoretical scan impedance of a dipole array without a ground plane is approximately equal to that of a rectangular waveguide array. As the dipole

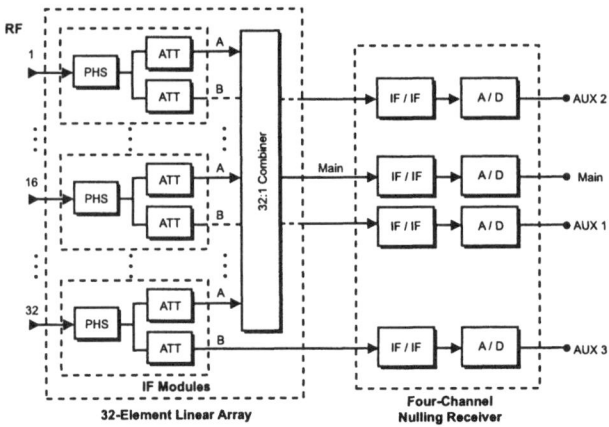

Figure 6.4 Block diagram for the test array configuration with the four-channel adaptive nulling receiver. © 1990 IEEE [3].

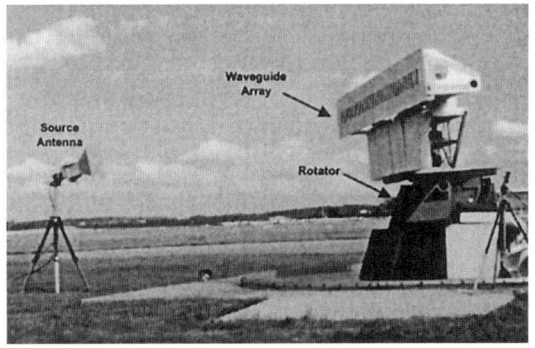

Figure 6.5 Photograph showing the focused near-field nulling test configuration. © 1990 IEEE [3].

array is the dual of the waveguide array, E- and H-planes are interchanged. Thin dipoles make the analysis considerably simpler here compared to analyzing rectangular waveguides, which are used in the prototype array. The theoretical array model is compared to the prototype linear array of waveguides in Figure 6.6. (Note: The finite rectangular waveguide array analysis provided in Chapter 16 would be another way to analyze the linear array of waveguides.) The spacing of the dipoles is the same as for the waveguide array. The near-field source dipole is copolarized with the dipole array and is constrained to lie on a line parallel to the array. This differs slightly from the experimental arrangement where the source antenna is fixed and the test antenna rotates, keeping the range constant.

Array mutual coupling effects (and source to array coupling effects) are taken into account using a sinusoidal-Galerkin moment method formulation. The adaptive nulling dipole array and near-field source dipole (CW calibration dipole and jammer dipole) are analyzed using the same method of moments formulation presented in Chapter 4. The analysis assumes a 1-MHz nulling bandwidth.

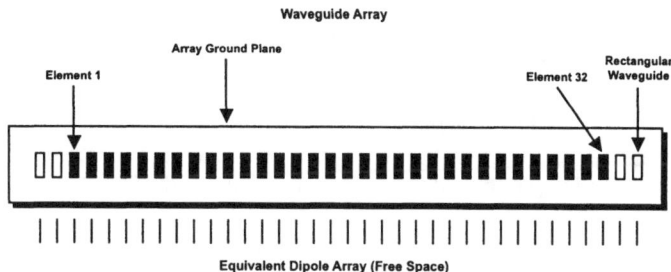

Figure 6.6 Theoretical array model. An array of linear dipoles in free space is used to model the rectangular waveguide array.

6.4 RESULTS

6.4.1 EQUALIZATION PERFORMANCE

Receiver channel tracking performance was investigated as a function of the number of equalizer taps (filter coefficients) allowed [5]. The equalizer results shown in Figure 6.7 indicate that 16 taps are sufficient for achieving 50-dB cancellation in the experimental nulling system. If channel equalization is not used, the channel tracking error varies between −30 dB and −10 dB. These tracking errors can be attributed primarily to the narrowband cavity filters. One of the cavity filters (Channel D in Figure 6.7) had a substantial deviation in channel characteristics from the other cavity filters. Figure 6.8 shows the frequency response of the four receiver channels before and after equalization using 16 taps. The significant amplitude difference between the channels that was evident in Figure 6.8(a) has been eliminated, as shown in Figure 6.8(b).

6.4.2 FOCUSED NEAR-FIELD NULLING PERFORMANCE

The phase shifters in the T/R modules for the waveguide array were calibrated, using a near-field CW radiating source at 1.3 GHz, to steer the receive main beam at $\theta = -30°$ with a focal distance equal to 1.35 active-array diameters. The quiescent near-field radiation pattern (−20 dB Chebyshev taper) was then

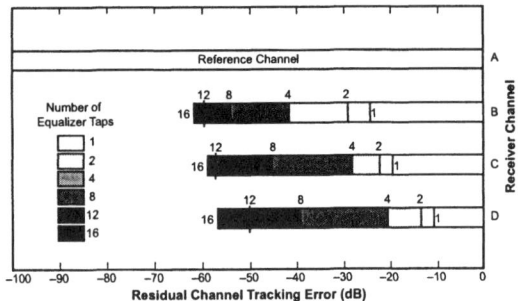

Figure 6.7 Measured FIR equalization performance in terms of residual channel tracking error when the number of equalizer taps is varied from 1 to 16 for the four-channel adaptive nulling receiver. © 1991 IEEE [5].

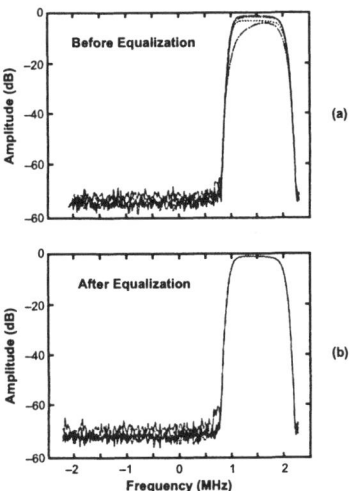

Figure 6.8 Measured frequency response for the four channels of the adaptive nulling receiver. (a) Before equalization, (b) after equalization with 16 taps. © 1991 IEEE [5].

measured using a noise source that had a bandwidth of 50 MHz. In Figure 6.9, the measured pattern is compared to the dipole array theoretical pattern, which is computed using a CW radiating signal, and good agreement is observed. The noise source was then used to illuminate the prototype array on a sidelobe close to broadside. Prior to adaptive nulling, the main channel output power was approximately 50 dB above the receiver noise level. Digitally equalized channel adaptive weights were computed, and were then frozen while the antenna was rotated in azimuth with the noise interference illuminating the array. The measured received power at the adaptive array output as a function of azimuth angle is shown in Figure 6.10. The simulated radiation pattern utilizes a CW probe source to compute the pattern, with adaptive weights that are computed based on nulling a wideband noise source over a 1-MHz nulling bandwidth. The adaptive cancellation of the noise source is approximately 50 dB in both the measurements and in the simulation. From theoretical simulations the same level of cancellation would be achieved in the far field.

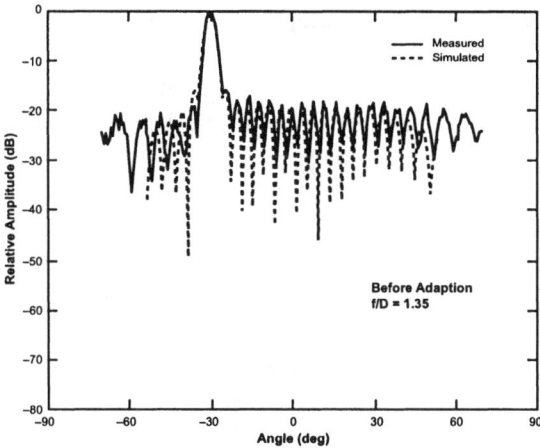

Figure 6.9 Measured and simulated focused near-field radiation patterns before adaption. © 1990 IEEE [3].

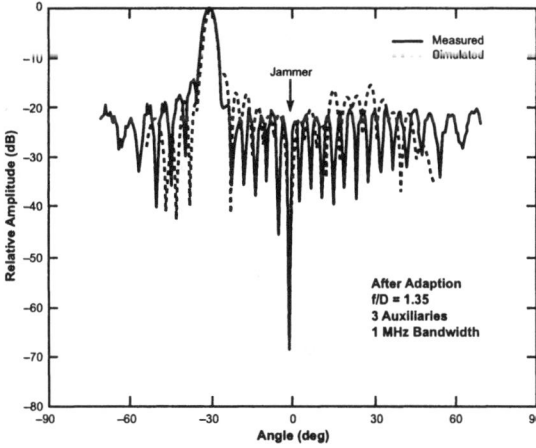

Figure 6.10 Measured and simulated focused near-field radiation patterns after digital nulling of one near-field jammer. Three auxiliaries are used in the nulling process and the nulling bandwidth is 1 MHz. © 1990 IEEE [3].

6.5 SUMMARY

In this chapter, focused near-field adaptive nulling has been demonstrated using a prototype linear array antenna and four-channel nulling receiver. The measured near-field focused array radiation patterns and adaptive cancellation for one near-field interference source were shown to be in good agreement with array simulations. The results indicate that adaptive antenna performance testing using focused near-field adaptive nulling, at approximately one to two aperture diameters range distance, is a viable alternative to conventional far-field adaptive nulling testing. Future experimental studies of focused near-field adaptive nulling could consider multiple jammers and the addition of simulated clutter and target sources.

References

[1] Fenn, A.J., "Evaluation of Adaptive Phased Array Far-Field Nulling Performance in the Near-Field Region," *IEEE Trans. Antennas Propag.*, Vol. 38, No. 2, 1990, pp. 173-185.

[2] Fenn, A.J., "Analysis of Phase-Focused Near-Field Testing for Multiphase-Center Adaptive Radar Sysems," *IEEE Trans. Antennas Propagat.*, Vol. 40, No. 8, August 1992, pp. 878-887.

[3] Fenn, A.J., et al., "Focused Near-Field Adaptive Nulling: Experimental Investigation," *1990 IEEE Antennas and Propagation Society International Symposium Digest,* Vol. 1, May 7-11, 1990, pp. 186-189.

[4] Scharfman, W.E., and G. August, "Pattern Measurements of Phased-Arrayed Antennas by Focusing into the Near Zone," in *Phased Array Antennas (Proc. of the 1970 Phased Array Antenna Symposium)*, A.A. Oliner and G.H. Knittel, (eds.), Dedham, MA: Artech House, 1972, pp. 344-350.

[5] Johnson, J.R., A.J. Fenn, H.M. Aumann, and F.G. Willwerth, "An Experimental Adaptive Nulling Receiver Utilizing the Sample Matrix Inversion Algorithm with Channel Equalization," *IEEE Trans. Microwave Theory Techniques*, Vol. 39, No. 5, 1991, pp. 798-808.

[6] Reed, I.S., J.D. Mallet and L.E. Brennnan, "Rapid Convergence in Adaptive Arrays," *IEEE Trans. Aerospace and Electronic Systems*, Vol. 10, 1974, pp. 853-863.

[7] Horowitz, L.L., et al., "Controlling Adaptive Antenna Arrays with the Sample Matrix Inversion Algorithm," *IEEE Trans. Aerospace and Electronic Systems*, Vol. 15, 1979, pp. 840-848.

[8] Horowitz, L.L., "Convergence Rate of the Extended SMI Algorithm for Narrowband Adaptive Arrays," *IEEE Trans. Aerospace and Electronic Systems*, Vol. 16, 1980, pp. 738-740.

[9] Manolakis, D.G., V.K. Ingle, and S.M. Kogon, *Statistical and Adaptive Signal Processing: Spectral Estimation, Signal Modeling, Adaptive Filtering, and Array Processing,* Norwood, MA: Artech House, 2005, pp. 659-669.

[10] Aumann, H.M., and F.G. Willwerth, "Intermediate Frequency Transmit/Receive Modules for Low-Sidelobe Phased Array Application," *Proc. of the 1988 IEEE National Radar Conference*, 1988, pp. 33-37.

[11] Diamond, B.L., and D.M. Bernella, "Correlation Between Dipole Array Calculations and Waveguide Array Measurements," *1966 IEEE AP-S International Symposium Digest*, 1966, pp. 372-379.

7

Experimental Testing of High-Resolution Nulling with a Multiple Beam Antenna

7.1 INTRODUCTION

The adaptive array antenna concepts and techniques discussed in the first six chapters of this book could be used in a variety of applications for radar and communications. Referring to the conceptual diagrams in Figures 1.1 and 1.2, in some adaptive nulling antenna applications, a multiple beam antenna (MBA) could be considered for producing multiple high gain beams arranged over a desired field of view. For example, future geosynchronous extremely high frequency (EHF) communications satellite uplink receiving antennas with high-resolution adaptive nulling capability could provide sufficient pattern gain to desired users while maintaining pattern nulls on interference sources (jammers) in close proximity to the users. High resolution or pattern discrimination between users and jammers can be achieved by using an adaptive antenna with a large electrical diameter that can produce narrow nulls [1-9]. However, for the case of a high-resolution single filled-aperture MBA, the required number of beams can be excessive for uniform pattern coverage of a given region. As described in this chapter, a distributed (sparse) multi-aperture multiple beam antenna with highly overlapped beams can provide high-resolution nulling. As discussed earlier, high-resolution nulling for a satellite system can be important for system users to maintain a communications link in the presence of jamming. Figure 7.1 depicts the field of view from a conceptual communications satellite where there are multiple users and a single jammer that is located close to a user terminal. For a satellite communications antenna at geosynchronous altitude (altitude 35,786 km above mean sea level), the Earth field of view is approximately 18°. An example angular separation between a jammer and user of about 0.1° requires a high-resolution null from geosynchronous altitude to cancel the jamming, without severely impacting the signal-to-noise ratio (SNR) for the user.

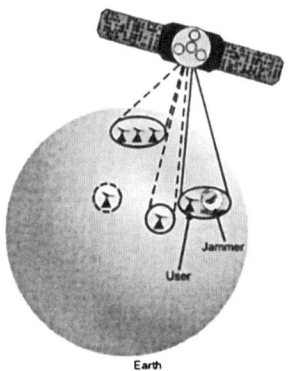

Figure 7.1 Example of a communications satellite with multiple users and a jammer operating in close proximity to a communications user.

Multiple beams are necessary to provide simultaneous coverage areas. These multiple beams would be repositioned rapidly to provide time-division multiple access communications capability. User terminals can be located within a given region or they may be located globally. In the case where jamming is present, the uplink communications performance can be significantly enhanced with adaptive nulling and, for example, 30 dB or more of adaptive nulling could be desired.

For satellite applications, lightweight adaptive nulling antennas are important. Millimeter-wave antennas can provide multiple-gigahertz wide bandwidth capability. In this chapter, the design and measured adaptive nulling performance of a prototype lightweight EHF (43.5 to 45.5 GHz) 127-beam multiaperture multiple beam antenna is investigated. Details of the MBA design and construction are given in the next section. The nulling weight network and nulling algorithm used for anechoic chamber tests are discussed. Radiation pattern measurements and nulling tests with one and two simulated jammers are described in Section 7.3.

7.2 MBA DESIGN AND CONSTRUCTION

The antenna described here achieves high-resolution adaptive nulling by means of a distributed (sparse) multiple aperture design. Figure 7.2 depicts the interlaced beam pattern of a seven-lens 127-beam multiple beam antenna

(Figure 7.3). With a thinned MBA approach, adjacent beams are highly overlapped and are generated from different apertures. The antenna concept uses seven space-fed lenses of 24 cm diameter, located at the center and around the periphery of a 122 cm circular structure, as shown in Figure 7.4. Six of these lenses are fed by 18 horns and one by 19 horns, producing 127 possible beam positions, as shown in Figure 7.3. A switching matrix comprised of two switch tree waveguide networks at each lens selects any two of its beams to feed its receivers, providing 14 receiving channels that are fed into the nulling processor. Based on the theory developed in Chapter 1, this multiple beam antenna has fourteen degrees of freedom for forming adaptive nulls. Fixed time delay lines in each channel are used to remove dispersion introduced by the distributed lenses. The nulling processor depicted in Figure 7.2 produces, for example, three independent user beams, in which adaptive nulls are used to maintain the desired user communications links in the presence of jamming. For this example, these user beams can be configured to cover a 1.5° area for many users while simultaneously servicing two independent users at arbitrary positions over the Earth field of view.

Adaptive antenna pattern nulling measurements with a four-channel RF nulling weight network and CW interference source are described. A nulling angular resolution of approximately 0.1° is demonstrated with the four-channel test bed.

The multiple beam antenna described in this chapter was designed to survive launch and operate in a space environment. A layout for the seven distributed lenses and the 127 beams of the multiple beam antenna prototype are shown in Figure 7.3. The lenses and the radiation beams are numbered 1

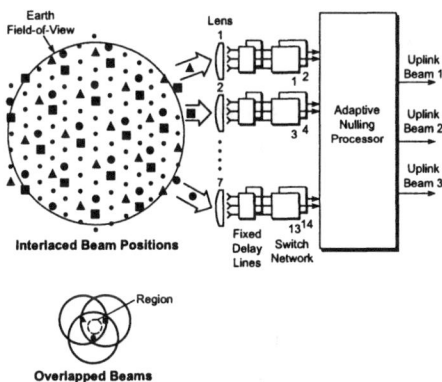

Figure 7.2 Interlaced beams and switch network for a seven-lens 127-beam adaptive nulling multiple beam antenna.

through 7. Four of the adjacent beams, from lenses 2, 4, 5, and 7, are used in the adaptive nulling measurements described in the next section.

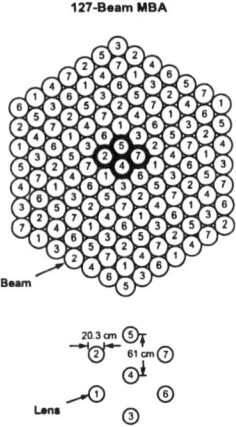

Figure 7.3 Layout for the 127 beams and the seven lenses in the lightweight multiple beam antenna. The four beams, indicated by the heavy-line circles, were used in anechoic chamber adaptive nulling measurements.

With the multiaperture MBA design, seven individual multiple beam antennas are located on a hexagonal lattice having a spacing of 61 cm between apertures, as depicted in the lightweight aluminum structure shown in Figure 7.4. The depth of the MBA lens structure is 60.8 cm and the depth of the switch tree bay is 15.2 cm. The 61 cm aperture spacing of the MBA lenses was chosen, by means of a computer simulation model, to provide the desired nulling resolution of 0.1° such that the adapted antenna gain recovers to within -15 dB of the peak beam gain prior to adaptive nulling. The center aperture is a 19-beam multiple beam antenna and the six surrounding MBA aperture antennas have 18 beams/feeds per aperture. The desired minimum gain for three independent beams covering a 1.5° area diameter is assumed to be 31 dBi. The peak gain of an aperture with area A can be expressed as

$$G = \eta \frac{4\pi A}{\lambda^2} \qquad (7.1)$$

where η is the aperture efficiency and λ is the wavelength.

The peak gain of each beam was desired to be approximately 37.0 dBi at the center frequency (44.5 GHz). From (7.1), to achieve this gain requires a lens with diameter 20.3 cm assuming a 60 percent aperture efficiency. The half-power beamwidth of each of the 127 beams is approximately 2.5° and the 6-dB beamwidth is approximately 3.4°, which satisfies the 1.5° area diameter.

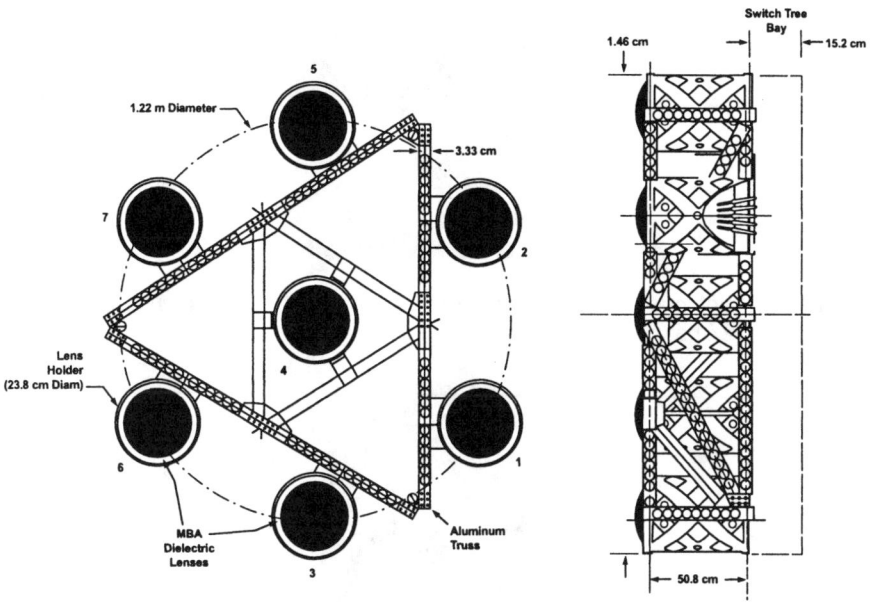

Figure 7.4 Diagram depicting a front and side view of the lightweight multiple beam antenna.

The 127 beams are highly overlapped (−1.3 dB relative to the beam peak) and spaced at 1.6° intervals to fully cover the Earth's field of view from geosynchronous altitude.

A photograph of one of seven of the fully assembled MBAs is shown in Figure 7.5, and a photograph of one of the MBA feed horn clusters is shown in Figure 7.6. A diagram illustrating the feed switch tree network is shown in Figure 7.7. A photograph of the switch tree assembly mounted on the back of the feed horn cluster plate is shown in Figure 7.8. A photograph of the fully assembled 127-beam antenna in an anechoic chamber is shown in Figure 7.9.

Each of the seven MBAs has a mass of approximately 2.8 kg and each is comprised of a right-hand circularly polarized (RHCP) conical horn feed cluster illuminating a 20.3-cm diameter Rexolite lens (relative dielectric constant ϵ_r=2.54). The front (Earth-side) and back (feed-side) surfaces of each lens are waffled to reduce mismatch losses. The feed-side lens surface is zoned into two flat surfaces to reduce the lens mass. The focal length to diameter (F/D) ratio for each lens is 1.5. In each MBA lens, the beams are offset from broadside by displacing the feeds, on a spherical surface, from the focal point of the lens. Final alignment of the individual multiple beam feeds within the lightweight mechanical structure, shown in Figure 7.4, is achieved

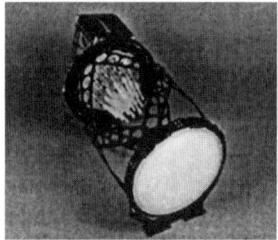

Figure 7.5 Photograph of the central 19-beam multiple beam antenna used in the 127-beam MBA.

Figure 7.6 Photograph of one of the multiple beam antenna feed horn clusters.

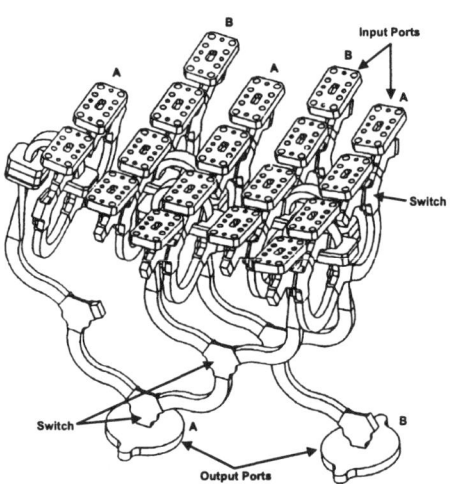

Figure 7.7 Three-dimensional view of an 18:2 switch tree used in the 127-beam multiple beam antenna.

with measurement of the beam peak positions and subsequent correction of each feed orientation by means of mechanical shims. The total mass of the

Figure 7.8 Photograph of one of the multiple beam antenna switch tree assemblies.

Figure 7.9 Photograph of the fully assembled 127-beam multiple beam antenna mounted in an anechoic chamber.

fully assembled 127-beam MBA, including the switch trees, is 25.3 kg.

A lightweight switch tree waveguide beamformer is used to select two receive beams from each distributed MBA aperture for a total of fourteen simultaneous receive beams. The RF switch structure and the switch drive electronics were manufactured by Electromagnetic Sciences, Inc. The switch trees utilize a 4:1 switch tree as the basic building block. The insertion loss for a 4:1 switch tree is approximately 0.5 dB and the isolation is approximately 50 dB to any off port. The switching speed is less than 1 μs. A lightweight switch tree structure is achieved by using flangeless waveguide interconnects, thin-wall electroformed waveguide, and ferrite switches. A sketch of an 18:2 switch tree that feeds one of the 18-beam MBAs is shown in Figure 7.7. This switch tree is composed of an 8:1 switch tree and a 10:1 switch tree. The total mass of the switch tree including waveguide, electronics, brackets, and other mounting hardware is 666g. The 18 waveguide input flanges are mounted on a thin aluminum spherical feed horn plate that has a mass of 286g.

The feed horns are made of electroformed copper with a gold finish and have a rectangular (WR22) to circular waveguide transition section (1.27 cm long), followed by a uniform circular waveguide section (0.686 cm diameter, 3.175 cm long) containing a dielectric-vane polarizer that generates the desired RHCP signal and a resistive card to reduce undesired left-hand circular polarization (LHCP) (cross-polarized) signals. To obtain the desired illumination of the MBA lens, a thin-wall Rexolite tube is mounted in the circular waveguide feed aperture of each conical horn. The effect of the dielectric tube is to narrow the beam pattern from the conical horn, which reduces the lens spillover and improves antenna efficiency. The conical feed horn diameter is 2.1 cm and the Rexolite tube with a wall thickness 0.254 mm and a diameter of 0.64 cm, extends 5.1 cm from the feed horn aperture. The total length of the feed horn, including the dielectric tube and brass flange, is 12.55 cm. The effective axial length of the conical horn is 3.9 cm. The mass of one of the feed horns is approximately 8g.

For anechoic chamber measurements, the test bed RF nulling weight network consisted of four rectangular waveguide transmission lines with four pairs of mechanically adjustable attenuators and phase shifters. Waveguide shims were used to accurately match the line lengths of the four rectangular waveguide transmission lines. The attenuators and phase shifters were set with an accuracy that enabled a cancellation of 40 dB in bench tests with a jammer simulator.

In the anechoic chamber tests, adaptive weight control was determined by the sample matrix inversion (SMI) algorithm, as discussed in Chapter 3, operating in an open-loop mode. The jammer covariance matrix for four beams was obtained by measuring the received signal at five equally spaced frequencies across the 2-GHz nulling bandwidth, and then computing the frequency-averaged cross-correlations between channels as in (3.7). The quiescent weight vector was chosen to have unity amplitude on one of the beams with the three remaining beam weights set to zero. The adaptive weights are computed from the product of the covariance matrix inverse and the quiescent weight vector, as given by (3.9). The adaptive weights were accurately applied by utilizing calibration data obtained from bench tests of the RF nulling weight network.

7.3 MEASURED RESULTS

Radiation pattern measurements of the feed horns were conducted in an anechoic chamber. A typical radiation pattern at 44.5 GHz of one of the feed horns, with Rexolite tube, is shown in Figure 7.10, where the half power beamwidth is observed to be 18°. The measured boresight gain of the feed horn is approximately 20.9 dBi at 44.5 GHz. The feed horns have a measured axial ratio at boresight of approximately 0.5 dB (maximum) over the 43.5- to 45.5-GHz band.

A compact range reflector system, with a 2m quiet zone, was used in the anechoic chamber measurements of the 127-beam MBA. A measured RHCP radiation pattern cut, over $\pm 4.5°$ for the center beam of the center lens is shown in Figure 7.11. The measured half-power beamwidth is 2.5° and the first sidelobe is less than approximately -24 dB.

Jammer angle of arrival was simulated by mechanically rotating the 127-beam MBA to the desired angle and by measuring the jammer received complex signal vector. The output signals from four of the MBA beams were fed into the four-channel RF nulling weight network. A 4×4 jammer

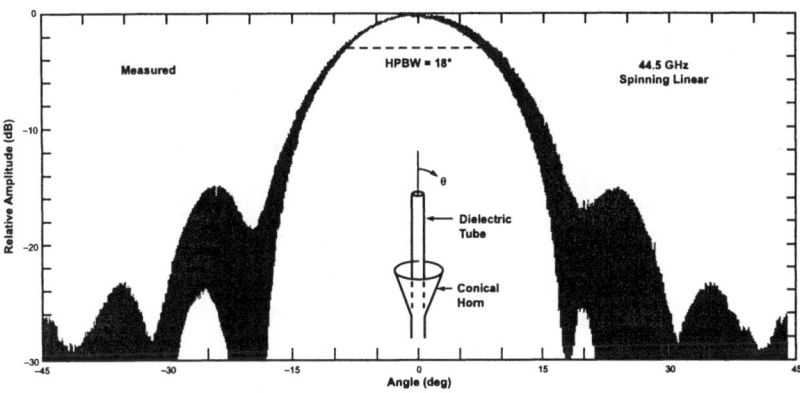

Figure 7.10 Measured radiation pattern for a feed horn used in the 127-beam MBA.

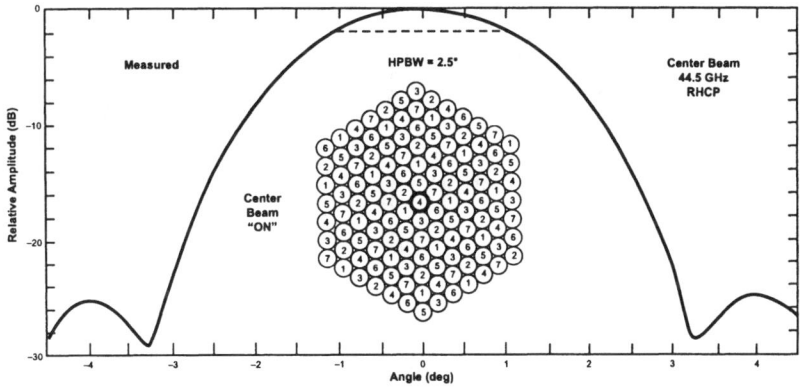

Figure 7.11 Measured radiation pattern for the center beam of the 127-beam MBA.

covariance matrix was computed from the signals measured in the four RF nulling channels.

In the nulling experiments, four adjacent beams were selected as depicted in Figure 7.3, by using the ferrite switches in the switch tree network. The MBA quiescent conditions have one beam (beam 2) turned on, as shown in the two-dimensional radiation pattern in Figure 7.12(a). A single simulated jammer was positioned at AZ= $-0.4°$, EL= $0.85°$ as indicated by the small square box with the jammer position indicated by the capital letter J. Figure 7.12(b) shows an enlarged view (square box region in Figure 7.12(a)) of the adapted radiation pattern, where a null is clearly formed in the direction of the jammer and the azimuth null width is on the order of $0.1°$. Figure 7.12(c) is an azimuth cut through the jammer elevation angle before and after adaptive nulling. The amplitude ripple in the adapted pattern is expected, due to interference between the highly overlapped beams; however, the minima in the ripple pattern are not deep enough to disrupt communications. The increase in gain by approximately 6 dB over the quiescent pattern is also expected and, again, is attributed to the highly overlapped beams. Figure 7.13 shows the null depth in the direction of the

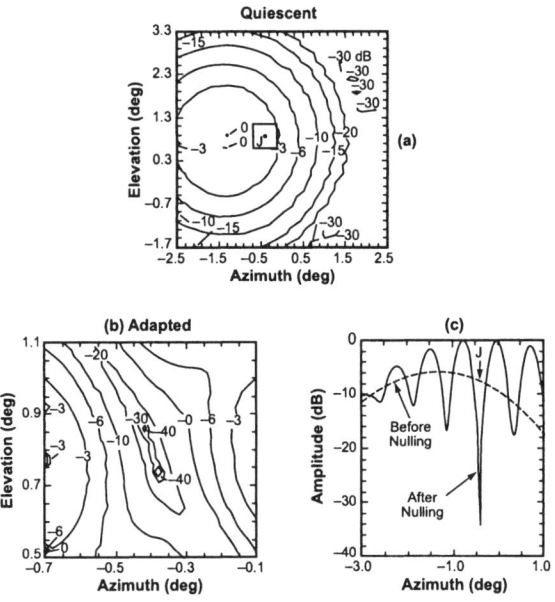

Figure 7.12 Comparison of two-dimensional and one-dimensional quiescent and adapted radiation patterns measured for the 127-beam MBA for a single jammer located at AZ= $-0.40°$, EL= $0.85°$. (a) Before adaption, (b) after adaption, and (c) azimuth cut before and after adaption.

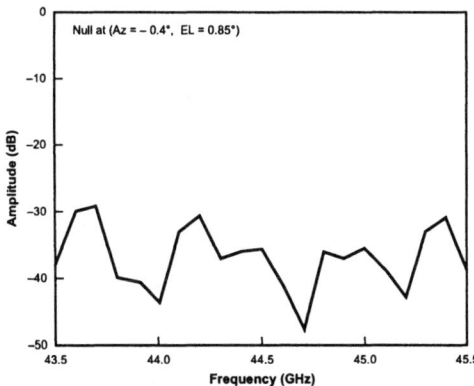

Figure 7.13 Measured adaptive null depth versus frequency achieved by the multiple beam antenna for a single jammer.

jammer as a function of frequency. In Figure 7.13, a cancellation of greater than 30 dB is observed over most of the 2 GHz nulling bandwidth. Figure 7.14 shows measured nulling results for the case of two jammers (located at AZ= −0.7°, EL= 0.4° and at AZ= 0.7°, EL= 0.7°). A cancellation of greater than 30 dB occurs in the direction of each jammer. For the interested reader, further details of this 127-beam MBA design and adaptive nulling results are discussed in [7-10].

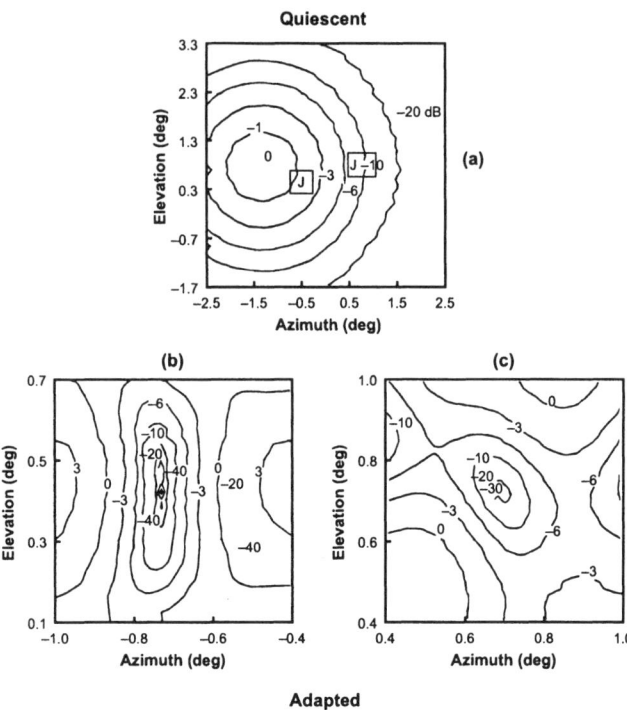

Figure 7.14 Adaptive nulling test results for two jammers. (a) Quiescent radiation pattern two-dimensional contours, (b) adaptive radiation pattern for jammer #1 (displayed with an expanded scale), and (c) adaptive radiation pattern for jammer #2.

7.4 SUMMARY

This chapter has described the design, construction, and measured performance of a lightweight high-resolution adaptive nulling EHF (43.5-45.5 GHz) multiple beam antenna. The 127-beam MBA consists of seven individual MBAs located on a distributed hexagonal lattice. The beam selection is achieved with a ferrite switch tree, which allows for a total of 14 simultaneous beams. Adaptive nulling measurements were performed with a four-channel RF nulling weight network with one and two jammers. A nulling resolution of $0.1°$ and a wideband cancellation of greater than 30 dB was experimentally demonstrated with this distributed aperture multiple beam antenna.

References

[1] Mayhan, J.T., "Nulling Limitations for a Multiple-Beam Antenna," *IEEE Trans. Antennas and Propagat.*, Vol. AP-24, No. 6, 1976, pp. 769-779.

[2] Mayhan, J.T., "Adaptive Nulling with Multiple-Beam Antennas," *IEEE Trans. Antennas and Propagat.*, Vol. 26, No. 2, 1978, pp. 267-273.

[3] Mayhan, J.T., "Area Coverage Adaptive Nulling from Geosynchronous Satellites: Phased Arrays Versus Multiple-Beam Antennas," *IEEE Trans. Antennas and Propagat.*, Vol. 34, No. 3, 1986, pp. 410-419.

[4] Johnson, R.C., and H. Jasik, *Antenna Engineering Handbook*, 2nd ed., New York: McGraw-Hill, 1984, Ch. 22, pp. 22-1 to 22-20.

[5] Cummings, W.C., P.C. Jain, and L.J. Ricardi, "Fundamental Performance Characteristics That Influence EHF MILSATCOM Systems," *IEEE Trans. on Communications*, Vol. 27, No. 10, October 1979, pp. 1423-1435.

[6] Tavormina, J.J., "Multiple-Beam Antennas for Military Satellite Communications," *Microwave System News & Communications Technology*, Vol. 18, No. 10, October 1988, pp. 20-24, 29-30.

[7] Fenn, A.J., et al., "High-Resolution Adaptive Nulling Performance for a Lightweight Agile EHF Multiple Beam Antenna," *1991 IEEE Military Communications Conference, MILCOM 91 Conference Record*, Vol. 2, 1991, pp. 678-682.

[8] Cummings, W.C., "An Adaptive Nulling Antenna for Military Satellite Communications," *The Lincoln Laboratory Journal*, Vol. 5, No. 2, 1992, pp. 173-194.

[9] Fenn, A.J., "High-Resolution Adaptive Nulling Performance for a Lightweight Agile EHF Multiple Beam Antenna," *Proc. of the 1994 Int. Symp. on Antennas*, 1994, pp. 190-195.

[10] Rispin, L.W., and D.S. Besse, "A Multiple-Aperture Multiple-Beam EHF Antenna for Satellite Communications: Antenna Measurement Results," Massachusetts Institute of Technology, Lincoln Laboratory Technical Report 957, August 13, 1992.

8

Phased Array Antennas: An Introduction

8.1 INTRODUCTION

Antenna systems are often required to have large apertures with significant radiated power levels, sensitive receive capability, and rapid beam scanning. While reflector antenna systems with single or multiple feeds can meet many requirements, phased array antenna systems with thousands of individual radiating antenna elements provide increased beam agility and graceful degradation [1-14]. Figure 8.1 depicts a phased array aperture that produces a narrow-beam (pencil beam) gain radiation pattern $G(\theta, \phi)$ that can be electronically scanned. In general, the phased array aperture can be composed of radiating and/or receiving antenna elements that are located on a planar or conformal surface. In this book, a heavy emphasis is on planar periodic phased arrays. Electronic scanning of the array antenna main beam is effected by means of phase shifters and/or time delays connected to individual array elements or to subarrays (groups) of elements. Chapters 1 through 7 have discussed array antennas and multiple beam antennas in the context of adaptive suppression of jamming and clutter. In this chapter, some of the basic characteristics of phased array antennas with an emphasis on large apertures are reviewed (Section 8.2). An example of phased array antenna measured characteristics is described in Section 8.3. Phased array analysis, design, and testing techniques for various types of array elements are described in Chapters 9 to 16.

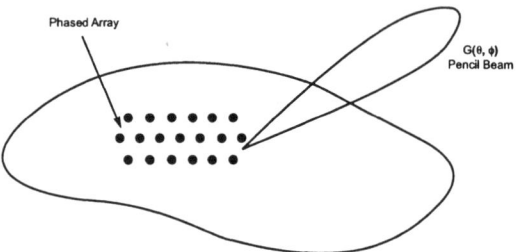

Figure 8.1 Large phased array aperture that produces an electronically steerable narrow-beam radiation pattern.

8.2 THEORETICAL BACKGROUND

A basic diagram of a transmitting phased array antenna is depicted, as a linear array, in Figure 8.2, An RF source has its signal divided into a number of channels by means of a power divider network. Each output path from the power divider is connected to a phase shifting device that applies a progressive (usually linear) phase shift from element to element such that the main beam of the array is scanned to a desired angle. In this phased array architecture example, prior to the divided and phase-shifted signal reaching each of the antenna radiating elements, amplification is effected such that a desired power level is reached. The phase-steered and amplified RF signal from each element is coherent and additive in the direction of the signal path. However, a percentage of the signal from each array element is electromagnetically coupled (curved arrow in Figure 8.2) into the surrounding array elements, and the coupled signal generally becomes weaker as the distance to the coupled element increases. This coupled signal is referred to as array mutual coupling.

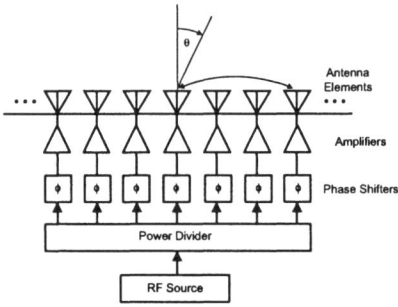

Figure 8.2 Block diagram for a transmitting phased array antenna.

As depicted in Figure 8.3, usually a phased array antenna is required to generate a radiation main beam at an angle θ_s that can be scanned over a certain angular sector. The challenge is to design the array so that the beam scanning can be accomplished in an efficient manner with desired radiation characteristics.

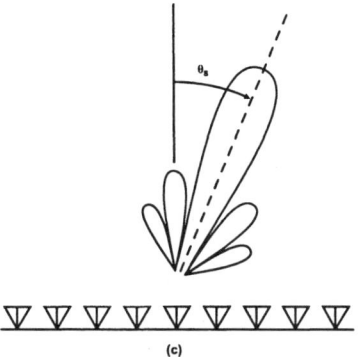

Figure 8.3 Example linear array antenna with the main beam steered to the angle θ_s.

A sketch of a phased array antenna aperture in standard spherical coordinates is shown in Figure 8.4. In the far field, depending on the array element design and antenna orientation, one or both of two orthogonal electric-field components (E_θ, E_ϕ) can be produced, and in the near field a radial component of the electric field (denoted E_r) can also exist (see Chapter 10).

Some of the key parameters that are involved in phased array antenna design are bandwidth, polarization, scan sector, antenna gain and reflection coefficient versus scan angle, and peak and average sidelobe levels. Additional design parameters involve volume, mass, power, thermal/cooling, complexity, and cost for the antenna array system, which are not addressed in this book. Transmit/receive (T/R) modules containing phase shifters to steer the main beam, amplifiers (power amplifiers for transmit and low noise amplifiers for receive) to provide the desired signal level and noise figure, and attenuators for generating low sidelobe radiation patterns are used in many phased arrays. A simplified block diagram for an example transmit/receive module is shown in Figure 8.5. In Figure 8.5, by means of switches and proper timing, the phase shifter is used for both transmit and receive. In this example, for the receive function, two independent phase centers or channels are formed in beamformers A and B (see Chapters 2, 6, and 11 for similar beamforming

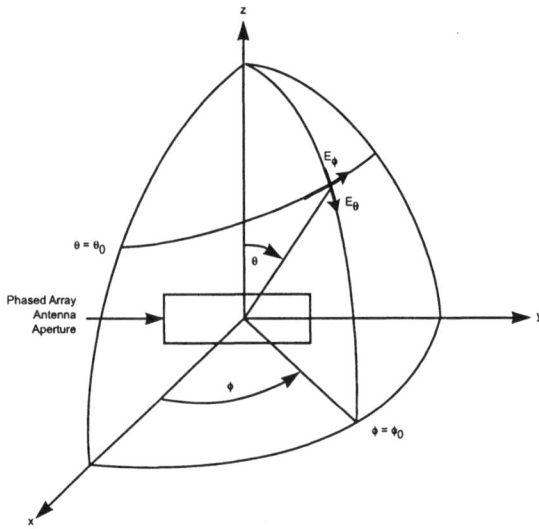

Figure 8.4 Phased array antenna aperture in the spherical coordinate system.

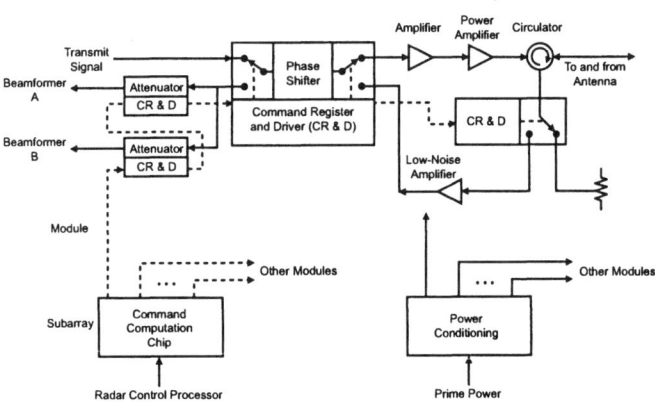

Figure 8.5 Example transmit/receive module block diagram for a phased array antenna.

examples). Many different architectures exist for T/R modules, which are designed based on the system requirements. Basic array theory will now be discussed.

Consider, for example, the simple case of an N-element array of isotropic radiating antenna elements with the $e^{j\omega t}$ time-harmonic dependence of the currents suppressed, where $\omega = 2\pi f$ is the radian frequency, t is time, and f is the frequency. The far-field array radiation pattern, or array factor

(AF) in this case, in terms of standard spherical coordinates can be expressed by superposition of the N radiating currents as [15]

$$\text{AF}(\theta, \phi) = \sum_{n=1}^{N} i_n e^{j\Psi_n} \tag{8.1}$$

where i_n is the complex current induced in the nth radiating element, and (x_n, y_n, z_n) are the coordinates of the nth array element, and where

$$\Psi_n = \beta \sin\theta (x_n \cos\phi + y_n \sin\phi) + \beta z_n \cos\theta \tag{8.2}$$

where $\beta = 2\pi/\lambda$ is the phase constant with λ denoting the wavelength. In order to electrically scan the main beam to a particular scan angle (θ_s, ϕ_s), the phasing of the element currents, denoted Ψ_{ns}, is simply set to be the negative value (phase conjugate) of Ψ_n evaluated at (θ_s, ϕ_s), that is,

$$\Psi_{ns} = -\beta \sin\theta_s (x_n \cos\phi_s + y_n \sin\phi_s) - \beta z_n \cos\theta_s \tag{8.3}$$

The array currents i_n can be expressed in complex form as

$$i_n = A_n e^{j\Psi_{ns}} \tag{8.4}$$

where A_n is the current amplitude of the nth array element. Equation (8.4) is general in the sense that amplitude and phase quantization and random errors of the transmit/receive modules can be readily included. Furthermore, (8.2) and (8.3) allow errors in element positioning $(\Delta x_n, \Delta y_n, \Delta z_n)$ to be taken into account. The array factor for the electrically scanned array case can be written as

$$\text{AF}(\theta, \phi) = \sum_{n=1}^{N} A_n e^{j(\Psi_n + \Psi_{ns})} \tag{8.5}$$

If the current amplitudes are assumed to be uniform with unity amplitude, that is, $A_n = 1$, it follows that the array factor evaluated at the scan angle (θ_s, ϕ_s) is computed as

$$\text{AF}(\theta_s, \phi_s) = \sum_{n=1}^{N} 1 = N \tag{8.6}$$

For a linear array along the x axis, both y_n and z_n are zero. Then with $\phi = 0°$ using (8.2) and (8.3), (8.5) reduces to

$$\text{AF}(\theta) = \sum_{n=1}^{N} A_n e^{j\beta x_n (\sin\theta - \sin\theta_s)} \tag{8.7}$$

As an example of the array factor for an ideal array, consider an eight-element linear array of isotropic array elements with $\lambda/2$ element spacing, as depicted in Figure 8.6.

For this eight-element array, using (8.7) the array factor for uniform illumination for scan angles of $\theta_s = 0°$ and $\theta_s = 45°$ are shown in Figure 8.7.

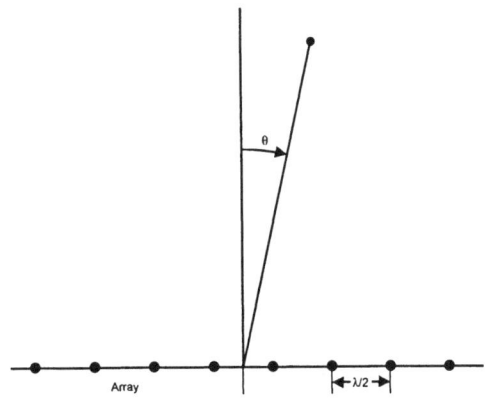

Figure 8.6 Example eight-element linear array antenna.

If the array elements are not isotropic radiators, but have a radiation pattern that depends on angle, the array radiation pattern $P(\theta, \phi)$ can be expressed as the product of the array factor and the element radiation pattern (denoted $p_e(\theta, \phi)$) as

$$P(\theta, \phi) = p_e(\theta, \phi) \text{AF} \tag{8.8}$$

The directivity of an antenna is a function of the radiation intensity normalized by the average radiation intensity as [15, pp. 34-35]

$$D(\theta, \phi) = \frac{U(\theta, \phi)}{U_{\text{ave}}} \tag{8.9}$$

where

$$U_{\text{ave}} = \frac{1}{4\pi} \int_{\theta=0}^{\pi} \int_{\phi=0}^{2\pi} U(\theta, \phi) \sin\theta \, d\theta \, d\phi \tag{8.10}$$

The antenna directivity, $D(\theta, \phi)$, relative to an isotropic radiator, depends only on the shape of the radiation pattern of the antenna and does not depend on power delivered to the antenna. The antenna directivity in decibels relative to an isotropic radiator (dBi) is computed by evaluating $10 \log_{10} D(\theta, \phi)$. To

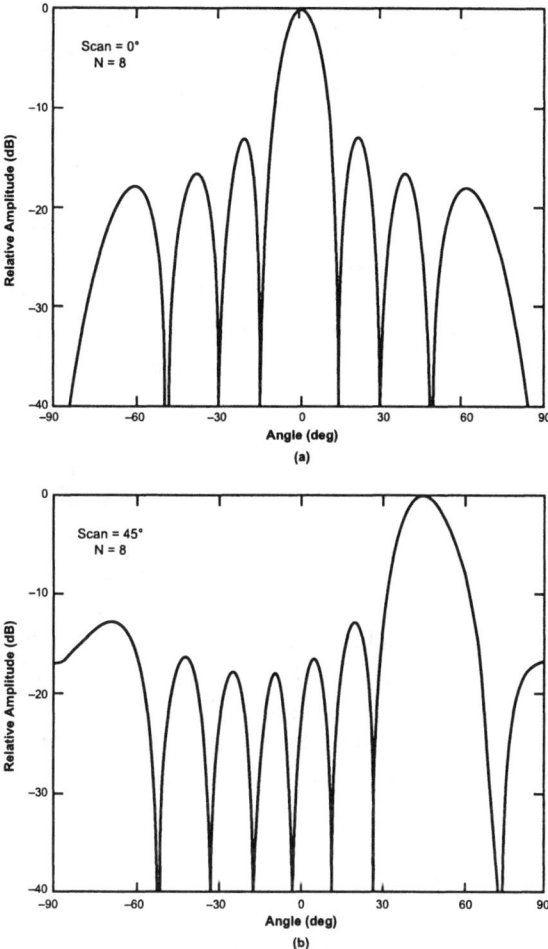

Figure 8.7 Calculated array factor for an eight-element linear array antenna. (a) Scan angle 0° (broadside), and (b) scan angle 45°.

evaluate (8.9), the radiation intensity in (8.10) can be computed from the antenna radiation pattern, denoted $P(\theta, \phi)$, as

$$U(\theta, \phi) = |P(\theta, \phi)|^2 = P(\theta, \phi) P^*(\theta, \phi) \qquad (8.11)$$

where * denotes conjugate. The antenna directivity does not take into account any losses in the antenna and its surroundings. The term antenna gain is used to take into account both the radiation pattern (directivity) as well as any losses.

The terminology in the literature regarding antenna gain and power can be confusing or vague in some instances, so a brief discussion is appropriate. Transmitter power, transmit power, available power, power accepted by the antenna, input power, power input to the antenna, and power radiated by the antenna are some of the terms found in the literature. In the case of gain, the terms directive gain (replaced by the term *directivity*), power gain, effective gain, gain loss due to impedance mismatch, gain loss due to polarization mismatch, gain loss due to mutual coupling, and gain loss due to power absorbed by the antenna or surrounding structures, are found in the literature. In any phased array antenna electromagnetic modeling analysis or measurements, it is important to clearly define how any of these terms are being used.

Consider a transmission line with a transmitter source of some power level attached at one end of the transmission line. At the other end of this transmission line, which will be the attachment point to the transmitting antenna, a power meter is attached and registers a power level of P_o watts. Thus, the amount of power connected to the antenna is accurately known. Now the transmitting antenna element in the array environment, with all mutual coupling effects included, can have a mismatch to the transmission line, which will result in reflected power into the transmission line (that is assumed to be terminated in a resistive load matched to the characteristic impedance of the transmission line) that will not radiate. This reflected power reduces the effective antenna gain and can be quantified as impedance mismatch loss, mismatch loss, or transmission loss. Also, since there is mutual coupling between array elements, some of the power radiated by the transmit element will be received and consumed (power loss) by the surrounding array elements. The net effect of the gain loss mechanisms will result in a reduced transmitted power density (watts per meter squared) at a near-field or far-field position from the array.

The antenna gain, relative to an isotropic radiator with the same input power, can be expressed as [15, p. 37]

$$G(\theta, \phi) = \frac{4\pi U(\theta, \phi)}{P_{in}} \tag{8.12}$$

where P_{in} is the input power, that is, the power at the end of the transmission line that will be connected to the antenna. Due to losses from antenna impedance mismatch and any losses due to components between the transmitter and the antenna terminals, the power transmitted by the antenna is generally less than the input power. For electrically small array elements, the finite resistive loss of the antenna element itself can significantly reduce both the efficiency and gain of the antenna [15, pp. 47-53]. The antenna gain in dBi is computed by evaluating $10 \log_{10} G(\theta, \phi)$.

Assuming that the antenna, with input impedance Z_{in}, is connected to a transmission line with characteristic impedance Z_o, the voltage reflection coefficient Γ is given by

$$\Gamma = \frac{Z_{in} - Z_o}{Z_{in} + Z_o} \quad (8.13)$$

The loss in antenna gain due to antenna input impedance mismatch alone is computed from the power transmission coefficient, denoted $|T|^2$,

$$|T|^2 = 1 - |\Gamma|^2 \quad (8.14)$$

The antenna gain and directivity are related by the following expression

$$G(\theta, \phi) = \epsilon D(\theta, \phi) \quad (8.15)$$

where ϵ is the antenna efficiency, which includes losses. For a 100% efficient antenna, $\epsilon = 1$ and the gain is equal to the directivity.

Array mutual coupling, or electromagnetic field coupling between array elements, as depicted in Figure 8.8, is an important effect that can strongly influence the radiation pattern of the individual array elements and the overall array patterns. Array mutual coupling is defined as the ratio of the coupled signal voltage in a surrounding array element relative to the input signal voltage in a reference element, when the surrounding elements are terminated in matched resistive loads. Letting S_{on} denote the complex mutual coupling between the reference element and the nth array element, then the scan reflection coefficient for a planar phased array is calculated as [1, p. 22]

$$\Gamma_o(\theta_s, \phi_s) = \sum_n S_{on} e^{\Phi_{no}(x_n, y_n, 0)} = \sum_n S_{on} e^{-j\beta \sin\theta_o (x_n \cos\phi_o + y_n \sin\phi_o)} \quad (8.16)$$

where (x_n, y_n) are the coordinates of the nth array element.

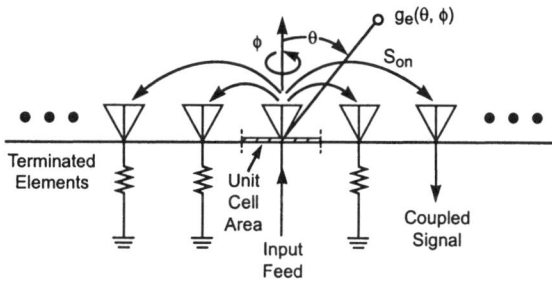

Figure 8.8 Diagram depicting an array with one element driven, with coupling to the surrounding elements.

The scanned array power transmission coefficient denoted $|T(\theta_s, \phi_s)|^2$, due to the scan reflection coefficient is expressed as

$$|T(\theta_s, \phi_s)|^2 = 1 - |\Gamma_o(\theta_s, \phi_s)|^2 \qquad (8.17)$$

The scan mismatch loss in dB can be computed from $10 \log_{10} |T(\theta_s, \phi_s)|^2$.

When a single antenna element is driven in a passively terminated array (also referred to as an embedded element), the fractional dissipated power (denoted F_p relative to the input power to the element) due to array mutual coupling is computed as the summation of the magnitude squared of the mutual coupling coefficients, that is [1, p. 22],

$$F_p = \sum_n |S_{on}|^2 \qquad (8.18)$$

This fractional dissipated power is the fractional power lost to the terminating loads in the transmission lines connected to the array elements. The array element gain $g_e(\theta, \phi)$ is then the product of the element directivity $D_e(\theta, \phi)$ and the quantity $(1 - F_p)$, that is,

$$g_e(\theta, \phi) = D_e(\theta, \phi)(1 - \sum_n |S_{on}|^2) \qquad (8.19)$$

The maximum gain (relative to an isotropic radiator) of an antenna aperture of arbitrary shape is given by the following expression

$$G_{max} = 4\pi A_{em}/\lambda^2 \qquad (8.20)$$

where A_{em} is the antenna maximum effective aperture area in square meters and λ is the wavelength in meters. As the frequency increases, the wavelength is reduced according to the following expression,

$$\lambda = c/f \qquad (8.21)$$

where c is the speed of light and f is the frequency. Thus, for a fixed aperture dimension in meters, ideally the gain will increase as the frequency increases. However, for a practical antenna aperture the surface tolerances and, in the case of a phased array, transmit/receive (T/R) module errors will introduce phase errors and will cause a reduction in antenna gain and an increase in sidelobe levels.

In decibels, the maximum gain of an antenna aperture is given by

$$G_{\text{max, dB}} = 10 \log_{10} G_{max} \qquad (8.22)$$

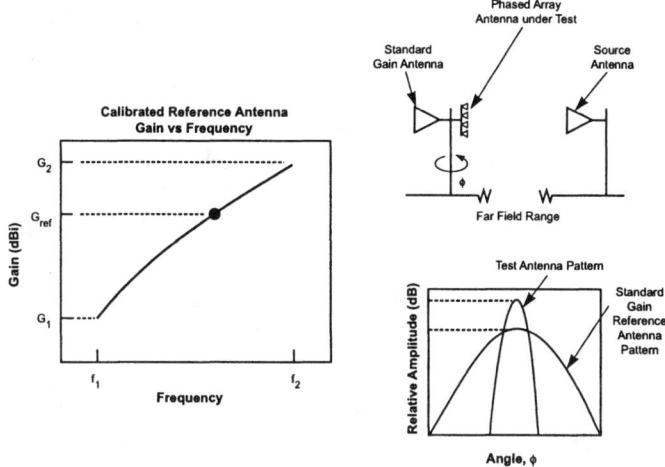

Figure 8.9 Measurement of phased array antenna gain in comparison to a standard gain antenna.

In measuring the gain of a phased array antenna, as depicted in Figure 8.9, a calibrated standard gain antenna can be used as a reference. In this gain measurement, the difference in gain (in decibels) between the antenna under test and the reference standard gain antenna is measured. The measured gain is usually expressed in dBi.

The element gain pattern for an infinite array (or very large finite array) is a function of the unit-cell area, the scan power transmission coefficient, and the projected aperture [1, pp. 23-25]. The projected aperture will vary as $\cos \theta$, where θ is the angle from broadside. For a rectangular grid array, the element gain pattern is given by

$$g_e(\theta, \phi) = \frac{4\pi d_x d_y}{\lambda^2} |T(\theta, \phi)|^2 \cos \theta \tag{8.23}$$

where d_x and d_y are the element spacings in the x and y directions, respectively. For an equilateral-triangle (hexagonal) array lattice, the element gain is expressed as

$$g_e(\theta, \phi) = 2\pi \sqrt{3} b^2 |T(\theta, \phi)|^2 \cos \theta \tag{8.24}$$

where b is the side of the equilateral triangle for the element spacing.

One of the fundamental challenges in designing a phased array is that significant portions of the microwave field transmitted by one element of the array can be received by the surrounding array antenna elements. This effect, which is known as array mutual coupling, can sometimes lead to surface waves and result in a substantial or total loss of transmitted or received signal, depending on the coherent combination of all of the mutual-coupling signals in the array. This total loss of signal is referred to in the literature as a blind spot as is depicted conceptually in Figure 8.10. A phased array blind spot can be described as either a null in the element gain pattern or as a peak tending toward unity in the scan array reflection coefficient. The amplitudes and phases of the array mutual-coupling signals depend primarily on the type (shape) of the radiating antenna elements, the spacing between the array elements, and the number of surrounding radiating elements. Numerous studies have investigated many different array-element designs, taking into account mutual-coupling effects [16-34]. A significant challenge in designing phased arrays is meeting requirements of scan volume and bandwidth while avoiding blind spots. The dashed curve in Figure 8.10(a) shows an idealized element-gain pattern, varying as the cosine (projected aperture) of the scan angle that covers the scan sector, with signal strength dropping outside of the sector. A dipole or other type of array element would attempt to provide the ideal element-gain pattern, but generally the element gain pattern has a larger taper (see Chapter 14 for an example) than the previous ideal cosine function. Figure 8.11 shows a photograph of one of the early L-band linearly polarized straight-arm dipole-phased-array test beds used in measuring array-element patterns, mutual coupling, and array scan impedance [11-14].

In some cases, it is desirable to provide dual-polarized or circularly polarized scan coverage with constant gain for any azimuthal (ϕ) angle, as

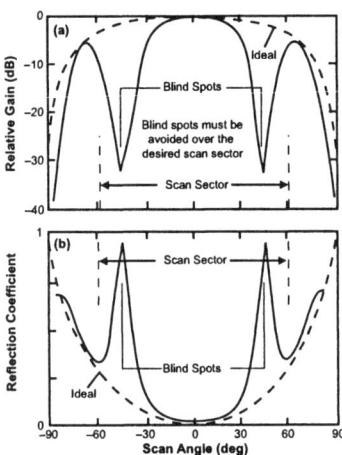

Figure 8.10 Conceptual diagram demonstrating blind spots for a phased array antenna. (a) Element gain pattern and (b) scan reflection coefficient.

Figure 8.11 Photograph showing an example 10 × 10 dipole array antenna.

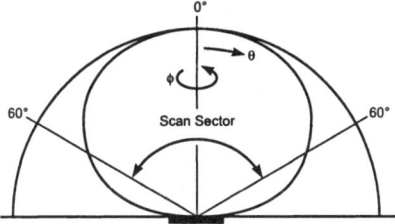

Figure 8.12 Conceptual polar diagram showing an element gain pattern, with peak radiation at broadside.

depicted in Figure 8.12, and this topic is addressed in Chapter 14 with crossed V-shaped dipole array elements.

When all of the phase shifters of an array are properly aligned, the array produces a main beam in the desired pointing direction. Generally, the corporate feed (for example, the power divider in Figure 8.2) is designed with minimal crosstalk between channels. However, once the signals have reached the radiating antenna elements, a significant amount of crosstalk (i.e., array mutual coupling) occurs. The amplitudes and phases of these mutual-coupling signals can significantly impact the performance of the phased array. If the array-element spacing is around one-half wavelength, substantial amounts of mutual coupling can occur. This coupling can manifest itself in undesired changes in the element's gain radiation pattern and its scan reflection coefficient. Unless care is taken in the design of the array, blind spots in the scan sector can occur. These blind spots are angles where the element gain pattern has a null and the reflection coefficient of the array has a peak close to unity, as depicted in Figure 8.10. At these blind spots the signal is significantly reduced in amplitude.

Sometimes it is desirable to place a blind spot in directions where it is undesirable to transmit or receive energy. For example, Figure 8.13 shows an example of a broadside-null radiation pattern that could be generated by either a monopole (see Chapter 9) or uniform current loop (see Chapter 13) antenna element. These types of elements could be useful when broadside radiation is undesirable, such as in reducing broadside clutter and jamming in the case of a spaceborne radar (see Chapters 2, 9, and 11). As the radar beam is steered away from 0° (broadside) toward 60°, the conventional broadside-peak-type element radiation pattern drops off, but the broadside-null-type element radiation pattern increases to a peak at about 45° to 50°.

Phased-array antennas require accurate calibration of their transmit and receive channels, so that the radar main beam can be pointed in the correct direction and the sidelobe levels of the radar antenna can be controlled. In practice, the phase shift through a channel is often affected by temperature and electronic drift; thus, methods for calibration of a fielded radar system are required. For example, airborne and space-based phased arrays containing many hundreds or thousands of transmit/receive channels require onboard techniques for in-flight calibration. One in-flight calibration technique involves the use of the inherent array mutual coupling to transmit

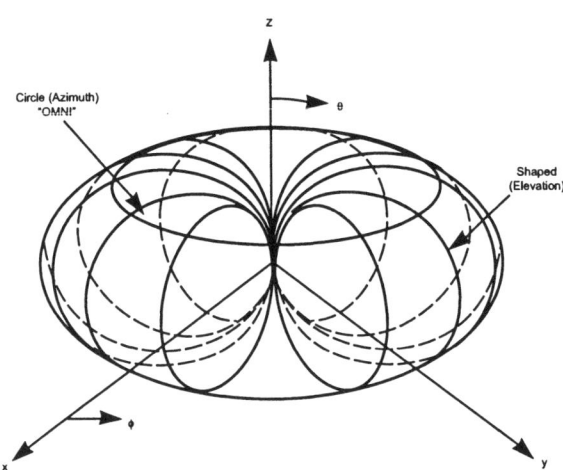

Figure 8.13 Artist depiction of a monopole-type radiation pattern for an element in a phased array antenna.

and receive signals between pairs of elements in the array, as described Aumann et al. [35]. The measured signals between all pairs of elements in the array allow a complete characterization of the relative amplitude and phase response of each channel in the array beamformer. Thus, the channel phase shifters and attenuators (illustrated in Figure 8.5) can be calibrated to generate any desired phase/amplitude distribution across the aperture of the array.

The effective isotropic radiated power, denoted EIRP, of an aperture is given by the product of the antenna gain G_{max} and the transmitter power P_t delivered to the antenna as

$$\text{EIRP} = P_t G_t \tag{8.25}$$

Note that the power transmitted by the antenna is usually less than the power delivered by the generator feeding the antenna. This difference in power is attributed to any RF losses between the generator and the antenna as well as any mismatch loss due to the reflection coefficient at the antenna terminals.

The power density P_d (with units of W/m^2) at a distance r from the aperture (one-way path) is given by

$$P_d = \frac{P_t G_t}{4\pi r^2} \tag{8.26}$$

Based on a receive antenna's maximum effective aperture, denoted A_{em}, the aperture collects incident power density and converts it to an output power. Thus, the received power P_r of an antenna aperture can be expressed as

$$P_r = P_d A_{em} \tag{8.27}$$

Furthermore, from (8.20), it follows that the receive antenna gain is expressed as

$$G_r = 4\pi A_{em}/\lambda^2 \tag{8.28}$$

and solving for A_{em} yields

$$A_{em} = G_r \lambda^2/(4\pi) \tag{8.29}$$

Thus, it follows that

$$P_r = \frac{P_t G_t G_r \lambda^2}{(4\pi r)^2} \tag{8.30}$$

Equation (8.30) can be rearranged to express the mutual coupling power ratio (P_r/P_t) between two antennas as

$$\frac{P_r}{P_t} = \frac{G_t G_r \lambda^2}{(4\pi r)^2} \tag{8.31}$$

If the overall antenna aperture is assumed to be an array of identical elements, the aperture can be divided into element apertures, which are also referred to as element unit cells (see Figure 8.8). For a large N-element array with an effective aperture A_{em}, the unit cell area is equal to A_{em}/N. If the embedded element gain patterns are either computed or measured, the array gain can be computed as follows. At range distance r, the embedded-element radiated power density, denoted P_{de}, due to the element transmit power (input power to the element) and embedded element gain is given by

$$P_{de} = \frac{P_{te} g_e(\theta, \phi)}{4\pi r^2} \tag{8.32}$$

and solving for g_e,

$$g_e = 4\pi r^2 \frac{P_{de}}{P_{te}} \tag{8.33}$$

In the far field, the plane wave power density of the element is given in terms of the array-element electric field intensity $E_e(\theta, \phi)$ as

$$P_{de} = \frac{|E_e(\theta, \phi)|^2}{120\pi} \tag{8.34}$$

Substituting (8.34) in (8.33) and simplifying yields

$$g_e(\theta, \phi) = \frac{r^2}{30} \frac{|E_e(\theta, \phi)|^2}{P_{te}} \tag{8.35}$$

Thus, it follows that

$$|E_e(\theta, \phi)| = \frac{\sqrt{30}}{r} \sqrt{g_e(\theta, \phi)} \sqrt{P_{te}} \tag{8.36}$$

The element gain pattern, $g_e(\theta, \phi)$ includes the effect of array mutual coupling and mismatch losses.

The transmit power of the nth element, is proportional to A_n^2, and using this relation in (8.36) yields

$$|E_{en}(\theta, \phi)| = \frac{\sqrt{30}}{r} \sqrt{g_{en}(\theta, \phi)} A_n \tag{8.37}$$

To compute the maximum gain of an array at a particular scan angle (θ_s, ϕ_s), by superposition the electric field contributions for the embedded elements can be summed as follows,

$$E_{\text{array, max}} = \frac{\sqrt{30}}{r} \sum_n \sqrt{g_{en}} A_n \tag{8.38}$$

Now, following along the lines of (8.35) the array gain can be expressed as

$$G_{\text{array, max}} = \frac{r^2}{30} \frac{|E_{\text{array, max}}|^2}{P_{\text{t, array}}} \quad (8.39)$$

where

$$P_{\text{t, array}} = \sum_{n=1}^{N} A_n^2 \quad (8.40)$$

Substituting (8.38) and (8.40) in (8.39) and simplifying yields [18, p. 55]

$$G_{\text{array, max}} = \frac{|\sum_{n=1}^{N} \sqrt{g_{en}} A_n|^2}{\sum_{n=1}^{N} A_n^2} \quad (8.41)$$

For the case of a large array with uniform array illumination (with unity amplitude at each array element), it can be assumed that all of the elements have the same element gain pattern (that is, $g_{en} = g_e$) and (8.41) reduces to

$$G_{\text{array, max}}(\theta_s, \phi_s) = g_e(\theta_s, \phi_s) N \quad (8.42)$$

That is, for uniform illumination the array gain is equal to the element gain times the number of array elements. If the array illumination is not uniform, a taper efficiency must be included in computing the array gain from the element gain.

To compute the array gain pattern from the element electric field versus angle (θ, ϕ), the element phase pattern $\Psi_{en}(\theta, \phi)$ (either calculated or measured) must be used, that is,

$$E_{en}(\theta, \phi) = |E_{en}(\theta, \phi)| e^{j\Psi_{en}(\theta, \phi)} \quad (8.43)$$

The array pattern E_{array} is then obtained by the superposition of the individual array element patterns with the desired array amplitude taper and with proper phasing of the array to steer the main beam in a desired direction. Thus,

$$E_{\text{array}}(\theta, \phi) = \sum_n E_{en}(\theta, \phi) e^{j\Psi_{ns}} \quad (8.44)$$

Or from (8.43)

$$E_{\text{array}}(\theta, \phi) = |E_{en}(\theta, \phi)| e^{j(\Psi_{en} + \Psi_{ns})} \quad (8.45)$$

and using (8.37) yields

$$E_{\text{array}}(\theta, \phi) = \frac{\sqrt{30}}{r} \sum_n \sqrt{g_{en}(\theta, \phi)} A_n e^{j(\Psi_{en} + \Psi_{ns})} \quad (8.46)$$

The array gain as a function of angle can then be computed from the electric field as

$$G_{\text{array}}(\theta, \phi) = \frac{r^2}{30} \frac{|E_{\text{array}}(\theta, \phi)|^2}{P_{\text{t, array}}} \qquad (8.47)$$

Substituting (8.40) and (8.46) in (8.47) yields the array gain in terms of the element gain and element phase functions and the array illumination, that is,

$$G_{\text{array}}(\theta, \phi) = \frac{|\sum_n \sqrt{g_{en}(\theta, \phi)} A_n e^{j(\Psi_{en} + \Psi_{ns})}|^2}{\sum_n A_n^2} \qquad (8.48)$$

In designing phased array antennas, the maximum electrical spacing d between the elements of a phased array is usually chosen to avoid grating lobes (multiple peaks) according to the following condition [4, p. 31]:

$$\frac{d}{\lambda_h} \leq \frac{1}{1 + \sin|\theta_s|} \qquad (8.49)$$

where λ_h is the wavelength at the highest frequency and θ_s is the scan angle from broadside. For many wide-angle scanning phased arrays, the electrical spacing is often chosen to be close to one-half wavelength.

8.3 EXAMPLE MEASUREMENTS OF ARRAY MUTUAL COUPLING AND ARRAY ELEMENT GAIN PATTERN

The previous section described the theory for characterizing phased array antennas by means of the element gain pattern and array mutual coupling. The center element far-field radiation pattern of a small planar array of radiating antenna elements can provide an estimate of the scan coverage and gain for a large phased array antenna [33]. An example of the measured element gain and array mutual coupling of a 49-element (7 row by 7 column) circular microstrip patch-element array operating at center frequency 1.3 GHz ($\lambda_o = 23.1$ cm) is discussed here. Some general references on the design of microstrip patch antennas and arrays are found in [36-38], and analysis and measurements of probe-fed circular microstrip patch arrays have been investigated [39].

The circular microstrip patch antenna elements were mounted on a square aluminum ground plane (0.3175 cm thick) with side dimension 1.22m as depicted in Figure 8.14. The elements were arranged in a square grid with element spacing 10.922 cm ($0.473\lambda_o$). The diameter of the circular microstrip patch antenna is 8.382 cm ($0.36\lambda_o$) and the thickness of the dielectric substrate

(Rexolite) is 0.1588 cm. Note: Rexolite has a dielectric constant of 2.55 and loss tangent of about 0.0004 at 1.3 GHz. The array elements were tuned to 1.3 GHz using a coaxial probe feed offset by 1.2 cm from a centrally located shorting pin. The surrounding microstrip patch elements were terminated in 50-ohm resistive loads during center element gain pattern measurements and array mutual coupling measurements.

The amplitude of the mutual coupling between the center element and the surrounding circular patch elements in Figure 8.14's 7×7 array were measured at 1.3 GHz, using the element numbering as shown in Figure 8.14. The array elements are numbered 1 to 7 in column 1, 8 to 14 in column 2, and so forth. Element number 25 is the center element of the array. The measured array mutual coupling data ($20 \log_{10} |S_{on}|$) rounded to the nearest decibel are presented as data superimposed on the array in Figure 8.15, and are plotted in Figure 8.16.

As discussed in the previous section, let $|S_{on}|$ represent the magnitude of the mutual coupling from the center element to the surrounding elements of an N-element array. Then the fractional power dissipated due to array mutual coupling is computed from (8.18). The element gain is the product of the directivity and the efficiency $(1 - F_p)$. A major contributor to the mutual coupling dissipated power is the coupling from the center element (element number 25) to the adjacent elements. For example, adjacent H-plane

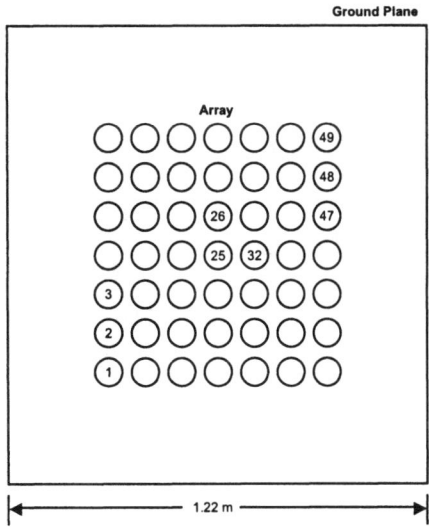

Figure 8.14 Diagram depicting a 7 row by 7 column circular microstrip patch array.

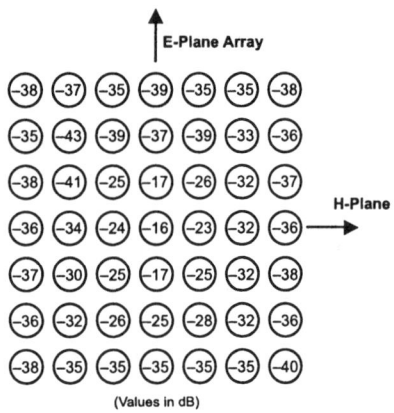

Figure 8.15 Measured array mutual coupling amplitudes (in decibels) superimposed on a diagram of the 7 row by 7 column circular microstrip patch array.

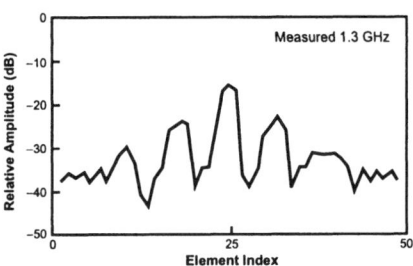

Figure 8.16 Measured mutual coupling for the center element of a 7 row by 7 column circular microstrip patch array.

coupling (element 25 to element 32) is −23 dB. Similarly, the adjacent E-plane coupling (element number 25 to element number 26) is −17 dB. The mutual coupling is observed to be stronger in the E-plane compared to the H-plane. The return loss of element 25 is −15.8 dB at 1.3 GHz. Using the previous measured mutual coupling data from Figure 8.15 in (8.18), the array element efficiency $(1 - F_p)$ is computed to be 0.89 or −0.49 dB. Thus, about one-half dB of loss due to mutual coupling occurs for this array example.

The E-plane (Figure 8.17(a)) and H-plane (Figure 8.17(b)) center-element gain patterns of the microstrip patch array were measured at 1.3 GHz in an anechoic chamber, using a range distance of 8.5 meters. By comparison with a standard gain antenna, the measured center-element peak gain on boresight is 3.6 dBi. At 60° from broadside the E-plane or H-plane element gain reduction is about −4.5 dB (relative to the broadside gain). Thus, based on the measured element gain pattern, good wide-angle scanning should be

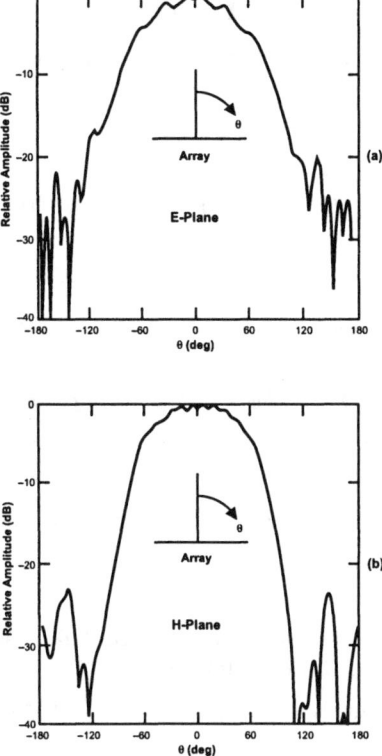

Figure 8.17 Measured radiation patterns for the center element of a 7 row by 7 column circular microstrip patch array. The element surrounding the center element are terminated in 50-ohm resistive loads. (a) E plane pattern and (b) H-plane pattern.

possible.

8.4 SUMMARY

This chapter has introduced some of the important characteristics of phased array antennas. Array element gain and mutual coupling have been described. In designing phased array antenna elements, the element must provide an element gain pattern that covers the desired scan sector. In the chapters that follow, phased array analyses and measurement techniques for several types of array antenna elements are presented.

References

[1] Amitay, N., V. Galindo, and C. P. Wu, *Theory and Analysis of Phased Array Antennas*, New York: Wiley, 1972, ch. 1.

[2] Oliner, A.A., and G.H. Knittel, (eds.), *Phased Array Antennas*, Dedham, MA: Artech House, 1972.

[3] Hansen, R.C., (ed.), *Significant Phased Array Papers*, Dedham, MA: Artech House, 1973.

[4] Mailloux, R.J., *Phased Array Handbook*, Norwood, MA: Artech House, 1994.

[5] Brookner, E., (ed.), *Practical Phased-Array Antenna Systems*, Norwood, MA: Artech House, 1991.

[6] Hansen, R.C., *Microwave Scanning Antennas, Vol. II: Array Theory and Practice*, Los Altos, CA: Peninsula Publishing, 1985.

[7] Hansen, R.C., *Phased Array Antennas*, New York: Wiley, 1998.

[8] Bhattacharyya, A.K., *Phased Array Antennas*, New York: Wiley, 2006.

[9] Brookner, E., "Phased-Array Radars," *Scientific American*, Vol. 252, No. 2, 1985, pp. 94-102.

[10] Fenn, A.J., et al., "The Development of Phased Array Radar Technology," *Lincoln Laboratory Journal*, Vol. 12, No. 2, 2000, pp. 321-340.

[11] Allen, J.L., et al., "Phased Array Radar Studies, 1 July 1959 to 1 July 1960," Technical Report 228, Lincoln Laboratory, August 12, 1960, DTIC No. AD-0249470.

[12] Allen, J.L., et al., "Phased Array Radar Studies, 1 July 1, 1960 to 1 July 1961," Technical Report 236, Lincoln Laboratory, November 13, 1961, DTIC No. AD-271724.

[13] Allen, J.L., et al., "Phased Array Radar Studies, July 1, 1961 to July 1, 1963," Technical Report 299, Lincoln Laboratory, February 20, 1963, DTIC No. AD-417572.

[14] Allen, J.L., et al., "Phased Array Radar Studies, July 1, 1963 to July 1, 1964," Technical Report 381, Lincoln Laboratory, March 31, 1965, DTIC No. AD-629363.

[15] Stutzman, W.L., and G.A. Thiele, *Antenna Theory and Design*, New York: Wiley, 1981.

[16] Fenn, A.J., "Theoretical and Experimental Study of Monopole Phased Array Antennas," *IEEE Trans. Antennas Propag.*, Vol. 34, No. 10, 1985, pp. 1118-1126.

[17] Fenn, A.J., "Arrays of Horizontally Polarized Loop-Fed Slotted Cylinder Antennas," *IEEE Trans. Antennas Propag.*, Vol. 33, No. 4, 1985, pp. 375-382.

[18] Allen, J.L., "Theory of Array Antennas (with Emphasis on Radar Applications)," Technical Report 323, Lincoln Laboratory, July 25, 1963, DTIC No. AD-422945.

[19] Diamond, B.L., "A Generalized Approach to the Analysis of Infinite Planar Phased Array Antennas," *Proc. IEEE*, Vol. 56, No. 11, 1968, pp. 1837-1851.

[20] Diamond, B.L., and G.H. Knittel, "A New Procedure for the Design of a Waveguide Element for a Phased-Array Antenna," in *Phased Array Antennas*, A.A. Oliner and G.H. Knittel, (eds.), Dedham, MA: Artech House, 1972, pp. 149-156.

[21] Tsandoulas, G.N., "Unidimensionally Scanned Phased Arrays," *IEEE Trans. Antennas Propagat.*, Vol. 28, No. 1, 1980, pp. 86-99.

[22] Chen, M.H., G.N. Tsandoulas, and F.G. Willwerth, "Modal Characteristics of Quadruple-Ridged Circular and Square Waveguides," *IEEE Trans. Microwave Theory Tech.*, Vol. 22, No. 8, 1974, pp. 801-804.

[23] Chen, M.H., and G.N. Tsandoulas, "A Wide-Band Square-Waveguide Array Polarizer," *IEEE Trans. Antennas Propag.*, Vol. 21, No. 3, 1973, pp. 389-391.

[24] Tsandoulas, G.N., "Wideband Limitations of Waveguide Arrays," *Microwave J.*, Vol. 15, No. 9, 1972, pp. 49-56.

[25] Tsandoulas, G.N., and G.H. Knittel, "The Analysis and Design of Dual-Polarization Square-Waveguide Phased Arrays," *IEEE Trans. Antennas Propagat.*, Vol. 21, No. 6, 1973, pp. 796-808.

[26] Knittel, G.H., "Relation of Radar Range Resolution and Signal-to-Noise Ratio to Phased-Array Bandwidth," *IEEE Trans. Antennas Propagat.*, Vol. 22, No. 3, 1974, pp. 418-426.

[27] Allen, J.L., "Gain and Impedance Variation in Scanned Dipole Arrays," *IRE Trans. Antennas Propagat.*, Vol. 10, No. 5, 1962, pp. 566-572.

[28] Allen, J.L., and W.P. Delaney, "On the Effect of Mutual Coupling on Unequally Spaced Dipole Arrays," *IRE Trans. Antennas Propagat.*, Vol. 10, No. 6, 1962, pp. 784-785.

[29] Allen, J.L., "On Array Element Impedance Variation with Spacing," *IRE Trans. Antennas Propagat.*, Vol. 12, No. 3, 1964, pp. 371-372.

[30] Allen, J.L., "On Surface-Wave Coupling Between Elements of Large Arrays," *IEEE Trans. Antennas Propagat.*, Vol. 13, No. 4, 1965, pp. 638-639.

[31] Allen, J.L., and B.L. Diamond, "Mutual Coupling in Array Antennas," Technical Report 424, Lincoln Laboratory, October 4, 1966, DTIC No. AD-648153.

[32] Allen, J.L., "A Simple Model for Mutual Coupling Effects on Patterns of Unequally Spaced Arrays," *IEEE Trans. Antennas Propag.*, Vol. 15, No. 4, 1967, pp. 530-533.

[33] Diamond, B.L., "Small Arrays – Their Analysis and Their Use for the Design of Array Elements," in *Phased Array Antennas*, A.A. Oliner and G.H. Knittel, (eds.), Dedham, MA: Artech House, 1972, pp. 127-131.

[34] Brookner, E., (ed.), *Radar Technology*, Norwood, MA: Artech House, 1991.

[35] Aumann, H.M., A.J. Fenn, and F.G. Willwerth, "Phased Array Antenna Calibration and Pattern Prediction Using Mutual Coupling Measurements," *IEEE Trans. Antennas Propagat.*, Vol. 37, No. 7, 1989, pp. 844-850.

[36] Bhartia, P., K.V.S. Rao, and R.S. Tomar, *Millimeter-Wave Microstrip and Printed Circuit Antennas*, Norwood, MA: Artech House, 1991.

[37] Bahl, I.J., and P. Bhartia, *Microstrip Antennas*, Dedham, MA: Artech House, 1980.

[38] James, J.R., and P.S. Hall, (eds.), *Handbook of Microstrip Antennas*, London: Peter Peregrinus, 1989.

[39] Deshpande, M.D., and M.C. Bailey, "Analysis of Finite Phased Arrays of Circular

Microstrip Patches," *IEEE Trans. Antennas Propagat.*, Vol. 37, No. 11, 1989, pp. 1355-1360.

9

Monopole Phased Array Antenna Design, Analysis, and Measurements

9.1 INTRODUCTION

As discussed in Chapter 8, periodic planar phased array antennas are generally designed to provide close to uniform element gain over a portion of a hemisphere. The ideal element gain pattern for many applications has a cosine variation from broadside. Peak gain occurs at or near broadside and wide-angle scanning out to approximately 60° is possible with many array element designs [1-4] such as dipoles, waveguides, and microstrip patch elements. The present chapter considers an alternate element design for wide-angle scanning phased arrays in which a pattern null rather than a pattern maximum is formed at broadside. For certain phased array antenna applications, maximum gain is desirable away from broadside with minimum gain or a null occurring at broadside. A two-dimensional periodic monopole array provides this type of pattern coverage [5].

This chapter investigates a monopole phased array antenna design as a possible candidate for a low-Earth-orbit space-based radar (SBR) [6, 7]. The pertinent orbit geometry for such a spaceborne sensor at low altitude in the range of about 1100 km to 1700 km is shown in Figure 9.1. As targets are desired to be detected from horizon to horizon, referring to Figure 2.9 and (2.16) this yields a maximum scanning cone of approximately ±60°. One of the important challenges of a space-based radar is to operate with a large amount of ground clutter, which can be addressed using a technique known as displaced phase center antenna (DPCA) [6, 7]. Clutter cancellation for phased arrays using the displaced phase center technique is described in Chapters 2,

5, and 11. In Figure 9.1 the minimum cone surrounding nadir is approximately ±30° and represents a clutter-dominated region, that is, the SBR would not attempt to scan the main beam in this region, because the greatly enhanced amount of high-grazing-angle reflected surface clutter prevents the realization of adequate radar performance. Since this region would not be used, the array antenna radiation pattern amplitude can be low in order to reduce the clutter return inside the approximate ±30° cone. This reduction in the nadir-region clutter can be achieved by using low sidelobe receive aperture amplitude distributions, sidelobe cancellation techniques, and/or by the array element design as discussed here.

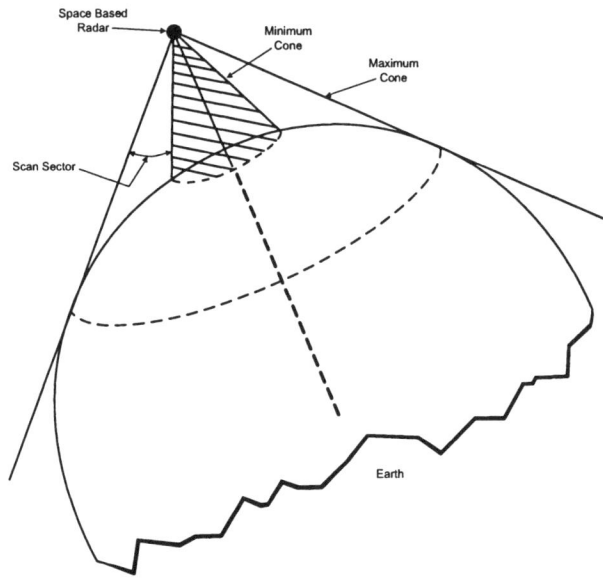

Figure 9.1 Low-altitude space-based radar phased array antenna scanning geometry. The antenna aperture is assumed to be nadir-pointed as the satellite orbits the Earth. The region within the minimum cone is referred to as the nadir hole (shaded region) where the surface clutter dominates over any moving radar targets. The desired scanning would be between the edge of the nadir hole and the horizon. © 1983 IEEE [6].

From the previous discussion, the SBR desired scanning region is defined by the region contained between the minimum (±30°) and maximum (±60°) cones. A conical scan or azimuthal scan is taken to be one where the main beam is steered about nadir along a constant cone angle. An elevation scan is defined as when the main beam is steered between two cones for a fixed azimuth angle. The satellite-to-Earth slant range increases as the main beam is steered away from nadir (elevation scan). To compensate for the increase in propagation path loss, the array element radiation pattern should provide more gain at wide angles than at angles close to nadir. Furthermore, the slant range is constant for a given elevation angle so the array elements should be omnidirectional in azimuth. This helps in part to constrain the array element design.

The SBR antenna bandwidth and polarization are important issues. For purposes of this chapter the desired bandwidth is assumed to be approximately 15 percent at L-band, centered at 1.3 GHz. This chapter also assumes that a single linear polarization (vertical) will provide adequate performance for omnidirectional pattern coverage. A simple monopole array element provides vertical polarization when the array face is horizontal. For horizontal polarization, a more complicated loop-fed slotted cylinder element (with a monopole type pattern) is a possible candidate and is described in Chapter 13. In cases where Faraday rotation [8] is significant, a dual-polarized element (such as the crossed V-dipole discussed in Chapter 14) could be required.

A displaced phase center antenna (DPCA) SBR monopole phased array system for canceling clutter was analyzed in Chapter 2, and a subscale array was measured and is described in Chapter 11. A low-sidelobe monopole phased array antenna was measured and is described in Chapter 12. An antenna for the spaceborne radar application could employ a large aperture consisting of tens of thousands of antenna elements. Since a large number of array elements is involved, a simple radiator design such as a monopole would be beneficial to the overall SBR system.

9.2 MONOPOLE PHASED ARRAYS

An antenna element that offers promise of good spaceborne radar system performance is a simple coaxial-fed cylindrical monopole with a ground plane. An array of monopoles is not new; however, they are usually found in small arrays (less than 100 elements). This element does not appear to have been used in any large array antennas (1000 elements or more) in the open literature. This is expected because phased array antennas are required generally to scan to broadside from the array plane. The monopole array cannot scan the main beam to the broadside angle, because the element has a deep pattern null there. However, a null at broadside for the SBR application was previously noted to be an asset for nadir clutter reduction.

A few theoretical and experimental studies of large arrays of monopoles have been published. Herper and Hessel [9] studied an infinite phased array of monopoles on an infinite ground plane. Schuman et al. [10] has also investigated the infinite array scan impedance for monopole arrays. In both studies, the array element reflection coefficient was found to have a minimum in the 50° to 60° scan region. In other words, the monopole array input impedance is well-matched to the feedline at wide scan angles. Monopole arrays have also been studied as a resonance absorber by Kurtze and Neumann [11].

In this chapter both finite and infinite array analyses and finite array measurements are used to assess the potential merits of a wide-angle scanning two-dimensional periodic monopole phased array antenna. The monopoles are assumed to be mounted on an infinite ground plane, which helps to simplify the analysis. The two array lattices of interest are square and hexagonal, both of which are included in the formulation. A sketch of a monopole array with a general skewed grid is shown in Figure 9.2.

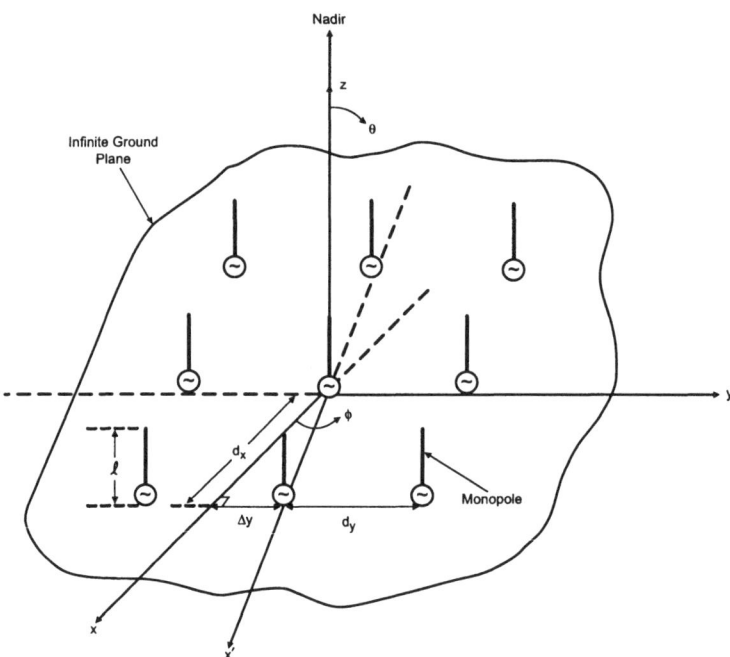

Figure 9.2 Monopole phased array antenna arranged in a two-dimensional skew-symmetric lattice. © 1985 IEEE [5].

The monopole array antenna elements analyzed here are assumed to have a wire radius less than or equal to approximately 0.01λ, which allows an analysis based on thin-wire techniques. Additionally, the monopoles are assumed to be approximately one-quarter wavelength long. In this case the current distribution on each monopole is nearly sinusoidal, and this assumption simplifies the analysis considerably. Section 9.3 discusses the formulation used to analyze finite phased arrays of monopoles. The method of moments is implemented to enforce the boundary conditions on each monopole. This method inherently includes the effects of array mutual coupling. By this approach the element currents, radiation patterns, gain, input impedance, and array mutual coupling are readily obtained. All of the elements can be excited (driven) to compute the array input impedance and patterns. A single driven element, in the presence of matched-loaded elements, is also analyzed readily to determine the element-gain pattern and mutual coupling.

In Section 9.4 an infinite array formulation for monopole arrays is given. The basic approach is to characterize the scanned array input impedance by a summation of plane waves. The element-gain pattern is then readily obtained from the active element (scanned array) transmission coefficient.

The motivation for investigating both finite and infinite array analyses is, in part, to have an independent check of the theory for large arrays. Also, for monopole elements it is relatively easy to build a small array to verify the finite array formulation. This is discussed for a 121-element array in Section 9.5. The finite array theory is used to investigate the element gain pattern and input impedance as a function of the array size. Edge effects and mutual coupling are also investigated. The infinite array analysis is particularly helpful here in showing that there is no occurrence of a blind spot in the desired scan sector. Blind spots are of concern for any large phased array design [1-4]. As demonstrated in Section 9.5, the finite array formulation can be used to analyze arrays of tens of thousands of elements. In Section 9.6 it is concluded that the monopole element is an attractive candidate for the low-altitude spaceborne radar application.

9.3 THEORY FOR ANALYSIS OF FINITE ARRAYS OF MONOPOLES

9.3.1 INTRODUCTION

In this section the method of moments is used to analyze a finite array of vertical monopoles on an infinite ground plane. Mutual coupling is inherently included by using this approach. This analysis allows computation of array element currents, radiation patterns, gain, input impedance, and mutual coupling.

Two array lattices are of interest here, square and equilateral triangle (hexagonal). Of the two lattices the square is most easily analyzed because of the large amount of symmetry in the mutual impedance matrix. From a practical standpoint of reducing the number of elements, the triangular lattice is often used in phased array antenna designs [12]. In the following analysis, both lattices are treated.

9.3.2 MATRIX EQUATIONS FOR THE ARRAY ELEMENT CURRENTS

Consider now a finite array of monopoles arranged in a skewed lattice as shown in Figure 9.2. The skewed lattice allows both square and hexagonal grids to be analyzed. By image theory the ground plane can be removed and monopole images are then introduced as shown in Figure 9.3. This application of image theory gives the nearly equivalent problem of analyzing a finite array of dipoles in free space. The primary differences are that the input impedance of the equivalent dipole is twice that of the monopole and the gain of the equivalent dipole is one-half that of the monopole. When these factors are properly taken into account, a complete electromagnetic description of a monopole array is possible.

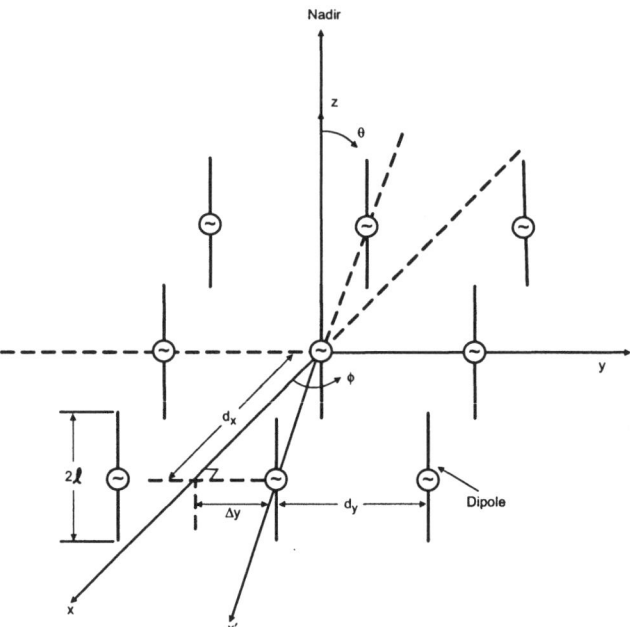

Figure 9.3 Application of image theory to a monopole phased array antenna, arranged in a two-dimensional skew-symmetric lattice, results in an equivalent dipole array for analysis. © 1985 IEEE [5].

The dipole image array approach has been taken here because the mutual impedance between dipoles in free space is readily computed. An equation for the mutual impedance between two parallel dipoles is given in a paper by Castello and Munk [13]. The dipoles are assumed to be electrically thin, that is, the wire radius is less than or equal to approximately 0.01λ.

The monopole elements are assumed to be approximately one-quarter wavelength long at the radar center frequency. Thus, the equivalent dipoles are approximately one-half wavelength long. The current distribution of a dipole near half-wavelength resonance is approximately sinusoidal. Thus, the electromagnetic properties of the equivalent dipole array can be accurately computed by assuming that the current distribution on each element is sinusoidal. The limitations imposed by this assumption will be discussed in the results section.

In the moment method formulation [14, pp. 306-374], the boundary condition that must be satisfied is that the tangential electric field E_m must be zero at the mth dipole element. That is,

$$E_m = E_m^s + E_m^i = 0, \quad m = 1, 2, \ldots, N \tag{9.1}$$

where E_m^i is the incident field at dipole m, E_m^s is the scattered field at dipole m, and N is the number of array elements.

The scattered field can be expressed as the superposition of the fields radiated by the dipole currents $i_n(z)$, $n = 1, 2, \ldots, N$. The dipole currents are assumed to be of the form

$$i_n(z) = i_n \frac{\sin \beta(L/2 - |z|)}{\sin \beta L/2} \tag{9.2}$$

where $\beta = 2\pi/\lambda = \omega\sqrt{\mu\epsilon}$ is the phase constant (angular wavenumber), L is the dipole length, and i_n is the complex terminal current of the dipole. The normalized sinusoidal function in (9.2) is referred to here as the basis function $b_n(z)$. The dipole terminal current i_n is a scan-dependent complex number that is to be determined.

The scattered electric field is written as

$$E_m^s(z) = \sum_{n=1}^{N} i_n \int_m G(\mathbf{r}, \mathbf{r}') b_n(z') dz' \tag{9.3}$$

where $G(\mathbf{r}, \mathbf{r}')$ is the free-space Green's function, $\mathbf{r}$ is the observation position vector on dipole m, and $\mathbf{r}'$ is the source position vector on dipole n.

Substituting (9.3) in (9.1) yields

$$E_m^i(z) = -\sum_{n=1}^{N} i_n \int_m G(\mathbf{r}, \mathbf{r}') b_n(z') dz' \tag{9.4}$$

where $m = 1, 2, \ldots, N$.

A Galerkin's formulation is obtained by the integral weighting of each of the previous equations with testing functions $t_m(z)$ that are equal to the basis function $b_m(z)$. Equation (9.4) now becomes

$$\int_m t_m(z)E_m^i(z)dz = -\sum_{n=1}^N i_n \int_m t_m(z) \int_n G(\mathbf{r},\mathbf{r}')b_n(z')dz'dz \qquad (9.5)$$

where $m = 1, 2, \ldots, N$, or

$$V_m = \sum_{n=1}^N i_n Z_{mn} \quad m = 1, 2, \ldots, N \qquad (9.6)$$

where

$$V_m = \int_m t_m(z)E_m^i(z)dz \qquad (9.7)$$

is the voltage excitation and

$$Z_{mn} = -\int_m \int_n t_m(z)G(\mathbf{r},\mathbf{r}')b_n(z')dz'dz \qquad (9.8)$$

is the open-circuit mutual impedance between elements m and n, expressed as the matrix $\mathbf{Z}$. These expressions can be combined in terms of the applied voltage excitation vector $(\mathbf{V})$, the impedance matrix $\mathbf{Z}$, and unknown current vector $(\mathbf{i})$ to give the matrix equation

$$\mathbf{V} = \mathbf{Z}\mathbf{i} \qquad (9.9)$$

that is,

$$\begin{aligned}
V_1 &= Z_{11}i_1 + Z_{12}i_2 + \cdots + Z_{1N}i_N \\
V_2 &= Z_{21}i_1 + Z_{22}i_2 + \cdots + Z_{2N}i_N \\
&\quad \cdot \\
&\quad \cdot \\
&\quad \cdot \\
V_N &= Z_{N1}i_1 + Z_{N2}i_2 + \cdots + Z_{NN}i_N
\end{aligned} \qquad (9.10)$$

The unknown current coefficients $i_n, n = 1, 2, \ldots, N$ are found by matrix inversion from

$$\mathbf{i} = \mathbf{Z}^{-1}\mathbf{V} \qquad (9.11)$$

The voltage source model chosen for each dipole is a delta gap (also referred to as a slice) generator. This model assumes that the impressed electric field exists only at the dipole terminals (that is, $z = 0$). In this case the nth excitation, V_n^g, can be defined to be one volt in amplitude. The required steering phase for an element in the mth row and qth column of the array is related to the main beam scan direction (θ_s, ϕ_s) by

$$\psi_x = \beta q d_x \sin \theta_s \cos \phi_s \qquad (9.12)$$
$$\psi_y = \beta (m d_y + q \Delta y) \sin \theta_s \sin \phi_s \qquad (9.13)$$

so

$$V_{qm}^g = e^{-j(\psi_x + \psi_y)} \qquad (9.14)$$

defines the generator voltages.

In the present form, (9.14) assumes a uniform amplitude excitation. It should be noted, however, that any amplitude taper can be used. This would be done by multiplying the qmth term of the voltage matrix by the appropriate amplitude coefficient A_{qm}.

The array elements are assumed to be excited by constant incident-power sources. This type of feeding is referred to as free excitation, because the complex voltage across the terminals of the array element can vary according to the scan conditions [15]. The equivalent circuit for the mth dipole port ($m = 1, 2, \ldots, N$) is shown in Figure 9.4.

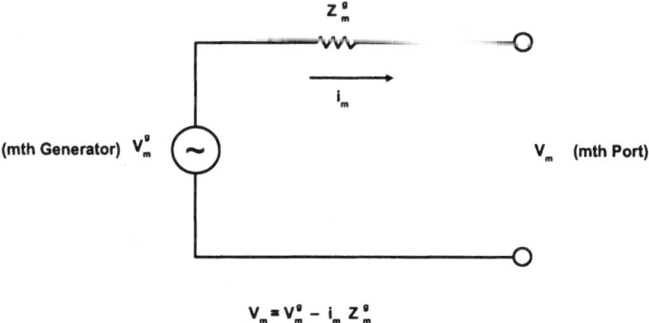

Figure 9.4 Equivalent circuit model for the mth array element port.

The effect of the generator (or load) impedance (Z_m^g) is included in the voltage excitation matrix. That is,

$$V_m = V_m^g - i_m Z_m^g \qquad (9.15)$$

then substituting (9.15) into (9.6) yields

$$\sum_{n=1}^{N} Z_{mn} i_n = V_m^g - i_m Z_m^g \tag{9.16}$$

or

$$\sum_{n=1}^{N} Z'_{mn} i_n = V_m^g \tag{9.17}$$

where

$$Z'_{mn} = Z_{mn} + Z_m^g \tag{9.18}$$

Thus, the effect of the generator impedance is to add Z_m^g to the diagonal elements of the impedance matrix.

For an electrically scanned phased array all generators are on, hence every V_m^g is nonzero. For a single-element excited array, only one V_m^g is nonzero, which can be assumed equal to one volt. For both types of arrays, the value of Z_m^g can be chosen to be 50 ohms, corresponding to the generator impedance which is the same as the characteristic impedance of the feedline (usually 50 ohms). The unknown currents $i_n, n = 1, 2, \ldots, N$ are determined from

$$\boldsymbol{i} = [\boldsymbol{Z'}]^{-1} \boldsymbol{V}^g \tag{9.19}$$

where

$$\boldsymbol{Z'} = \boldsymbol{Z} + \boldsymbol{I} Z_m^g \tag{9.20}$$

where $\boldsymbol{I}$ represents the identity matrix. The current coefficients i_n determined from (9.19) are subtituted in (9.2) for computation of the array radiation pattern as given later in (9.27).

The mutual impedance as defined by (9.8) is a function of the dipole length and spacing. However, since the monopoles in the array are all the same length, in matrix form the mutual impedance varies according to the array lattice. When the array lattice is rectangular (or square) the open-ciruit mutual impedance matrix has block-Toeplitz symmetry. A block-Toeplitz matrix has the form in which the blocks along the main diagonal are equal and the off diagonal blocks are also equal. A numerically efficient software subroutine is available for solution of block-Toeplitz systems of equations [14, pp. 585-590; 16]. For this computer program, storage and computation time are significantly reduced when compared to other computer codes.

9.3.3 ARRAY INPUT IMPEDANCE, PATTERNS, AND GAIN

The open-circuit mutual impedances defined by (9.8) are independent of the array generator voltages. Hence, both fully excited arrays (all elements excited) and single-element excited arrays (the remainder terminated with matched resistive loads equal to the generator impedance) are readily analyzed.

The input impedance, denoted Z_{in}, of the mth dipole array element is computed from

$$Z_{in} = \frac{V_m}{i_m} \qquad (9.21)$$

where V_m is given by (9.15) and i_m is computed from (9.19). For a monopole above a ground plane, the voltage across the feed terminals is one half the voltage for the equivalent dipole in free space. This voltage relation occurs because the feed gap distance of the monopole is one-half that of the equivalent dipole. Assuming the same terminal current, the input impedance of a monopole over a ground plane is one-half that of a dipole in free space with twice the length of the monopole [14, p. 93], or

$$Z_{in_{monopole}} = \frac{1}{2} Z_{in_{dipole}} \qquad (9.22)$$

The far-zone electric field pattern of a linear dipole with sinusoidal current distribution is found by evaluating the following equation [14, p. 84]

$$E_\theta = j\omega\mu_o \frac{e^{-j\beta r}}{4\pi r} \sin\theta \int_{-L/2}^{L/2} i(z') e^{j\beta z' \cos\theta} dz' \qquad (9.23)$$

where μ_o is the permeability of free space and

$$i(z') = i_t \frac{\sin\beta(L/2 - |z'|)}{\sin\beta L/2} \qquad (9.24)$$

where i_t is the complex terminal current of $i(z')$.

Substituting (9.24) in (9.23) and evaluating the single integral gives

$$E_\theta = \frac{j\omega\mu_o}{\beta} \frac{e^{-j\beta r}}{4\pi r} \frac{2i_t}{\sin\beta L/2} \frac{\cos(\beta L/2 \cos\theta) - \cos(\beta L/2)}{\sin\theta} \qquad (9.25)$$

which is the field radiated by a single isolated dipole in free space.

To evaluate the field radiated by an array of dipoles having arbitrary complex currents $i_1, i_2, \ldots, i_N$, the element pattern and array factor (AF) can be used. For either a single-element excited array or a fully excited array, the current coefficients are obtained from (9.19). The array factor for isotropic point sources is given by

$$\text{AF} = \sum_{n=1}^{N} i_n e^{j\beta \sin\theta(x_n \cos\phi + y_n \sin\phi)} \qquad (9.26)$$

where (x_n, y_n) are the coordinates of the nth element. It is important to note that the currents used in the earlier array factor expression can include the effect of mutual coupling between array elements, when the currents are determined by (9.19). The currents can also be ideal currents in which case the array factor would not include mutual-coupling effects. In the case where all of the elements can be assumed to have the same pattern, the complete array radiated field $E_\theta^a(\theta, \phi)$ is given by the product of the element pattern, denoted $E_\theta^e(\theta)$, which is the single dipole pattern in (9.25), and the array factor (AF), that is,

$$E_\theta^a(\theta, \phi) = E_\theta^e(\theta) \mathrm{AF} = \frac{j60}{\sin \beta L/2} \frac{e^{-j\beta r}}{r} \frac{\cos(\beta L/2 \cos \theta) - \cos(\beta L/2)}{\sin \theta} \mathrm{AF} \quad (9.27)$$

Equation (9.27) gives the pattern of the array depending on whether one element or all the elements are driven. Note that the element pattern given by $E_\theta^e(\theta)$ can be an embedded element pattern determined by the method of moments for a particular large finite array size. To calculate the pattern of an array of tens of thousands of elements, the embedded element pattern for a smaller, but still large, array can be used in combination with the ideal array factor (calculated with currents that do not include mutual-coupling effects).

The antenna gain pattern, with respect to an isotropic radiator with the same input power, versus angle, denoted $G(\theta, \phi)$, not to be confused with the Green's function, can be computed according to

$$G(\theta, \phi) = \frac{4\pi U(\theta, \phi)}{P_{in}} \quad (9.28)$$

where $U(\theta, \phi)$ is the radiation intensity and P_{in} is the input power. The radiation intensity is defined to be

$$U(\theta, \phi) = \frac{1}{2} Re(\boldsymbol{E} \times \boldsymbol{H}^*) \cdot r^2 \hat{r} \quad (9.29)$$

where $\boldsymbol{H}$ is the magnetic field, Re means real, and $*$ means complex conjugate.

In the far-zone region of a z-directed dipole, the only nonzero field components are E_θ and H_ϕ. For a plane wave, the E_θ and H_ϕ components are related by [14, p. 22]

$$H_\phi = \frac{E_\theta}{\eta_o} \quad (9.30)$$

where $\eta_o = 120\pi$ ohms is the impedance of free space. The radiation intensity can now be expressed as

$$U(\theta, \phi) = \frac{1}{2} \frac{|E_\theta|^2}{\eta_o} r^2 \quad (9.31)$$

The input power is given by

$$P_{in} = \frac{1}{2} \sum_{n=1}^{N_a} |i_n|^2 R_{in} \qquad (9.32)$$

where $R_{in} = \text{Real}(Z_{in})$ and N_a is the number of excited elements. Thus, the gain of the dipole array is given by

$$G(\theta, \phi) = \frac{4\pi \frac{1}{2} |E_\theta|^2 r^2}{\eta_o \frac{1}{2} \sum_{n=1}^{N_a} |i_n|^2 R_{in}} \qquad (9.33)$$

which simplifies to

$$G(\theta, \phi) = \frac{|E_\theta|^2 r^2}{30 \sum_{n=1}^{N_a} |i_n|^2 R_{in}} \qquad (9.34)$$

Equation (9.34) is the gain of an array of dipoles in free space, and since the input power to a monopole array fed against an infinite ground plane radiates over one-half the space of the equivalent dipole, the monopole array gain is twice this value, or

$$G(\theta, \phi)_{monopole} = 2G(\theta, \phi)_{dipole} \qquad (9.35)$$

It is important to note that the antenna gain given by (9.34) does not include the loss due to impedance mismatch. This additional loss is taken into account by defining the voltage reflection coefficient

$$\Gamma = \frac{Z_{in} - Z_c}{Z_{in} + Z_c} \qquad (9.36)$$

where Z_c is the characteristic impedance of the feed line. The impedance mismatch loss is given by

$$\text{Mismatch Loss} = 10 \log_{10}(1 - |\Gamma|^2) \text{ dB} \qquad (9.37)$$

The impedance mismatch loss is included in the element gain pattern simulations given in Section 9.5.

The mutual impedance as given by (9.8) is readily computed for sinusoidal current functions. However, mutual impedance is not convenient for array measurements because it requires the elements to be open circuited. A more convenient measurement is the complex mutual coupling that requires that all elements, except the two under test, be terminated in matched loads. Mutual coupling is different from mutual impedance in that it compares like quantities, (for example, voltage to voltage). In other words, mutual coupling is dimensionless. A network analyzer measurement of the scattering parameter, denoted S21, is a convenient method to quantify mutual coupling in an array.

For mutual coupling measurements, one array element is connected to a transmitter and a second array element is connected to a receiver. The ratio of the received voltage to the reference transmitting voltage (referenced to the terminals of the transmitting and receiving array elements) is the complex mutual coupling (real and imaginary or amplitude and phase). The power in decibels for S21 is obtained by taking $20 \log_{10} |S21|$. It is of interest to be able to compute the theoretical mutual coupling scattering matrix for an array, which will later be used to compare against measured values of S21.

The scattering matrix of coupling coefficients is computed from the normalized array mutual impedance matrix, denoted z by using the following relation [17],

$$S = [z - I][z + I]^{-1} \tag{9.38}$$

where

$$z = \frac{Z}{Z_c} \tag{9.39}$$

In the scattering matrix S, the $S_{k,n}$th element represents the ratio of the signal received at the nth array element to the signal transmitted by the kth array element. Thus, $S_{k,n}$ is a normalized complex quantity.

Of particular interest is the case in which the mutual coupling occurs between the center element of a finite array and its surrounding elements. The scanned array case is defined here as corresponding to all of the elements in the array being excited with a complex voltage source. The reflection coefficient versus scan angle of the center element of the array can be determined by the appropriate summation of the calculated or measured mutual coupling coefficients. If the center element is chosen as a reference element, then the scan reflection coefficient is given by [1, 18]

$$\Gamma_o(\theta_s, \phi_s) = \sum_n S_{on} e^{-j\beta \sin \theta_s (x_n \cos \phi_s + y_n \sin \phi_s)} \tag{9.40}$$

where (x_n, y_n) are the coordinates of the nth array element. Equation (9.40) is used subsequently in this chapter, with measured values of S_{on}, to compute the scan reflection coefficient of a monopole phased array.

This section has addressed the problem of analyzing finite arrays of thin-wire monopoles on an infinite ground plane. Although large arrays of thousands of elements can be analyzed with this approach, it is also of interest to analyze infinite arrays of monopoles as is done in the next section. The infinite array analysis provides a check of the finite array formulation for large arrays and is particularly useful for verifying that there are no blind spots over the desired scan region.

9.4 THEORY FOR ANALYSIS OF INFINITE ARRAYS OF MONOPOLES

9.4.1 INTRODUCTION

The general theory for infinite array analysis used in this section was developed by Munk et al. [19-21]. The infinite array is formulated as a periodic surface with arbitrarily shaped identical linear elements. An important part of the theory is that the array elements do not have to lie within the plane of the array, which allows for the analysis of three-dimensional periodic surfaces such as an array of thin vertical monopoles on a ground plane (refer to Figure 9.2).

By applying image theory, an array of vertical monopoles with a skewed lattice on a ground plane can be analyzed as a periodic surface in free space. Each monopole element has a vertical image in the ground plane. The ground plane can be removed and a free-space periodic dipole array results, as shown in Figure 9.3. The input impedance of a reference element of the equivalent dipole array is twice that of the reference element of the monopole array. Additionally, the gain of the dipole array is one-half that of the monopole array.

The basic approach of the general theory of periodic surfaces is first to determine the field radiated by an infinite periodic array of Hertzian dipoles with infinitesimal length denoted dp. The radiated field of this Hertzian dipole array is expressed as an infinite summation of plane waves, both propagating and evanescent. Next, the transmitting current distribution (assumed to be piecewise-sinusoidal for the monopole elements) is imposed on the Hertzian element, and the total radiated field is computed by integration. A so-called exterior element (identical in shape to the reference element of the infinite array) is exposed to the field of the infinite array, and the induced voltage is determined by a second integration. The scan input impedance of a reference element of the infinite array is found by displacing the exterior element one wire radius from the reference element and computing the ratio of the induced voltage to the terminal current.

9.4.2 DERIVATION OF THE NEAR-ZONE RADIATED ELECTRIC FIELD

The general theory of periodic surfaces described by Munk is valid for arbitrary linear elements in either free-space or layered dielectric media. It is applied here to the free-space array of thin dipoles shown in Figure 9.3. The equivalent dipoles here are assumed to be arranged in a skewed lattice. In Figure 9.5, the element in the qth column and mth row has position $(x = qd_x, y = md_y + q\Delta y, z = 0)$, where d_x and d_y are the array element spacings in the x- and y-directions, respectively. The dipole elements are assumed to be oriented in the z-direction, and the feed terminals of each dipole are at $z = 0$.

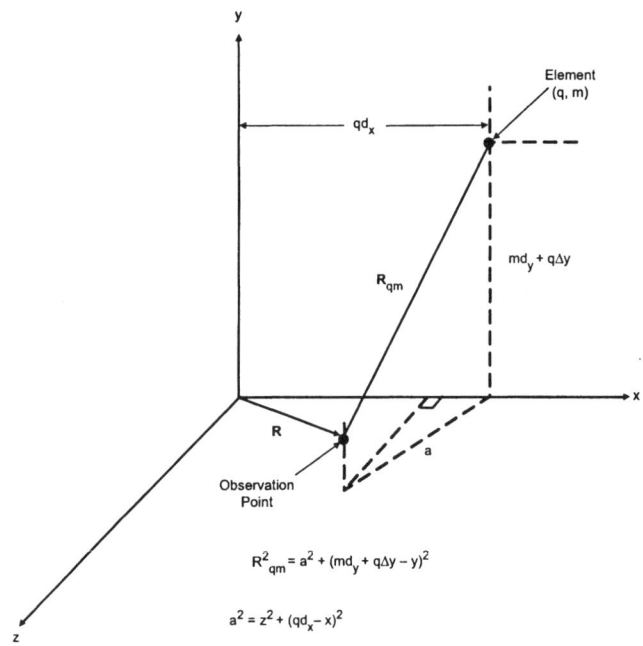

Figure 9.5 Geometry for the qmth array element and observation point in the rectangular coordinate system.

Let $\mathbf{R} = \hat{x}x + \hat{y}y + \hat{z}z$ be the observation position vector from the origin, and let R_{qm} be the distance from the qmth array element to the observation point. From geometry, it follows that

$$R_{qm}^2 = a^2 + (md_y + q\Delta y - y)^2 \tag{9.41}$$

$$a^2 = z^2 + (qd_x - x)^2 \tag{9.42}$$

The differential magnetic vector potential due to a Hertzian dipole with current i_{qm} and length dp is given by [20, p. 81]

$$dA_{qm} = \hat{p} \frac{\mu_o i_{qm} dp}{4\pi} \frac{e^{-j\beta R_{qm}}}{R_{qm}} \tag{9.43}$$

where $\hat{p} = \hat{z}$ is the dipole orientation, μ_o is the permeability of free space, and $\beta = 2\pi/\lambda$ is the phase propagation constant with λ the wavelength.

The total differential magnetic vector potential due to the entire infinite array of Hertzian dipoles is found by superposition of (9.43), that is,

$$dA = \sum_{q=-\infty}^{\infty} \sum_{m=-\infty}^{\infty} dA_{qm} \tag{9.44}$$

or

$$dA = \hat{p} \frac{\mu_o i_{qm} dp}{4\pi} \sum_{q=-\infty}^{\infty} \sum_{m=-\infty}^{\infty} i_{qm} \frac{e^{-j\beta R_{qm}}}{R_{qm}} \tag{9.45}$$

To scan the main beam of the array in the direction (θ_s, ϕ_s) (where the angles (θ, ϕ) are standard spherical coordinates (see Figure 9.2), the current excitation must be periodic. By Floquet's theorem [1, pp. 37-40],

$$i_{qm} = i(p) e^{-j\beta q d_x s_x} e^{-j\beta(md_y + q\Delta y) s_y} \tag{9.46}$$

where $i(p)$ is the transmitting current distribution on each element of the array and

$$s_x = \sin\theta_s \cos\phi_s \tag{9.47}$$

$$s_y = \sin\theta_s \sin\phi_s \tag{9.48}$$

Substituting (9.46) into (9.45) yields

$$dA = \hat{p} \frac{\mu_o i(p) dp}{4\pi} \sum_{q=-\infty}^{\infty} e^{-j\beta q d_x s_x} \sum_{m=-\infty}^{\infty} e^{-j\beta(md_y + q\Delta y) s_y} \frac{e^{-j\beta R_{qm}}}{R_{qm}} \tag{9.49}$$

or

$$dA = \sum_{q=-\infty}^{\infty} e^{-j\beta q(d_x s_x + \Delta y s_y)} dA_q \tag{9.50}$$

where using (9.41) for R_{qm},

$$dA_q = \hat{p} \frac{\mu_o i(p) dp}{4\pi} \sum_{m=-\infty}^{\infty} e^{-j\beta m d_y s_y} \frac{e^{-j\beta \sqrt{a^2 + (md_y + q\Delta y - y)^2}}}{\sqrt{a^2 + (md_y + q\Delta y - y)^2}} \tag{9.51}$$

Equation (9.51) represents the differential magnetic vector potential for a single column of Hertzian elements and is the superposition of an infinite number of spherical waves emanating from the array. This summation is slowly convergent, because the terms are damped by the $1/R_{qm}$ factor.

An equivalent expression for which the convergence is more rapid can be obtained by applying the Poisson sum formula [22, pp. 47-49]:

$$\sum_{m=-\infty}^{\infty} e^{jm\omega_o t} F(m\omega_o) = T \sum_{n=-\infty}^{\infty} f(t+nT) \qquad (9.52)$$

where $F(\omega)$ is the Fourier transform of $f(t)$; that is, $F(\omega) = \mathcal{F}[f(t)]$ and

$$T = \frac{2\pi}{\omega_o} \qquad (9.53)$$

Comparing (9.51) with the left side of (9.52), it is observed that

$$F(m\omega_o) = \frac{e^{-j\beta\sqrt{a^2+(md_y)^2}}}{\sqrt{a^2+(md_y)^2}} \qquad (9.54)$$

where

$$\omega_o = d_y \qquad (9.55)$$
$$t = -\beta s_y \qquad (9.56)$$

and

$$T = \frac{2\pi}{d_y} \qquad (9.57)$$

From Bateman's table of integral transforms, the following Fourier transform pair is obtained [23, Section 1.13, formula (42) for $\nu = 1/2$]:

$$\frac{e^{-j\beta\sqrt{a^2+\omega^2}}}{\sqrt{a^2+\omega^2}} = \mathcal{F}[\frac{H_o^{(2)}(a\sqrt{\beta^2-t^2})}{2j}] \qquad (9.58)$$

where $H_o^{(2)}$ is the zero-order Hankel function of the second kind. By using the frequency shift theorem [22, p. 15]

$$F(\omega - \omega_1) = \mathcal{F}[e^{j\omega_1 t} f(t)], \qquad (9.59)$$

(9.58) now becomes

$$\frac{e^{-j\beta\sqrt{a^2+(\omega-\omega_1)^2}}}{\sqrt{a^2+(\omega-\omega_1)^2}} = \mathcal{F}[e^{j\omega_1 t} \frac{H_o^{(2)}(a\sqrt{\beta^2-t^2})}{2j}] \qquad (9.60)$$

or
$$F_1(\omega) = \mathcal{F}[f_1(t)] \tag{9.61}$$

Using (9.52) and (9.60) in (9.51) with $\omega_1 = y - q\Delta y$ yields

$$dA_q = \hat{p}\frac{\mu_o i(p)dp}{4}\frac{2}{d_y}\sum_{n=-\infty}^{\infty} e^{-j(y-q\Delta y)(\beta s_y - n\frac{2\pi}{d_y})}\frac{1}{2j}$$
$$\cdot H_o^{(2)}[a\sqrt{\beta^2 - (-\beta s_y + n\frac{2\pi}{d_y})^2}] \tag{9.62}$$

but,

$$(y - q\Delta y)(\beta s_y - n\frac{2\pi}{d_y}) = \beta y(s_y - \frac{n\lambda}{d_y}) - q\beta(\Delta y s_y - \frac{\Delta y n\lambda}{d_y}) \tag{9.63}$$

so

$$dA_q = \hat{p}\frac{\mu_o i(p)dp}{4jd_y}\sum_{n=-\infty}^{\infty} e^{-j\beta y(s_y - \frac{n\lambda}{d_y})} e^{j\beta q(\Delta y s_y - \frac{\Delta y n\lambda}{d_y})} H_o^{(2)}(\beta\rho a) \tag{9.64}$$

where

$$\rho = \sqrt{1 - (s_y - \frac{n\lambda}{d_y})^2} \tag{9.65}$$

Equation (9.64) is the differential magnetic vector potential for the qth column of Hertzian elements, expressed as an infinite sum of outgoing cylindrical waves. Now, substituting (9.64) in (9.50) yields

$$dA = \hat{p}\frac{\mu_o i(p)dp}{4jd_y}\sum_{n=-\infty}^{\infty} e^{-j\beta y(s_y - \frac{n\lambda}{d_y})} \sum_{q=-\infty}^{\infty} e^{-j\beta q(d_x s_x + \Delta y s_y)}$$
$$\cdot e^{j\beta q(\Delta y s_y - \frac{n\Delta y \lambda}{d_y})} H_o^{(2)}(\beta\rho a) \tag{9.66}$$

where from (9.42),
$$a = \sqrt{z^2 + (qd_x - x)^2} \tag{9.67}$$

Simplifying (9.66) further by observing that the exponential terms involving $\Delta y s_y$ cancel, yields,

$$dA = \hat{p}\frac{\mu_o i(p)dp}{4jd_y}\sum_{n=-\infty}^{\infty} e^{-j\beta y(s_y - \frac{n\lambda}{d_y})} \sum_{q=-\infty}^{\infty} e^{-j\beta q d_x(s_x + \frac{n\Delta y \lambda}{d_x d_y})} H_o^{(2)}(\beta\rho a)$$
$$\tag{9.68}$$

or
$$dA = \hat{p}\frac{\mu_o i(p)dp}{4jd_y}\sum_{n=-\infty}^{\infty}e^{-j\beta y(s_y-\frac{n\lambda}{d_y})}\sum_{q=-\infty}^{\infty}dA_q \qquad (9.69)$$

where
$$\sum_{q=-\infty}^{\infty}dA_q = \sum_{q=-\infty}^{\infty}e^{-j\beta q d_x(s_x+\frac{n\Delta y\lambda}{d_x d_y})}H_o^{(2)}(\beta\rho a) \qquad (9.70)$$

Equation (9.69) can be transformed to a faster converging series by means of the Poisson sum formula

$$\sum_{q=-\infty}^{\infty}e^{jq\omega_o t}F(q\omega_o) = T\sum_{k=-\infty}^{\infty}f(t+kT) \qquad (9.71)$$

where
$$\omega_o = d_x \qquad (9.72)$$

$$t = -\beta(s_x + \frac{n\Delta y\lambda}{d_x d_y}) \qquad (9.73)$$

and
$$T = \frac{2\pi}{d_x} \qquad (9.74)$$

From Bateman [23, Section 1.13, formula (42) for $\nu = 0$], the following Fourier transform is obtained:

$$H_o^{(2)}(\beta\rho\sqrt{z^2+\omega^2}) = \mathcal{F}[\frac{e^{-jz\sqrt{(\beta\rho)^2-t^2}}}{\pi\sqrt{(\beta\rho)^2-t^2}}], \quad -\infty < t < \infty \qquad (9.75)$$

Applying the frequency shift theorem given by (9.59) to (9.75) yields

$$H_o^{(2)}(\beta\rho\sqrt{z^2+(\omega-\omega_2)^2}) = \mathcal{F}[e^{j\omega_2 t}\frac{e^{-jz\sqrt{(\beta\rho)^2-t^2}}}{\pi\sqrt{(\beta\rho)^2-t^2}}] \qquad (9.76)$$

where $\omega_2 = x$. Using (9.69), (9.71), (9.72), (9.73), (9.74), and (9.76) yields

$$dA = \hat{p}\frac{\mu_o i(p)dp}{4jd_y}\sum_{n=-\infty}^{\infty}e^{-j\beta y(s_y-\frac{n\lambda}{d_y})}\frac{2\pi}{d_x}\sum_{k=-\infty}^{\infty}e^{-jx[-\beta(s_x-\frac{n\Delta y\lambda}{d_x d_y})+k\frac{2\pi}{d_x}]}$$

$$\cdot \frac{e^{-jz\sqrt{(\beta\rho)^2-(-\beta(s_x+\frac{n\Delta y\lambda}{d_x d_y})+k\frac{2\pi}{d_x})^2}}}{\pi\sqrt{(\beta\rho)^2-(-\beta(s_x+\frac{n\Delta y\lambda}{d_x d_y})+k\frac{2\pi}{d_x})^2}} \qquad (9.77)$$

Substituting (9.65) in (9.77) yields

$$dA = \hat{p}\frac{\mu_0 i(p)dp}{2j\beta d_x d_y}\sum_{k=-\infty}^{\infty}\sum_{n=-\infty}^{\infty} e^{-j\beta y(s_y-\frac{n\lambda}{d_y})}e^{-j\beta x[s_x-k\frac{\lambda}{d_x}+\frac{n\Delta y\lambda}{d_x d_y}]}$$

$$\cdot \frac{e^{-j\beta z\sqrt{1-((s_x+\frac{n\Delta y\lambda}{d_x d_y})-k\frac{\lambda}{d_x})^2-(s_y-\frac{n\lambda}{d_y})^2}}}{\sqrt{1-((s_x+\frac{n\Delta y\lambda}{d_x d_y})-k\frac{\lambda}{d_x})^2-(s_y-\frac{n\lambda}{d_y})^2}} \quad (9.78)$$

In (9.78), $-n$ and $-k$ can be substituted for n and k, respectively, without changing the result. It then follows that

$$dA = \hat{p}\frac{\mu_0 i(p)dp}{2j\beta d_x d_y}\sum_{k=-\infty}^{\infty}\sum_{n=-\infty}^{\infty}\frac{e^{-j\beta \boldsymbol{R}\cdot\hat{\boldsymbol{r}}_\pm}}{r_z} \quad (9.79)$$

where

$$\boldsymbol{R} = \hat{x}x + \hat{y}y + \hat{z}z \quad (9.80)$$

$$\hat{\boldsymbol{r}}_\pm = \hat{x}(s_x + k\frac{\lambda}{d_x} - \frac{n\Delta y\lambda}{d_x d_y}) + \hat{y}(s_y + \frac{n\lambda}{d_y}) \pm \hat{z}r_z \quad (9.81)$$

where the plus (+) sign is chosen for $z > 0$ and the minus ($-$) sign is chosen for $z < 0$, and

$$r_z = \sqrt{1 - ((s_x + k\frac{\lambda}{d_r} - \frac{n\Delta y\lambda}{d_r d_y}))^2 - (s_y + \frac{n\lambda}{d_y})^2} \quad (9.82)$$

An important note is that when k and n are such that the square root in (9.82) is imaginary, the $-j$ value is chosen. Thus, evanescent waves are attenuated exponentially as the observation point recedes from the array.

The differential magnetic field now can be found by applying the relation [24, p. 205]

$$d\boldsymbol{H}(x,y,z) = \frac{1}{\mu_0}\nabla \times [d\boldsymbol{A}(x,y,z)] \quad (9.83)$$

to (9.79) which gives

$$d\boldsymbol{H} = \frac{i(p)dp}{2j\beta d_x d_y}\sum_{k=-\infty}^{\infty}\sum_{n=-\infty}^{\infty}\nabla \times \hat{p}\frac{e^{-j\beta \boldsymbol{R}\cdot\hat{\boldsymbol{r}}_\pm}}{r_z} \quad z \geq 0 \quad (9.84)$$

The curl operation in (9.84) is performed by use of the vector identity:

$$\nabla \times (\boldsymbol{A}\phi) = \phi\nabla \times \boldsymbol{A} - \boldsymbol{A} \times \nabla\phi \quad (9.85)$$

which gives

$$dH(x,y,z) = \frac{i(p)dp}{2d_x d_y} \sum_{k=-\infty}^{\infty} \sum_{n=-\infty}^{\infty} \frac{e^{-j\beta \mathbf{R}\cdot\hat{r}_\pm}}{r_z} \hat{p} \times \hat{r}_\pm \qquad (9.86)$$

where r_+ is used for $z > 0$ and r_- is used for $z < 0$.

The differential electric field in free space is found from the differential magnetic field by using the Maxwell equation

$$\mathbf{E} = \frac{1}{j\omega\epsilon_o} \nabla \times \mathbf{H} \qquad (9.87)$$

together with (9.86), such that

$$d\mathbf{E}(x,y,z) = \frac{\eta_o i(p)dp}{2d_x d_y} \sum_{k=-\infty}^{\infty} \sum_{n=-\infty}^{\infty} \frac{e^{-j\beta \mathbf{R}\cdot\hat{r}_\pm}}{r_z} [\hat{p} \times \hat{r}_\pm] \times \hat{r}_\pm \qquad (9.88)$$

where $\eta_o = 120\pi$ ohms is the impedance of free space. The general expression for the differential electric field due to a Hertzian dipole given by (9.88) is a plane wave summation that includes propagating and evanescent terms.

To compute the field radiated by the equivalent dipole array, it is required to integrate $d\mathbf{E}(x,y,z)$ over the dipole length. For planar dipole arrays, this integration is usually relatively simple; however, this is not the case when the elements protrude from the plane of the array. Note: The plane of the monopole array (or the equivalent dipole array) is taken to be at the ground plane. The integration must be performed carefully when the observation point is located between the tips of the dipole elements. This is due in part to $r_\pm$ in (9.88).

Equation (9.88) is valid for a reference Hertzian dipole located at the origin, that is, $z = 0$. Assuming that the current distribution $I(p)$ has only a z variation, it is necessary to shift the reference Hertzian element by

$$\mathbf{R}' = \hat{z}z' \qquad (9.89)$$

as shown in Figure 9.6.

At the observation point (x, y, z), the new observation position vector is

$$\mathbf{R}'' = \mathbf{R} - \mathbf{R}' \qquad (9.90)$$

In terms of the new observation position vector, $\mathbf{R}''$, the differential electric field of the Hertzian array is

$$d\mathbf{E}^{(a)}(\mathbf{R}'') = \frac{\eta_o}{2d_x d_y} i(z')dz' \sum_{k=-\infty}^{\infty} \sum_{n=-\infty}^{\infty} \frac{e^{-j\beta \mathbf{R}''\cdot\hat{r}_\pm}}{r_z} e_\pm^{(a)} \qquad (9.91)$$

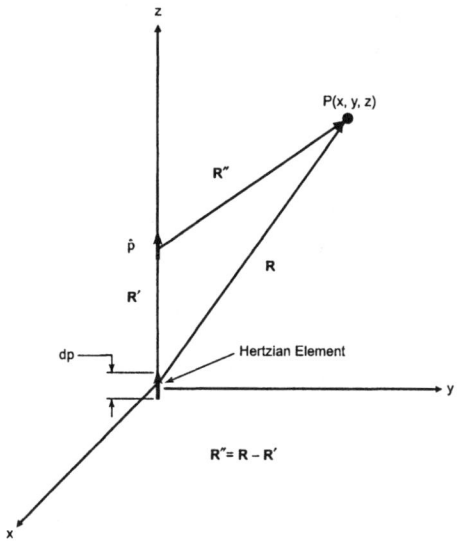

Figure 9.6 Shifting the reference Hertzian dipole current element to R'.

where
$$e_{\pm}^{(a)} = [\hat{p}^{(a)} \times \hat{r_{\pm}}] \times \hat{r_{\pm}} \tag{9.92}$$
is the polarization vector of the radiated field. Using (9.80) and (9.89) in (9.90) yields
$$R'' = R - R' = \hat{x}x + \hat{y}y + \hat{z}(z - z') \tag{9.93}$$
Substituting (9.93) into (9.91) and integrating over the length $(2l)$ of the reference dipole gives
$$E^{(a)} = \frac{\eta_o}{2d_x d_y} \sum_{k=-\infty}^{\infty} \sum_{n=-\infty}^{\infty} \frac{1}{r_z} \int_{z'=-l}^{z'=l} i(z') e_{\pm}^{(a)} e^{-j\beta((\hat{x}x+\hat{y}y+\hat{z}(z-z'))\cdot \hat{r_{\pm}})} dz' \tag{9.94}$$
where now
$$\hat{r_{\pm}} = \hat{x}(s_x + k\frac{\lambda}{d_x} - \frac{n\Delta y \lambda}{d_x d_y}) + \hat{y}(s_y + \frac{n\lambda}{d_y}) \pm \hat{z}r_z \tag{9.95}$$

In (9.95), the plus (+) sign is chosen for $z - z' > 0$ and the minus (−) sign is chosen for $z - z' < 0$. The factor
$$e^{-j\beta(\hat{x}x+\hat{y}y)\cdot \hat{r_{\pm}}} = e^{-j\beta(x(s_x+k\frac{\lambda}{d_x}-\frac{n\Delta y \lambda}{d_x d_y})+y(s_y+\frac{n\lambda}{d_y}))} \tag{9.96}$$
is independent of z' and has no sign ambiguity; thus, it can be removed from the integral of (9.94). However, other factors in the integrand have a sign

(denoted by sgn) that depends on the sign of $z - z'$ and must be studied carefully. From (9.81), it follows that

$$\hat{z} \cdot \hat{r}_{\pm} = r_z, \quad z - z' > 0 \qquad (9.97)$$

$$\hat{z} \cdot \hat{r}_{\pm} = -r_z, \quad z - z' < 0 \qquad (9.98)$$

Furthermore, $\text{sgn}(z - z')$ is positive for $z - z' > 0$ and $\text{sgn}(z - z')$ is negative for $z - z' < 0$ so it follows that

$$(z - z')\hat{z} \cdot \hat{r}_{\pm} = |z - z'|r_z \qquad (9.99)$$

Using (9.99), (9.94) reduces to

$$\boldsymbol{E}^{(a)}(x, y, z) = \frac{\eta_o}{2 d_x d_y} \sum_{k=-\infty}^{\infty} \sum_{n=-\infty}^{\infty} e^{-j\beta x (s_x + k\frac{\lambda}{d_x} - \frac{n \Delta y \lambda}{d_x d_y})} e^{-j\beta y (s_y + \frac{n\lambda}{d_y})}$$

$$\cdot \frac{1}{r_z} \int_{-l}^{l} i(z') e^{-j\beta |z-z'| r_z} \boldsymbol{e}_{\pm}^{(a)} dz' \qquad (9.100)$$

Equation (9.100) is the electric field radiated by the infinite array of vertical dipoles. This expression is valid in both the far field and in the near field, because it is a plane wave expansion that includes both propagating and evanescent terms.

In the next section, (9.100) is used to find the voltage induced at the terminals of an exterior element. The scan input impedance is then computed by evaluating the induced voltage for the exterior element at one wire radius away from the reference element and then dividing by the terminal current. The transmitting current distribution on the reference element is assumed to be piecewise-sinusoidal, that is,

$$i(z) = \sin \beta (l - |z|) \quad -l < z < l \qquad (9.101)$$

9.4.3 DERIVATION OF THE INDUCED VOLTAGE AT THE REFERENCE ELEMENT

In this section, a plane wave expansion is obtained for the voltage induced across the terminals of an exterior element that is assumed to be illuminated by the field from the infinite array. This plane wave expansion is used in the next section to compute the input impedance of a reference element of the array.

From Schelkunoff [25], the voltage induced across the terminals of a linear element is given by

$$V = \frac{1}{I^t} \int \boldsymbol{E} \cdot \hat{\boldsymbol{p}}\, i(p) dp \qquad (9.102)$$

where E is the incident electric field, $i(p)$ is the transmitting current distribution, $\hat{p}$ is the orientation of the linear element, and I^t is the terminal current for $i(p)$.

Assuming that the incident field is given by (9.100) (that is, the field from the equivalent dipole array) and the transmitting current distribution is given by (9.101), the voltage induced across the terminals of an exterior z-directed dipole is given by

$$V_{ea} = \frac{\eta_0}{2d_x d_y I^e} \sum_{k=-\infty}^{\infty} \sum_{n=-\infty}^{\infty} e^{-j\beta x(s_x + k\frac{\lambda}{d_x} - \frac{n\Delta y \lambda}{d_x d_y})} e^{-j\beta y(s_y + \frac{n\lambda}{d_y})} \frac{1}{r_z}$$

$$\cdot \int_{z=-l}^{z=l} \sin\beta(l - |z|) \int_{z'=-l}^{z'=l} \sin\beta(l - |z'|) e^{-j\beta|z-z'|r_z} e_{\pm}^{(a)} \cdot \hat{z} dz' dz \quad (9.103)$$

where $I^e = \sin\beta l$ is the terminal current of the exterior element. By using the vector identity

$$(A \times B) \times C = (C \cdot A)B - (C \cdot B)A \quad (9.104)$$

(9.92) can be expressed as

$$e_{\pm}^{(a)} = (\hat{r}_{\pm} \cdot \hat{z})\hat{r}_{\pm} - \hat{z} \quad (9.105)$$

Now, from (9.81), (9.97), (9.98), and (9.105) it follows that

$$e_{\pm}^{(a)} \cdot \hat{z} = r_z^2 - 1 \quad (9.106)$$

which is valid for all values of $z - z'$. Thus, the induced voltage can be written as

$$V_{ea} = \frac{\eta_0}{2d_x d_y I^e} \sum_{k=-\infty}^{\infty} \sum_{n=-\infty}^{\infty} e^{-j\beta x(s_x + k\frac{\lambda}{d_x} - \frac{n\Delta y \lambda}{d_x d_y})} e^{-j\beta y(s_y + \frac{n\lambda}{d_y})} \frac{(r_z^2 - 1)}{r_z}$$

$$\cdot \int_{z=-l}^{z=l} \sin\beta(l - |z|) \int_{z'=-l}^{z'=l} \sin\beta(l - |z'|) e^{-j\beta|z-z'|r_z} dz' dz \quad (9.107)$$

The double integrals (with respect to z and z' in (9.107)) can be regarded as the combined pattern functions (denoted by P_{ea}) of the reference element of the array and the exterior element. This double integration is evaluated as follows: By symmetry, the integral on z can be simplified as

$$P_{ea} = 2\int_{z=0}^{l} \sin\beta(l - z) \int_{z'=-l}^{l} \sin\beta(l - |z'|) e^{-j\beta|z-z'|r_z} dz' dz \quad (9.108)$$

Performing the integration on the variable z' yields (details are left to the reader)

$$P_{ea} = \frac{4}{\beta(1-r_z^2)} \int_0^l \sin\beta(l-z)[jr_z \sin\beta(l-z) - e^{-j\beta zr_z}\cos\beta l \\ + e^{-j\beta lr_z}\cos\beta zr_z]dz \quad (9.109)$$

Performing the integration on the variable z yields (details are left to the reader)

$$P_{ea} = \frac{2jr_z}{\beta^2(1-r_z^2)}(\beta l + \frac{\sin 2\beta l}{2}) \\ + \frac{4\cos\beta l}{\beta^2(1-r_z^2)^2}(\cos\beta l - e^{-j\beta lr_z} - jr_z \sin\beta l) \\ + \frac{4e^{-j\beta lr_z}}{\beta^2(1-r_z^2)^2}(\cos\beta lr_z - \cos\beta l) \quad (9.110)$$

Thus, the induced voltage is now expressed in terms of P_{ea} as

$$V_{ea} = \frac{\eta_0}{2d_x d_y I^e} \sum_{k=-\infty}^{\infty}\sum_{n=-\infty}^{\infty} e^{-j\beta x(s_x + k\frac{\lambda}{d_x} - \frac{n\Delta y\lambda}{d_x d_y})} e^{-j\beta y(s_y + \frac{n\lambda}{d_y})} \\ \cdot \frac{(r_z^2 - 1)}{r_z} P_{ea} \quad (9.111)$$

The expression given by (9.111) is the voltage induced across the terminals of an exterior dipole due to the field from the equivalent infinite dipole array. For input impedance calculations, the induced voltage is evaluated at a one-wire radius r_w displacement from the reference dipole element of the array. That is, x and y are computed by

$$x = r_w \cos\phi \quad (9.112)$$

$$y = r_w \sin\phi \quad (9.113)$$

where ϕ is the azimuth observation angle about the reference element axis. For thin dipoles ($r_w < 0.001\lambda$), the input impedance is independent of ϕ. When $r_w > 0.001\lambda$, it is generally necessary to average the input impedance over a series of azimuth angles.

9.4.4 SCAN INPUT IMPEDANCE

From Schelkunoff [25], the mutual impedance between an infinite array and an exterior element can be expressed as

$$Z_{ea} = -\frac{V_{ea}}{I^a} \qquad (9.114)$$

where V_{ea} is the induced voltage on the exterior element and $I^a = \sin\beta l$ is the terminal current for the reference element of the array. The scan input impedance of the array is evaluated by utilizing the induced voltage at a distance of one wire radius away from the reference element of the infinite array as in (9.111).

Substituting (9.111) into (9.114) and noting that the input impedance of the monopole array is one-half that of the equivalent dipole array results in the following equation for the scanned monopole array input impedance:

$$Z_{in}(\theta_s, \phi_s) = \frac{\eta_o}{d_x d_y \sin^2\beta l} \sum_{k=-\infty}^{\infty} \sum_{n=-\infty}^{\infty} e^{-j\beta x(s_x + k\frac{\lambda}{d_x} - \frac{n\Delta y\lambda}{d_x d_y})}$$
$$\cdot e^{-j\beta y(s_y + \frac{n\lambda}{d_y})} \frac{(r_z^2 - 1)}{r_z} P_{ea} \qquad (9.115)$$

where s_x and s_y are defined in (9.47) and (9.48), Δy is the shift in element position (see Figure 9.2), d_x and d_y are the interelement spacings, l is the monopole length, x and y are defined by (9.112) and (9.113), and r_z is defined by (9.82).

The scanned array reflection coefficient is obtained from the scanned array input impedance by the use of the following equation

$$\Gamma(\theta_s, \phi_s) = \frac{Z_{in}(\theta_s, \phi_s) - Z_c}{Z_{in}(\theta_s, \phi_s) + Z_c} \qquad (9.116)$$

where (θ_s, ϕ_s) is the scan direction, and Z_c is the characteristic impedance of the feed line.

9.4.5 THE ELEMENT-GAIN PATTERN FOR AN INFINITE ARRAY

In Section (9.2), expressions for computing the gain pattern of a monopole element in a finite array were given by (9.34) and (9.35). The element gain pattern for the infinite array is a function of the unit-cell area, the active transmission coefficient squared, and the projected aperture [6]. For a rectangular grid array, the element gain is given by

$$g_e(\theta, \phi) = \frac{4\pi d_x d_y}{\lambda^2} |T(\theta, \phi)|^2 \cos\theta \qquad (9.117)$$

where

$$|T(\theta,\phi)|^2 = 1 - |\Gamma(\theta,\phi)|^2 \tag{9.118}$$

is the active (scan) array power transmission coefficient, and $\Gamma(\theta,\phi)$ is given by (9.116). For an equilateral-triangle (hexagonal) array lattice, the element gain is expressed as

$$g_e(\theta,\phi) = 2\pi\sqrt{3}b^2|T(\theta,\phi)|^2\cos\theta \tag{9.119}$$

where b is the side of the equilateral triangle.

9.5 RESULTS

9.5.1 THE EFFECTS OF ARRAY SIZE FOR IDEAL ONE-QUARTER WAVELENGTH MONOPOLES

To better understand the electromagnetic behavior of a large phased array of monopoles, it is of interest to show some results for finite arrays of various sizes. The examples given are for one-quarter wavelength monopoles with one-half wavelength spacing on a square grid with an infinite ground plane. The monopoles are assumed to be electrically thin, with a wire radius of 0.001λ. The finite array formulation is utilized to compute the element gain pattern and input impedance. For example, a five row by five column passively terminated monopole array, with the center element driven, is depicted in Figure 9.7. Finite arrays up to size 25 rows by 25 columns are treated here. Additionally, the infinite array analysis is used for comparison and good agreement is shown.

The radiation pattern of an isolated monopole element and the same element embedded in arrays of various sizes is now investigated in Figure 9.8. Consider first the radiation pattern (E_θ component) of a single monopole antenna on an infinite ground plane. As shown in Figure 9.8 (solid curve) there is a null on axis ($\theta = 0°$) and the elevation pattern peak occurs as expected at $\theta = 90°$. The elevation pattern ($\phi = 0°$) for this array is superimposed with the single element pattern as well as other monopole array sizes, as shown in Figure 9.8. For the 5×5 array, the principal plane peak gain occurs at $\theta = 38°$. Successive elevation radiation patterns for the center monopole of 9×9, 11×11, and 25×25 arrays are included in Figure 9.8. The peak element gain occurs close to $\theta = 50°$ as the array size grows. For the 11×11 array, the element gain pattern has only small differences when compared to the results obtained for the 25×25 array. The element gain pattern for the infinite array of monopoles is also shown. The infinite array element peak gain is 2.7 dBi and occurs at $\theta = 50°$ – the infinite array element gain compares quite closely to the element gain pattern peak computed for the 25 × 25 array, which is 2.9 dBi. For the 11 × 11 array the azimuthal pattern variation at $\theta = 55°$ is approximately ±0.3 dB as shown in Figure 9.9.

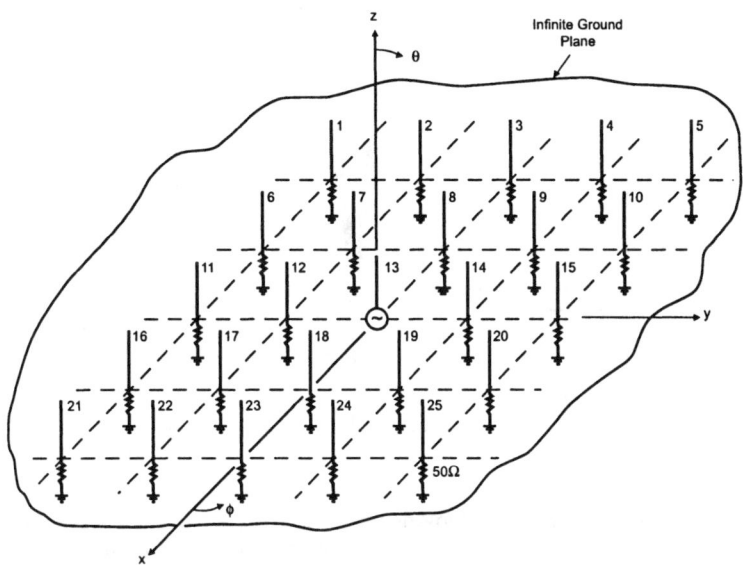

Figure 9.7 Example 5 × 5 monopole array with center element excited and surrounding elements passively terminated. © 1985 IEEE [5].

Also, the center element input impedance versus array size is summarized in Table 9.1. The input impedance of the $\lambda/4$ monopole with $\lambda/2$ element spacing tends to be almost purely resistive as the number of surrounding loaded monopoles increases, and the resistance is insensitive to array size.

The characteristics of the element gain patterns, scan reflection coefficient, and mutual coupling of elements in a large monopole array is of general interest. Consider, for example, the 11 × 11 array layout shown in Figure 9.10. Mutual coupling from the central element to the surrounding elements in the center row is depicted by the curved arrows. The elements in the central row are numbered from column −5 to column 5. The center row will be referred to as row number 0, and the element locations can be referred to as (column number, row number).

The calculated principal plane element gain patterns for the elements along the center row are shown in Figure 9.11. Approximately 2.2 dB of asymmetry occurs between the two (left and right) peak values of the edge element pattern. The edge-adjacent element (−4, 0) has about 1.2 dB of

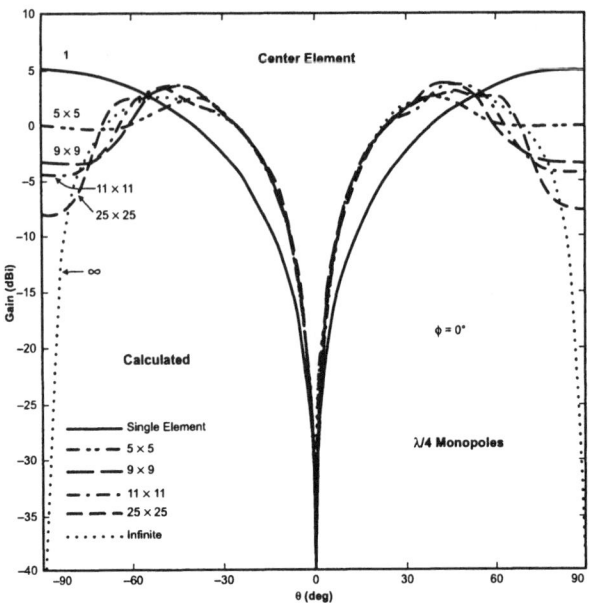

Figure 9.8 Theoretical center element gain pattern ($\phi = 0°$ cut) as a function of monopole array size. Monopole length is $\lambda/4$, wire radius is 0.001λ, and element spacing is $\lambda/2$. © 1985 IEEE [5].

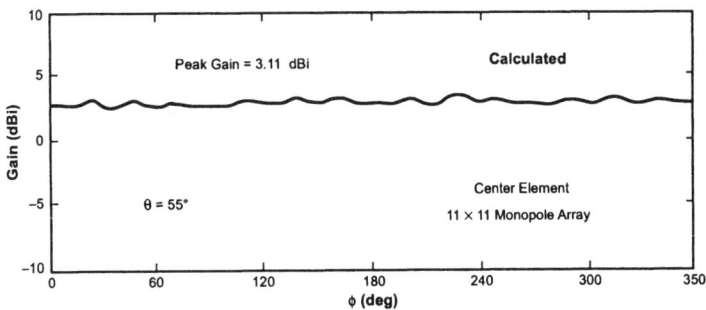

Figure 9.9 Conical gain pattern cut for the center monopole of an 11×11 array. Monopole length is $\lambda/4$, wire radius is 0.001λ, and element spacing is $\lambda/2$. © 1985 IEEE [5].

pattern asymmetry. For monopoles two or more elements away from the edge element, the radiation pattern peak asymmetry is less than 0.5 dB.

Next, the calculated input impedance of a few elements of the 121-element array is given in Table 9.2: the center element, a corner element, and an edge element of the center row. The real part of the input impedance is seen

Table 9.1
Theoretical Input Impedance for the Center Element of Passively Terminated Monopole Arrays

Array Size		No. of	Input Impedance
No. Rows	No. Columns	Elements	(ohms)
1	1	1	36.5 + j 21.0
3	3	9	38.0 + j 5.9
5	5	25	38.8 + j 3.2
7	7	49	38.7 + j 2.4
9	9	81	38.6 + j 2.1
11	11	121	38.5 + j 2.0
25	25	625	38.4 + j 1.8

Monopole length is $\lambda/4$, wire radius is $0.001\ \lambda$, and element spacing is $\lambda/2$ on a square grid.

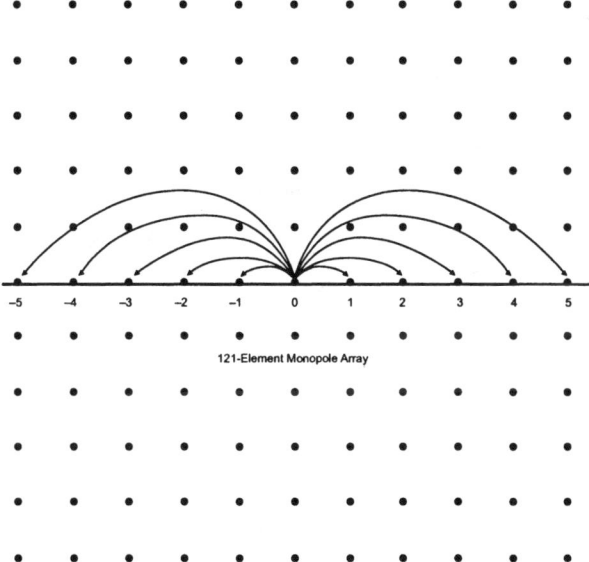

Figure 9.10 11 × 11 monopole array numbering convention for the center row. Mutual coupling between the center element and the surrounding elements in the center row is indicated by curved arrows. © 1985 IEEE [5].

to be relatively insensitive to element position in the array and the imaginary part varies by less than 20 ohms.

The calculated scan reflection coefficient of the monopole array versus scan angle (θ_s) is now considered. The magnitude of the reflection coefficient for the 11×11 array (center element) and infinite array is shown in Figure 9.12 for the principal plane scan, and good agreement between these two cases is evident. The scan reflection coefficient minimum is seen to occur in the vicinity of $\theta_s = 60°$, as desired for wide-angle scanning.

One of the conclusions that can be made from this computer simulation data is that a large array of monopoles has good wide-angle scanning capability. Based on the theoretical element gain pattern, main beam scanning with low gain loss is possible from approximately 30° to 60° from broadside. In reference to the infinite array data and 25×25 array data, it is apparent that the center element in an 11 row by 11 column array behaves very similar to that in a much larger array. The next section investigates a prototype 11×11 monopole array.

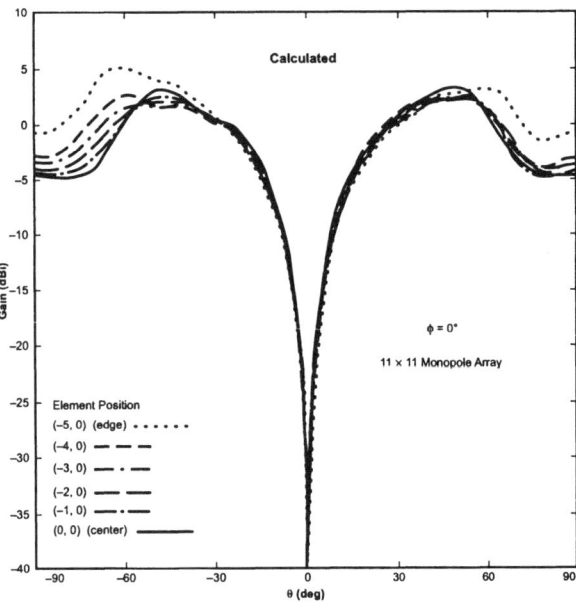

Figure 9.11 Theoretical element gain patterns as a function of element position along the center row of an 11×11 monopole array on an infinite ground plane. Monopole length is $\lambda/4$, wire radius is 0.001λ, and element spacing is $\lambda/2$. © 1985 IEEE [5].

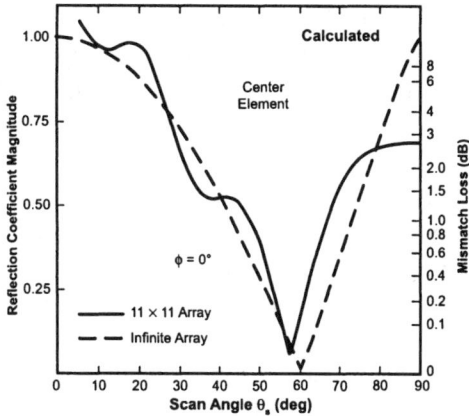

Figure 9.12 Theoretical center element reflection coefficient magnitude as a function of scan angle (θ_s is variable, $\phi_s = 0°$) for infinite and finite arrays of monopoles. Monopole length is $\lambda/4$, wire radius is 0.001λ, and element spacing is $\lambda/2$. © 1985 IEEE [5].

9.5.2 121-ELEMENT SQUARE GRID MONOPOLE ARRAY: EXPERIMENT AND THEORY

Based on the thin-wire monopole array data given in the previous section, a 121-element square grid array is considered well representative for demonstrating the monopole element performance in a large array. To verify the theory, a prototype array was designed, fabricated, and tested over the desired frequency range 1.2 to 1.4 GHz. Assuming an interelement spacing of 10.67 cm this yields a 1.07m by 1.07m array. At a center frequency of 1.3 GHz the element spacing is 0.462λ. The diameter of the monopole elements was chosen to be 0.3175 cm, which is 0.0148λ at the high frequency (1.4 GHz);

Table 9.2
Theoretical Input Impedance at Various Element Positions in an 11 × 11 Monopole Array

Element Position on 11 × 11 Array	Input Impedance (ohms)
Center Element (0,0)	38.5 + j 2.0
Edge Element of Center Row (−5,0)	39.7 + j 7.0
Corner Element (−5,−5)	39.7 + j 11.7
Isolated Element (reference)	36.5 + j 21.0

Monopole length is $\lambda/4$, wire radius is $0.001\ \lambda$, and element spacing is $\lambda/2$ on a square grid.

hence, the thin-wire formulation is applicable. The monopole element length can be optimized theoretically by computing the center-element scanned array reflection coefficient over the desired scan sector and frequency bandwidth. After running a number of cases, a length of 6.35 cm ($0.275\lambda_o$) was deemed appropriate.

Each monopole element was constructed by attaching a 0.3175 cm diameter brass rod to the center pin of a type-N panel connector. A photograph of a typical monopole array element installed on a ground plane is shown in Figure 9.13. A 1.22m × 1.22m square sheet of aluminum (0.3175 cm thick) was used for the 121-element array ground plane. The element mounting holes were machined such that the base region of each monopole is flush with the ground plane. A sketch of the overall array configuration is given in Figure 9.14 and a photograph of the assembled 121-element array on a test positioner in an anechoic chamber is shown in Figure 9.15. For far-field pattern measurements, the center element is driven and the surrounding elements are terminated in 50 Ω resistive loads.

The measured center element gain elevation pattern ($\phi = 0°$) at the center frequency 1.3 GHz is given in Figure 9.16. Included in this figure is the corresponding finite array and infinite array theoretical element gain patterns, which compare closely to the measured data. Figure 9.17 shows a measured conical radiation pattern cut (ϕ variable) at 1.3 GHz for $\theta = 55°$. The E_θ (principal) polarized pattern is seen to be nearly omnidirectional as expected. The cross-polarized component (E_ϕ) is down by more than 30 dB. These patterns confirm that wide-angle scanning is practical with a monopole element.

Next, referring to Figure 9.10, to measure the complex mutual coupling in the 11 × 11 array, the center element was connected to a CW transmitter and the relative received voltage was measured at each of the surrounding elements. Except for the transmitting element and the receiving element, all elements were terminated in 50-ohm resistive loads. This measurement yields the coupling coefficient (also referred to as the scattering parameter S_{21}). Figure 9.18(a) shows a plot of the amplitude of the coupling coefficients along the center row at 1.3 GHz. Figure 9.18(b) is the corresponding phase received at each element. The measured amplitude is in good agreement with the theory. However, there is a noticeable shift in phase between the measured and theoretical data. This shift is likely due in part to two approximations in the analysis. One is that a delta gap model is used for the feed region rather than the actual coaxial aperture. Second, the analysis assumes a one mode sinusoidal current distribution on the monopole, which is not exact. It is useful to note that the received amplitude is down by 40 dB at the edge elements. Thus, based on the mutual coupling amplitude data, little change in the element gain and scan reflection coefficient or scan input impedance would be expected by increasing the array size. The passive center element reflection coefficient versus frequency is shown in Figure 9.19, and there is generally good agreement between the measurements and moment method simulations for the 11 × 11 array.

Using the measured coupling coefficients in (9.40), the scan reflection coefficient and input impedance as a function of scan angle can be computed using the relation given in (9.36). This computation for the scan input impedance and scan reflection coefficient is done for the principal plane scan ($\phi_s = 0°$) in Figures 9.20 and 9.21 at 1.3 GHz. In Figures 9.20 and 9.21, the three curves are measured data, finite array theory, and infinite array

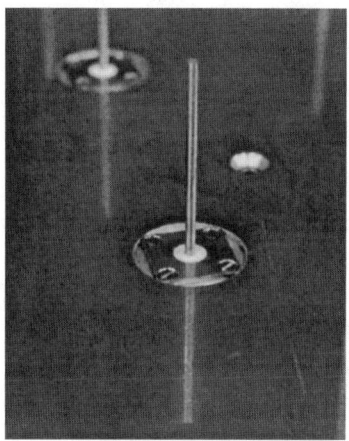

Figure 9.13 Photograph of a cylindrical monopole array element mounted on a conducting ground plane.

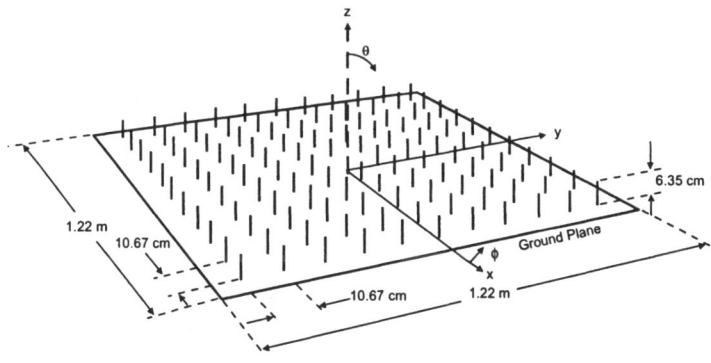

Figure 9.14 121-element monopole array layout. © 1985 IEEE [5].

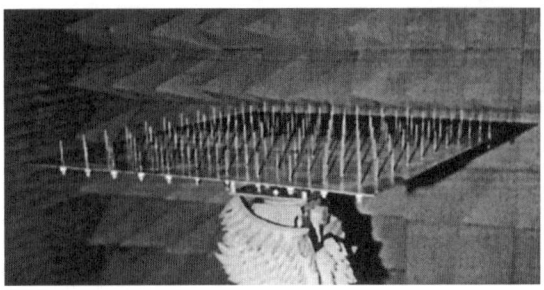

Figure 9.15 Photograph of a 121-element monopole array on an antenna positioner in an anechoic chamber. © 1985 IEEE [5].

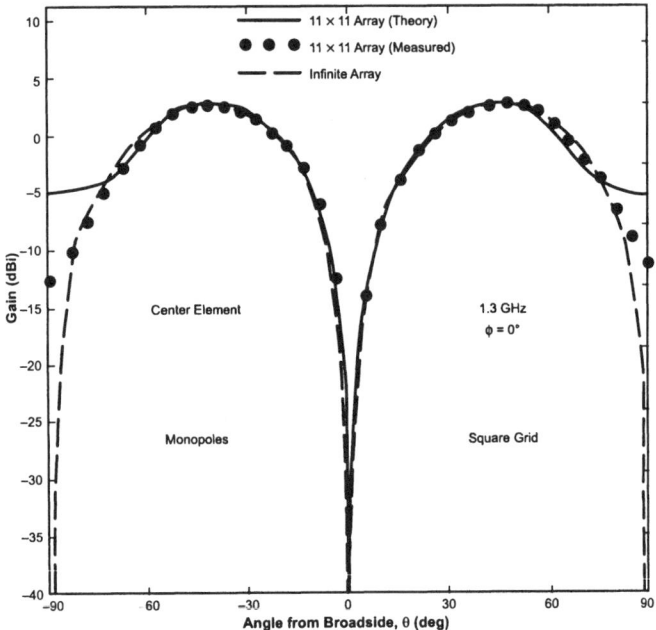

Figure 9.16 Measured and calculated center element gain patterns at 1.3 GHz for the 121-element monopole array compared with the element gain for an infinite array. Elevation cut at $\phi = 0°$. Monopole length is 6.35 cm, wire radius is 1.588 mm, and element spacing is 10.67 cm. © 1985 IEEE [5].

theory – there is generally good agreement between theory and experiment. In Figure 9.21 the reflection coefficient has a minimum at about $\theta = 52°$. Optimization of the input impedance of the monopole array can be achieved by applying standard impedance matching techniques [26]. The optimization would be with respect to a selected bandwidth and scan sector.

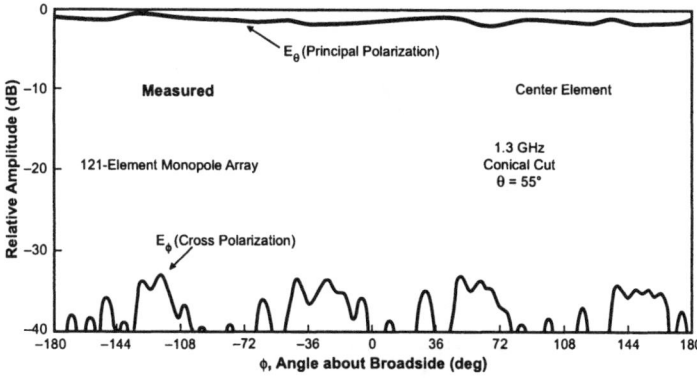

Figure 9.17 Measured center element radiation pattern at 1.3 GHz for the 121-element monopole array. Conical cut at $\theta = 55°$. Monopole length is 6.35 cm, wire radius is 1.588 mm, and element spacing is 10.67 cm. © 1985 IEEE [5].

The scanned radiation pattern of a large array of thousands of array elements, including mutual coupling effects, can be calculated in an approximate manner by first computing the element gain pattern in a large array and then multiplying it by the array factor given by (9.26). For example, using the element gain radiation pattern for a 25×25 (625 elements) monopole array (similar to the patterns shown in Figure 9.16, the radiation pattern of a large array with 150 columns and 75 rows (11,250 elements)) can be calculated for a 55° scan angle and is shown in Figure 9.22. The sidelobes are very low in the vicinity of $\theta = 0°$ as expected, and these low sidelobes would reject clutter in the nadir region, as depicted in Figure 9.1. The peak gain of this 11,250 element array is 42.4 dBi – the array gain was estimated by taking the gain of the element at $\theta_s = 55°$ and adding $10 \log_{10} N_a$, where $N_a = 11,250$ is the number of array elements.

Based on the results presented in this section, an approximate model for the embedded monopole array element gain pattern versus θ is just

$$g_e(\theta) = 2\sin^2(2\theta). \qquad (9.120)$$

The approximate gain model for an embedded monopole as given in (9.120) has a peak gain of 3.0 dBi, the peak occurs at $\theta = 45°$, and gain nulls occur at $\theta = -90°, 0°, 90°$. In Figure 9.23, the approximate embedded element gain pattern ((9.120) normalized to unity, $g'_e(\theta) = \sin^2(2\theta)$) is in good agreement with the normalized element gain computed for an infinite array with a square grid with 10.67 cm element spacing and monopoles with length 6.35 cm, and wire radius 1.588 mm.

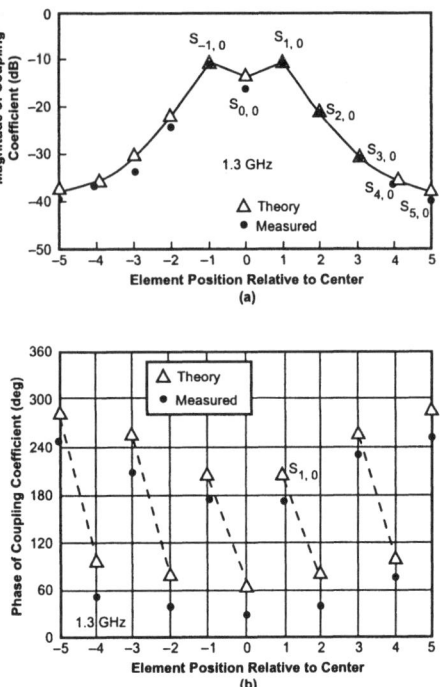

Figure 9.18 Measured and calculated mutual coupling coefficient as a function of element position for the 121-element monopole array. The center element is transmitting and the elements along the center row in Figure 9.10 are receiving. Monopole length is 6.35 cm, wire radius is 1.588 mm, and element spacing is 10.67 cm. © 1985 IEEE [5].

9.5.3 HEXAGONAL LATTICE INFINITE ARRAY RESULTS

The results of the present analysis are compared now to calculated and measured data in the literature for hexagonal lattice infinite arrays. For example, Herper and Hessel [9] performed a similar infinite array analysis (combination unit cell and variational impedance technique), which also assumed a sinusoidal current distribution. They used the waveguide simulator technique [3] to experimentally verify the theoretical array scan input impedance for a few scan angles. The array parameters used were monopole length = 0.25λ, monopole diameter = 0.02λ, and element spacing = 0.55λ at the center frequency. Their results are reproduced in Figure 9.24, along with a comparison of the present theory. The 45° scan angle data are all in good agreement. For the 65° scan angle, the agreement is not as good, but this is likely attributed to the single sinusoid current distribution assumption and the

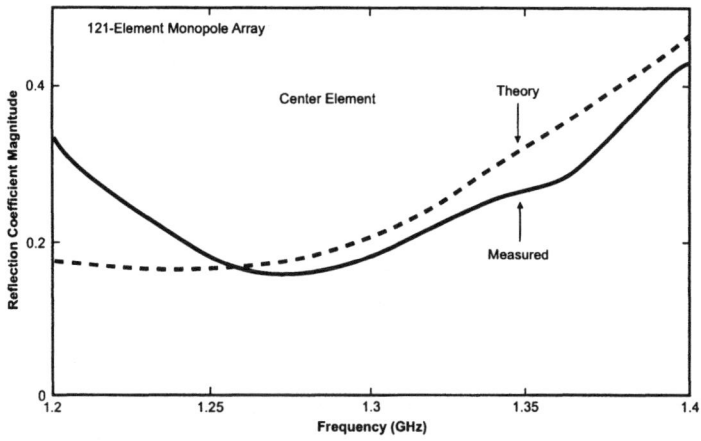

Figure 9.19 Measured and calculated center element passive reflection coefficient versus frequency for the 11 × 11 array. Monopole length is 6.35 cm, wire radius is 1.588 mm, and element spacing is 10.67 cm. © 1985 IEEE [5].

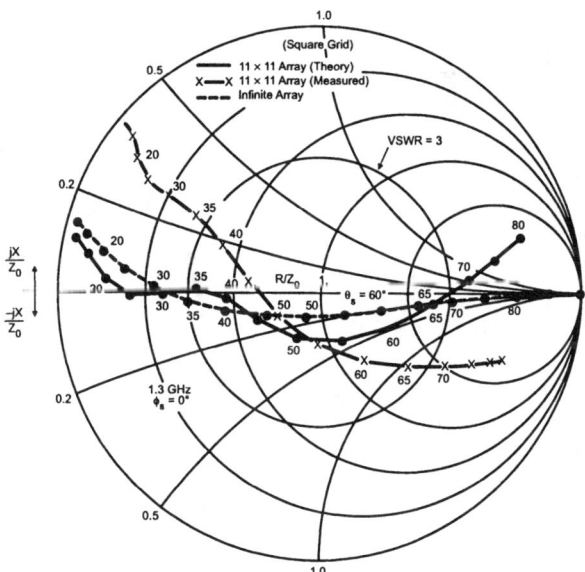

Figure 9.20 Measured and calculated scanned array input impedance at 1.3 GHz as a function of scan angle for the 121-element monopole array and for an infinite array. Monopole length is 6.35 cm, wire radius is 1.588 mm, and element spacing is 10.67 cm on a square grid. © 1985 IEEE [5].

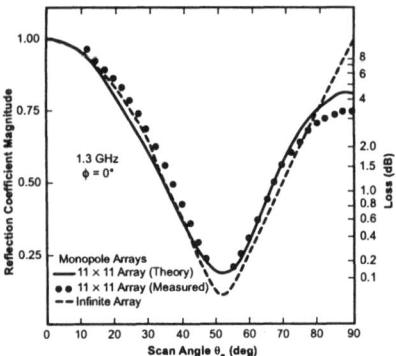

Figure 9.21 Measured and calculated scanned array reflection coefficient as a function of scan angle for the 121-element monopole array and for an infinite array. Monopole length is 6.35 cm, wire radius is 1.588 mm, and element spacing is 10.67 cm on a square grid. © 1985 IEEE [5].

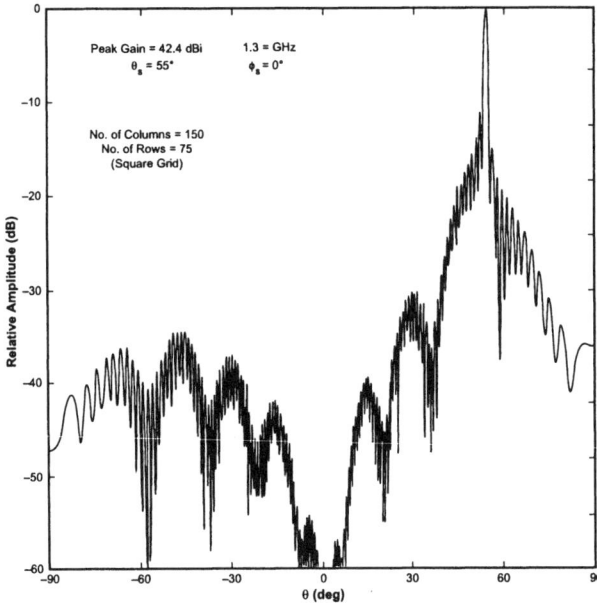

Figure 9.22 Simulated scanned ($\theta_s = 55°, \phi_s = 0°$) radiation pattern for an 11,250-element monopole phased array antenna. Monopole length is 6.35 cm, wire radius is 1.588 mm, and element spacing is 10.67 cm (square grid). There are 150 columns aligned in the $\phi = 0°$ direction, and there are 75 rows.

delta-gap model previously mentioned. Even with the simple model, however, the correct scan angle dependence and frequency behavior are predicted.

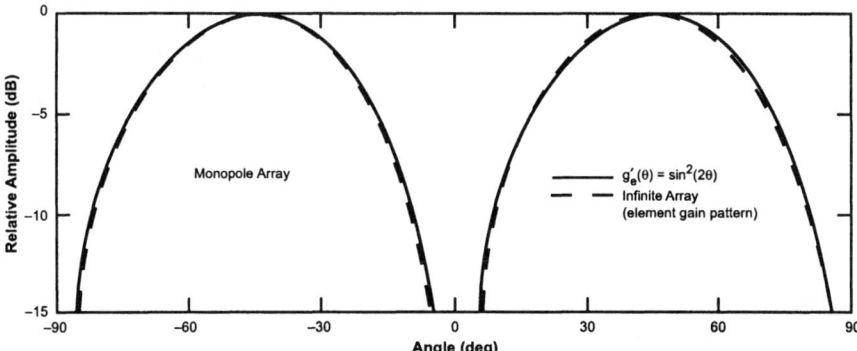

Figure 9.23 Comparison of approximate embedded monopole element gain model ((9.120) normalized to unity) with calculated element gain for an infinite phased array of monopoles. For the infinte array simulation, the monopole length is 6.35 cm, wire radius is 1.588 mm, and element spacing is 10.67 cm (square grid), and the frequency is 1.3 GHz.

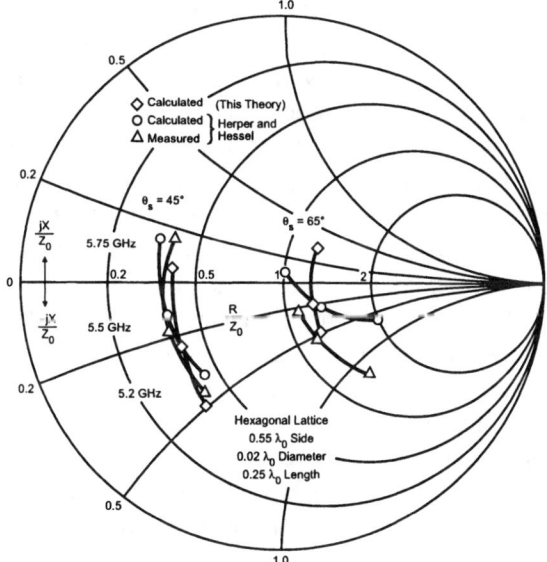

Figure 9.24 Calculated and measured array scan input impedance as a function of scan angle (θ_s) for infinite arrays of monopoles with hexagonal lattice. At center frequency, monopole length is $0.25\lambda_o$, wire radius is $0.01\lambda_o$, and element spacing is $0.55\lambda_o$. © 1985 IEEE [5].

9.6 SUMMARY

This chapter has described the theory and experimental results for two-dimensional periodic monopole phased array antennas. A sinusoidal-Galerkin's version of the method of moments was used to analyze finite arrays. An infinite array plane-wave representation of the scan impedance was given for monopoles with sinusoidal current, arranged on a general skewed grid. Element gain patterns and active input impedance were computed both for finite and infinite arrays of thin cylindrical monopoles. The effects of the array size and element position on the element gain pattern and input impedance were shown.

Good agreement for element gain patterns, mutual coupling, and input impedance was obtained between theory and measurements for a 121-element monopole array with square lattice. The center element gain pattern indicates good pattern coverage at wide angles from broadside. The radiation pattern is vertically polarized and nearly omnidirectional at a fixed angle from broadside. Wide angle scanning out to $60°$ with peak gain occuring near $50°$ from broadside is possible with this antenna configuration. Natural sidelobe suppression would occur in the vicinity of broadside, because of the monopole element pattern null. Good agreement between the present infinite array analysis and calculated and measured data for hexagonal lattices was shown.

Finally, there is the issue of possible blind spot occurrence in thin-monopole phased arrays. As is well known, blind spots are often due to the presence of higher-order modes within the array unit cell [6]. The earlier analysis has assumed that the transmitting current distribution is a single piecewise-sinusoidal function. The assumption of a symmetric current distribution on the equivalent dipole is valid because the monopole image current is always symmetric with the current on the monopole. The current vector direction is essentially parallel to the orientation of the thin monopole and the azimuthal (phi) component of current is negligible. Thus, higher-order modes are not provided for in the model. However, since the monopole length is close to $\lambda/4$, the fundamental resonance, it is unlikely that higher-order modes would exist to any appreciable degree. Based on these considerations and the presented theoretical and experimental results, blind spots are not expected to occur for phased arrays of thin monopoles. Monopole phased arrays are further discussed in detail in Chapters 10, 11, and 12.

9.7 PROBLEM SET

9.1 Derive (9.25)
9.2 Derive (9.109).
9.3 Derive (9.110).
9.4 Write software to evaluate the input impedance given by (9.115), and reflection coefficient as given by (9.116) for an infinite array of monopoles with parameters as in the example shown in Figure 9.12. Hint: Vary the

number of terms in the summation on k and n until the input impedance is converged.

References

[1] Amitay, N., V. Galindo, and C.P. Wu, *Theory and Analysis of Phased Array Antennas*, New York: Wiley, 1972, p. 22.

[2] Mailloux, R.J., "Phased Array Theory and Technology," *Proc. IEEE*, Vol. 70, 1982, pp. 246-291.

[3] Mailloux, R.J., *Phased Array Antenna Handbook*, Norwood, MA: Artech House, 1994.

[4] Hansen, R.C., *Phased Array Antennas*, New York: Wiley, 1998.

[5] Fenn, A.J., "Theoretical and Experimental Study of Monopole Phased Array Antennas," *IEEE Trans. Antennas Propagat.*, Vol. 33, No. 10, 1985, pp. 1118-1126.

[6] Kelly, E.J., and G.N. Tsandoulas, "A Displaced Phase Center Antenna Concept for Space Based Radar Applications," *IEEE Eascon*, Sept. 1983, pp. 141-148.

[7] Tsandoulas, G.N., "Space-Based Radar," *Science*, July 17, 1987, pp. 257-262.

[8] Skolnik, M.I., *Introduction to Radar Systems*, New York: McGraw-Hill, 1962, pp. 605-607.

[9] Herper, J.C., and A. Hessel, "Performance of $\lambda/4$ Monopole in a Phased Array," *IEEE Antennas and Propagat. Society, 1975 Symp. Digest*, pp. 301-304.

[10] Schuman, H.K., D.R. Pflug, and L.D. Thompson, "Infinite Planar Arrays of Arbitrarily Bent Thin Wire Radiators," *IEEE Trans. Antenna Propagat.*, Vol. 32, No. 3, 1984, pp. 364-377.

[11] Kurtze, G., and E.G. Neumann, "Absorption and Transmission of Electromagnetic Waves, Phase B: Development of 3-Dimensional Dipole Absorbers," III Physikalisches Institut der Universitat Gottingen, RADC TN-59-375 B, Technical Report Contract No. AF 61 (052)-154, September 30, 1959, AD No. 231020.

[12] Sharp, E.D., "A Triangular Arrangement of Planar-Array Elements that Reduces the Number Needed," *IRE Trans. Antennas Propagat.*, Vol. 9, No. 2, 1961, pp. 126-129.

[13] Castello, D., and B.A. Munk, "Table of Mutual Impedance of Identical Dipoles in Echelon," Report 2382-1, The Ohio State University ElectroScience Laboratory, Department of Electrical Engineering: prepared under Contract F33615-67-C-1507 for Air Force Avionics Laboratory, Wright-Patterson Air Force Base, Ohio, (AD 822013), October 1968.

[14] Stutzman, W.L., and G.A. Thiele, *Antenna Theory and Design*, New York: Wiley, 1981, pp. 306-374.

[15] Hansen, R.C., (ed.), *Microwave Scanning Antennas, Vol. II: Array Theory and Practice*, New York: Academic, 1966, pp. 213-216.

[16] Sinnott, D.H., "An Improved Algorithm for Matrix Analysis of Linear Antenna Arrays," Australian Defense Scientific Service, Weapons Research Establishment, Adelaide, South Australia, WRE-TECH. NOTE-1066(AP), 1974.

[17] Montgomery, C.G., R.H. Dicke, and E.M. Purcell, *Principles of Microwave Circuits*, Massachusetts Inst. Technol., Radiation Lab. Series, Vol. 8, New York: McGraw-Hill, 1948. pp.147-149.

[18] Brennecke, N.R., and W.N. Moule, "Uses of Fences to Optimize Operating Impedance of Phased Arrays Using an Improved Measuring Technique," *IEEE Antennas Propagat. Soc. Int. Symp. Dig.*, 1964, pp. 134-142.

[19] Munk, B.A., and G.A. Burrell. "Plane-Wave Expansion for Arrays of Arbitrarily Oriented Piecewise Linear Elements and its Application to Determining the Impedance of a Single Linear Antenna in a Lossy Half-Space," *IEEE Trans. Antennas Propagat.*, Vol. 27, No. 3, 1979, pp. 331-343.

[20] Munk, B.A., *Frequency Selective Surfaces: Theory and Design*, New York: Wiley, 2000.

[21] Munk, B.A., *Finite Antenna Arrays and FSS*, New York: Wiley, 2003.

[22] Papoulis, A., *The Fourier Integral and Its Applications*, New York: McGraw-Hill, 1962.

[23] Bateman, H., *Tables of Integral Transforms, Vol. I*, New York: McGraw-Hill, 1954,

[24] Kraus, J.D., *Antennas*, 2nd ed., New York: McGraw-Hill, 1988.

[25] Schelkunoff, S.A., *Antennas: Theory and Practice*, New York: Wiley, 1952, p. 298, Equation (100) and p. 366, Equation (26).

[26] Thomas, R.L., *A Practical Introduction to Impedance Matching*, Dedham, MA: Artech House, 1976.

10

Monopole Phased Array Field Characteristics in the Focused Near-Field Region

10.1 INTRODUCTION

The focused near-field adaptive nulling performance [1-4] of phased array antennas has been investigated in Chapters 3 through 5 as a testing methodology for evaluating the far-field performance of adaptive arrays. Isotropic point source arrays were investigated in Chapter 3, and practical arrays including monopole and dipole elements with mutual coupling effects were analyzed in Chapters 4 and 5. It was shown that the adaptive array characteristics in the focused near-field region are equivalent to conventional far-field characteristics. Experimental verification [4] of focused near-field adaptive nulling was described in Chapter 6. In the focused near-field nulling technique, the phased array antenna is focused at one to two aperture diameters, and radiating test sources are positioned on the focal plane. Near-field main beam focusing is achieved with calibration and phase control at each active (driven) element of the array, producing an antenna near-field radiation pattern equal to a conventional far-field pattern [5]. As part of the near-field radiation pattern formulation in [1, 2, 4], it was assumed that the radially polarized component of the electric field is negligible compared to the principal component. The principal electric-field component was computed from the tangential electric-field component while ignoring the radial component. The radially polarized electric field has been considered by several investigators for determining the cross-polarization level in reflector antenna systems [6, 7]. This author has investigated the radially polarized amplitude of a phased array focused in the near field [8]. The purpose of the present chapter is to show by computer simulation that, for a near-field focused phased array antenna composed of monopole elements, the assumption of a negligible radial component is valid [8]. The near-field focusing distance here is assumed to be in the range of one to two aperture diameters of the antenna under test.

The next section gives the design for a monopole phased array antenna used in the computer simulations. Section 10.3 describes the near-field formulation used to calculate the electric field components of the monopole phased array antenna. The results are presented in Section 10.4, and Section 10.5 contains a summary.

10.2 MONOPOLE PHASED ARRAY ANTENNA DESIGN

Consider Figure 10.1 which shows the phased array antenna to be analyzed in this chapter. The array consists of thin monopole elements having length 0.264λ with wire radius 0.007λ and having element spacing 0.473λ at center frequency 1.3 GHz. This array design has been used in evaluating planar near-field measurement techniques for low-sidelobe antenna applications [9] as discussed in Chapter 12. A monopole phased array antenna [10, 11] is useful in achieving wide-angle scanning as described in Chapter 9. The array is assumed to have 180 monopole elements arranged in 5 rows and 36 columns on a square lattice. To reduce edge effects, two guard bands of passively terminated (50-ohm loaded) elements (unshaded elements in Figure 10.1) are assumed to surround the driven portion (shaded elements) of the array, which forms a linear array of 32 active elements in the center row. With 10.922 cm interelement spacing, the length of the active array is 3.4m. Near field distances $z_o = 3.4$m (one aperture diameter) to $z_o = 6.8$m (two aperture diameters) are considered here.

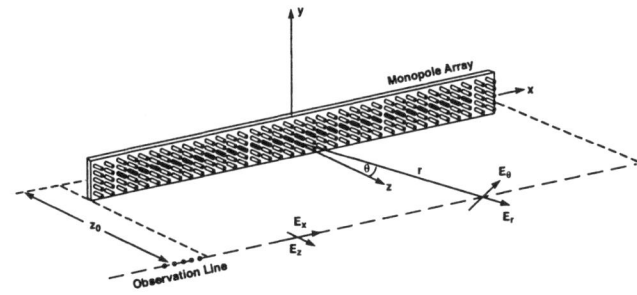

Figure 10.1 Monopole array geometry. The shaded elements are driven and the unshaded elements are passively terminated. © 1992 IEEE [8].

10.3 THEORETICAL FORMULATION FOR THE ELECTRIC FIELD COMPONENTS

In standard spherical coordinates, the far-field principal polarization of a monopole array, with $\hat{z}$-directed elements, is the E_θ or vertically polarized component. At a range distance of infinity, the radial component of the electric field, E_r, is zero; however, in the near field E_r is nonzero. For a thin monopole, the E_ϕ component of the electric field is theoretically zero both in the near field and in the far field and is ignored in this analysis.

The computation of the near-zone radial and vertical electric field components is accomplished by using expressions given in [12]. The method of moments [13] is used to include the effects of mutual coupling between the monopole array elements. The monopoles and infinite ground plane are replaced by an equivalent array of dipoles in free space as depicted in Figure 10.2. A piecewise-sinusoidal current distribution expressed as

$$i_n(z) = i_n^{term} \frac{\sin[k(l - |z|)]}{\sin kl} \tag{10.1}$$

is assumed on the nth array element. In (10.1), i_n^{term} is the terminal current for the nth element, and $k = 2\pi/\lambda$ is the angular wavenumber, with λ the wavelength, l is the monopole length (dipole half length), and $|z| \leq l$.

The near-zone electric field components for a center-fed linear dipole antenna are given in cylindrical coordinates as [12]

$$E_{nz} = \frac{j30 i_n^{term}}{\sin kl} (2 \frac{e^{-jkr_o}}{r_o} \cos kl - \frac{e^{-jkr_1}}{r_1} - \frac{e^{-jkr_2}}{r_2}) \tag{10.2}$$

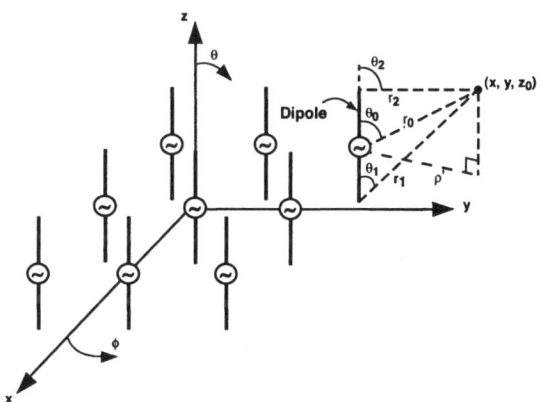

Figure 10.2 Near-field geometry for equivalent dipole array antenna. © 1992 IEEE [8].

$$E_{n\rho'} = \frac{j30i_n^{term}}{\rho' \sin kl}(e^{-jkr_1}\cos\theta_1 + e^{-jkr_2}\cos\theta_2 - 2\cos kl e^{-jkr_o}\cos\theta_o) \quad (10.3)$$

where $j = \sqrt{-1}$ and the parameters r_o, r_1, r_2, ρ' and $\theta_o, \theta_1, \theta_2$ are defined in Figure 10.2. The rectangular coordinate system electric-field components E_{nx} and E_{ny} are readily computed as

$$E_{nx} = E_{n\rho'}\cos\phi' \quad (10.4)$$

$$E_{ny} = E_{n\rho'}\sin\phi' \quad (10.5)$$

where $\phi' = \tan^{-1}((y-y_n)/(x-x_n))$ is the angle between $\hat{x}$ and $\hat{\rho}'$ with the coordinates (x_n, y_n) denoting the feed point of array element n. To compute the antenna near field, including array mutual-coupling effects, the array terminal currents are computed in the following manner:

Let Z represent the mutual impedance matrix for the equivalent free-space dipole array, such that

$$Z = Z^{o.c.} + Z_L I \quad (10.6)$$

where $Z^{o.c.}$ is the open-circuit mutual impedance matrix for the array, I is the identity matrix, and Z_L is the load impedance at each element. The moment method mutual impedance matrix is evaluated using computer subroutines developed in [14].

Define v as the voltage excitation matrix of the array. Then the array element terminal currents, i, are related to the voltage excitation matrix and impedance matrix by

$$v = Z \cdot i \quad (10.7)$$

The nth element of the voltage excitation matrix for a phased array antenna is given by

$$v_n = A_n e^{j\psi_n} \quad (10.8)$$

where A_n is the amplitude illumination and ψ_n is the phase delay which scans the main beam to the near-field position (z_o, θ_s, ϕ_s).

Using (10.6) in (10.7), the array terminal currents are found by solving the system of equations written in matrix form as

$$v = [Z^{o.c.} + Z_L I] \cdot i \quad (10.9)$$

The impedance matrix in (10.9) is of block-Toeplitz form for which special-purpose computer subroutines are used in solving for the unknown currents

[13, 15]. Having computed the array terminal currents from (10.9) and the nth-element near field by using (10.2) to (10.5), the array near-zone field including mutual coupling effects is expressed by summing the contributions from all N array elements as

$$E_x(x, y, z_o) = \sum_{n=1}^{N} E_{nx} \qquad (10.10)$$

$$E_y(x, y, z_o) = \sum_{n=1}^{N} E_{ny} \qquad (10.11)$$

$$E_z(x, y, z_o) = \sum_{n=1}^{N} E_{nz} \qquad (10.12)$$

After computing the rectangular components of the electric field, the equations for computing the exact near-field spherical components in the $\phi = 0°$ cut are given by a standard rectangular-to-spherical coordinates conversion,

$$E_r = E_x \sin\theta + E_z \cos\theta \qquad (10.13)$$

$$E_\theta = E_x \cos\theta - E_z \sin\theta \qquad (10.14)$$

Note that in the $\phi = 0°$ cut, for a linear array of $\hat{z}$ monopoles, the E_y component is theoretically zero and is ignored. If it is assumed that the radial component is zero (far-field (FF) assumption), then substituting $E_r^{FF} = 0$ in (10.13) yields

$$E_z^{FF} = -E_x \tan\theta \qquad (10.15)$$

Therefore, the E_z component is dependent only on the E_x component when the radial component is zero. From (10.15) it is observed that E_z^{FF} is in phase with E_x for $\theta < 0°$ and is 180° out of phase for $\theta > 0°$. Substituting (10.15) in (10.14) and simplifying yields

$$E_\theta^{FF} = \frac{E_x}{\cos\theta} \qquad (10.16)$$

for the case of a zero radial component in the far field (or a negligible radial component in the near field). Thus, (10.16) is an approximate expression for the vertically polarized near-field component whereas (10.14) is exact. Both (10.14) and (10.16) are applied in the next section to the case of a near-field focused phased array antenna with variable focal range and variable scan angle.

Table 10.1
Relative Radial and Normal Components for 32-Element Monopole Array with Focusing Distance z_o. Scan Angle is $\theta_s = -30°$

$z_o(m)$	E_r^{max}/E_θ^{max}	E_z^{max}/E_x^{max}
3.4	−19.1 dB	−4.5 dB
5.1	−22.2 dB	−4.6 dB
6.8	−24.5 dB	−4.7 dB

10.4 RESULTS

As an example of the principal-plane ($y = 0$) radial and vertical electric field components radiated by the 32-active element monopole array shown in Figure 10.1, consider first a $-30°$ scan angle from broadside and a Taylor $\bar{n} = 10$, 40 dB illumination function. The computed E_x and E_z components (from (10.10) and (10.12), respectively) on the observation line at one aperture diameter distance ($z_o = 3.4$m) are shown in Figure 10.3(a). The peak E_z (normal) component is down by -4.5 dB compared to the peak E_x (tangential) component. Using the conversion from rectangular to spherical coordinates ((10.13) and (10.14)), the radial and vertical field components are computed and are shown in Figure 10.3(b). Here, it is seen that the maximum radial component is down by -19.1 dB. Note that the near-field data in Figure 10.3(b) have been presented as a function of angle with respect to the phase center ($x = 0, y = 0, z = 0$) of the linear array. The corresponding data for near-field focusing and near-field observation at 1.5 aperture diameters ($z_o = 5.1$m distance) are shown in Figure 10.4.

In Figure 10.4(a), the peak E_z component is down by -4.6 dB compared to the peak E_x component. In Figure 10.4(b), the maximum radial component is -22.2 dB relative to the peak E_θ. When the array is focused at two array diameters ($z_o = 6.8$m), the rectangular and spherical components are shown in Figure 10.5.

In Figure 10.5(a), the peak E_z component is down by -4.7 dB compared to E_x, and in Figure 10.5(b), the radial component has dropped to -24.5 dB below the maximum vertical component. These ratios are conveniently summarized in Table 10.1. Thus, as would be expected for a given scan angle, the relative radial component decreases as the focusing range increases. In Figure 10.6(a) (one diameter distance) and Figure 10.6(b) (two diameters distance) a comparison is made between the exact near-field calculation of E_θ by (10.14) and the approximate near-zone E_θ obtained by (10.16).

Here it is demonstrated that there is little difference between the two methods of computation and so the radial component is demonstrated to be

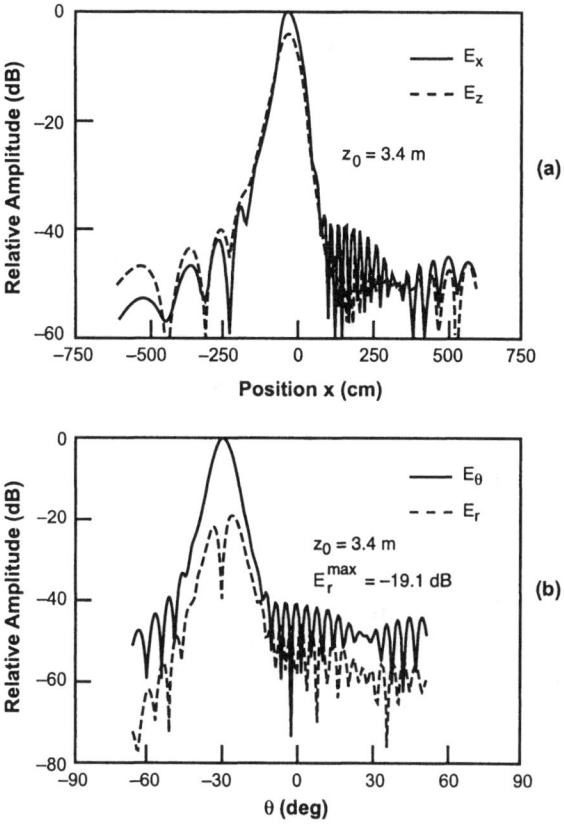

Figure 10.3 Simulated near-field amplitude for monopole array focused at one aperture diameter. The scan angle is $\theta_s = -30°$. (a) Tangential E_x and normal E_z electric field components and (b) radial E_r and vertical E_θ electric field components. © 1992 IEEE [8].

negligible. Next, in Figure 10.7 at one aperture diameter, scan angles of -40, -45, and $-50°$ are considered, for which the radial component is down by -20.7 dB, -21.8 dB, and -22.9 dB, respectively. Thus, in Figure 10.7, as the scan angle increases away from broadside, the near-field focal range increases, and the computed maximum radial component decreases in amplitude as expected. It is observed that the relative peak E_z (normal) component at the scan angles -40, -45, and $-50°$ has values -1.8 dB, -0.4 dB, and 1.0 dB, relative to the tangential component, as shown in Figure 10.8. These ratios are summarized in Table 10.2.

Expanded scale plots of the rectangular components in amplitude and phase across the main beam region for the $-50°$ scan angle at one diameter

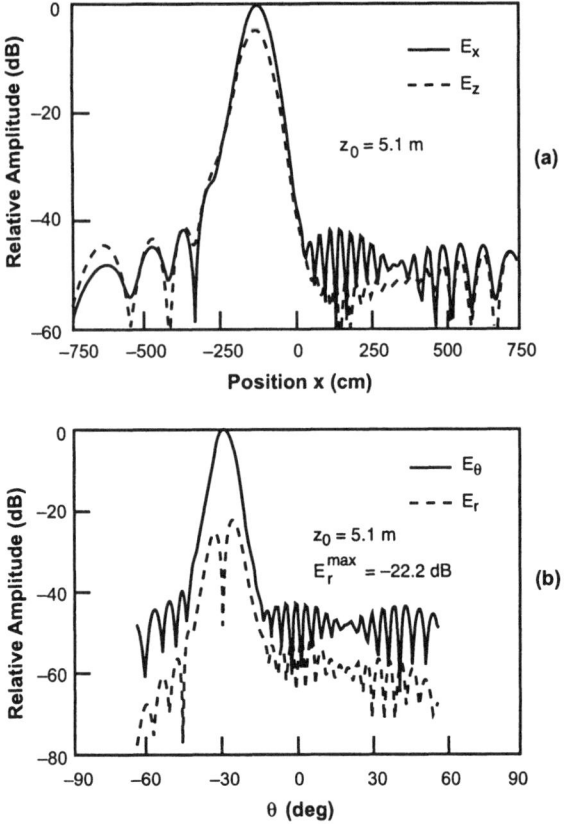

Figure 10.4 Simulated near-field amplitude for monopole array focused at 1.5 aperture diameters. The scan angle is $\theta_s = -30°$. (a) Tangential E_x and normal E_z electric field components and (b) radial E_r and vertical E_θ electric field components.

Table 10.2
Relative Radial and Normal Components for 32-Element Monopole Array with Scan Angle θ_s; Near-Field Focusing Distance and Observation Distance is $z_o = 3.4$m.

θ_s	E_r^{max}/E_θ^{max}	E_z^{max}/E_x^{max}
$-30°$	-19.1 dB	-4.5 dB
$-40°$	-20.7 dB	-1.8 dB
$-45°$	-21.8 dB	-0.4 dB
$-50°$	-22.9 dB	1.0 dB

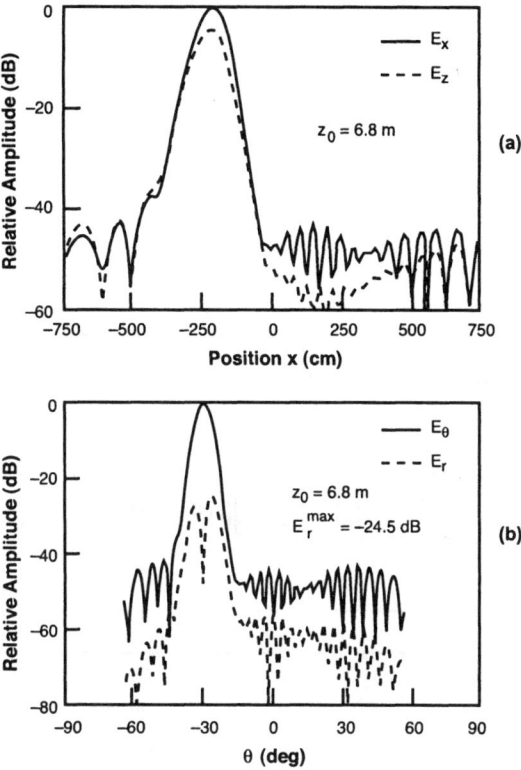

Figure 10.5 Simulated near-field amplitude for monopole array focused at two aperture diameters. The scan angle is $\theta_s = -30°$. (a) Tangential E_x and normal E_z electric field components and (b) radial E_r and vertical E_θ electric field components.

distance are shown in Figure 10.9.

As predicted earlier, the E_x and E_z components are nearly in phase (Figure 10.9(b)) for this negative scan angle. For positive scan angles E_x and E_z components would be nearly 180° out of phase in the main beam region.

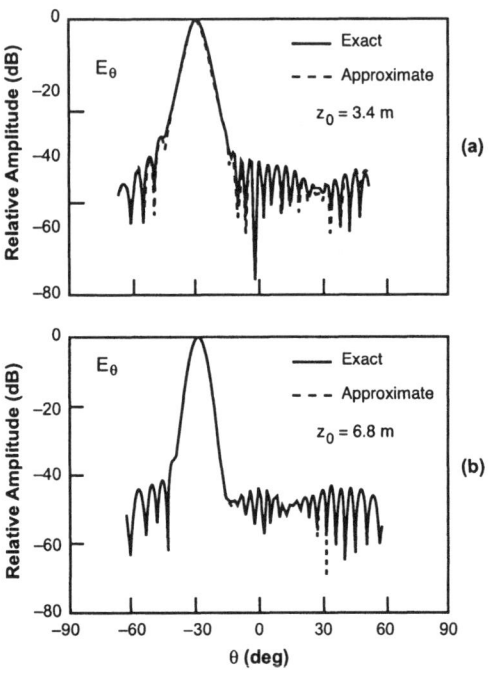

Figure 10.6 Comparison of exact and approximate near-field E_θ component at (a) one aperture diameter distance, and (b) two aperture diameters distance. © 1992 IEEE [8].

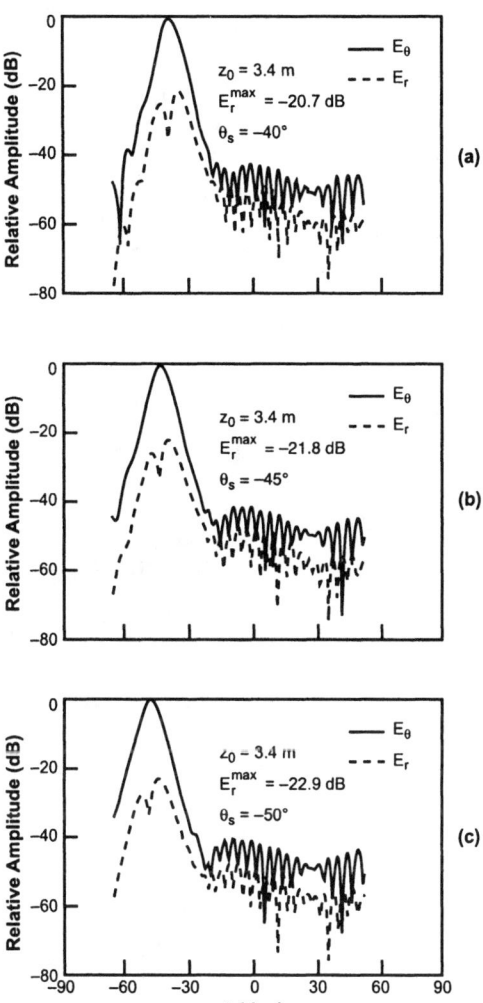

Figure 10.7 Simulated near-zone radial and vertical electric field components. The monopole array is focused at one aperture diameter and the scan angle is variable. (a) $\theta_s = -40°$ (b) $\theta_s = -45°$, and (c) $\theta_s = -50°$. © 1992 IEEE [8].

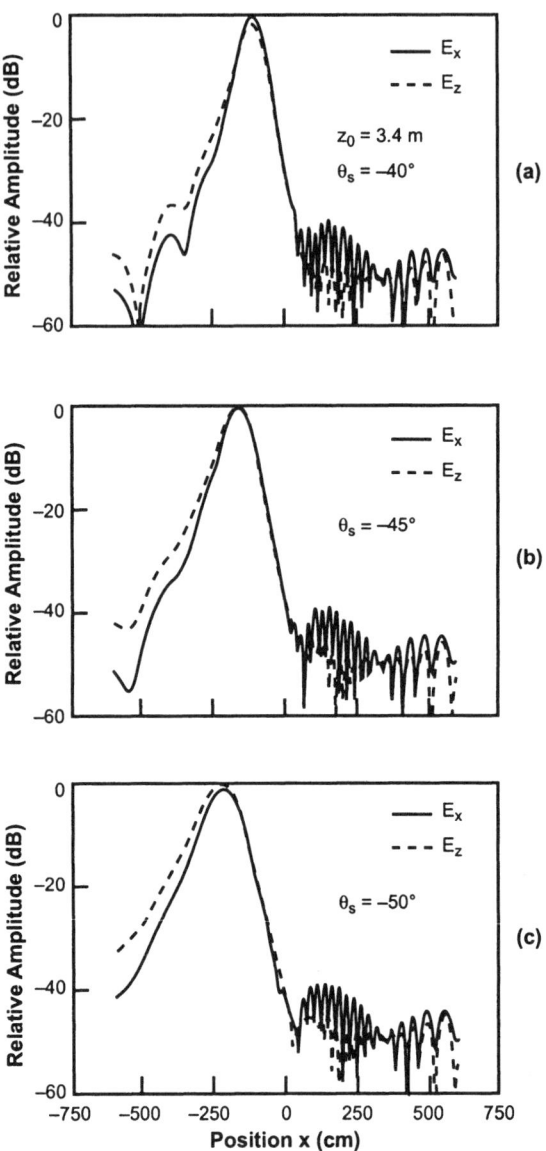

Figure 10.8 Simulated near-field amplitude for tangential and normal electric field components E_x and E_z. The monopole array is focused at one aperture diameter and the scan angle is (a) $\theta_s = -40°$, (b) $\theta_s = -45°$, and (c) $\theta_s = -50°$. © 1992 IEEE [8].

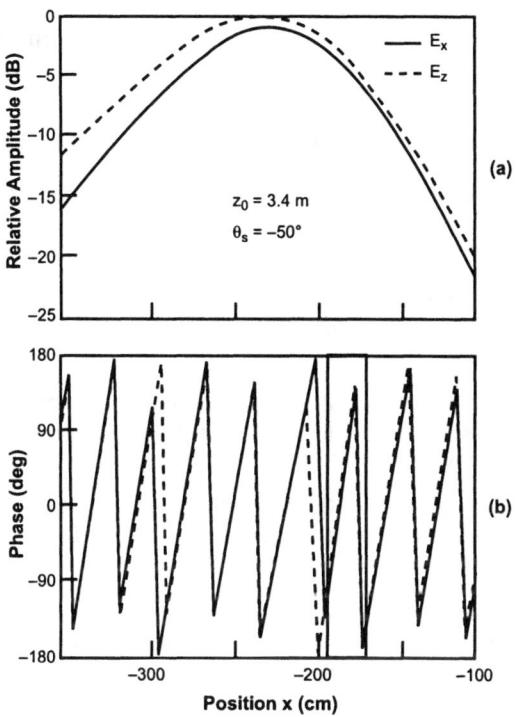

Figure 10.9 Expanded scale display of the rectangular components E_x and E_z in amplitude and phase across the main beam region for $-50°$ scan angle at one aperture diameter distance. (a) Amplitude and (b) phase. © 1992 IEEE [8].

10.5 SUMMARY

This chapter has shown moment method simulations of the principal and radial spherical components of the electric field for a monopole phased array antenna focused at a near-field range of one to two aperture diameters. For this low-sidelobe linear array, a negligible radial component exists in the focused near-field region. This fact means that only the computed or measured tangential electric field component is necessary in computing the focused near-field principal spherical component. A similar conclusion needs to be validated for other types of phased array radiating elements.

References

[1] Fenn, A.J., "Moment Method Analysis of Near Field Adaptive Nulling," *IEE Sixth International Conference on Antennas and Propagat., ICAP 89,* April 4-7, 1989, pp. 295-301.

[2] Fenn, A.J., "Analysis of Phase-Focused Near-Field Testing for Multiphase-Center Adaptive Radar Systems," *IEEE Trans. Antennas and Propagat.*, Vol. 40, No. 8, 1992, pp. 878-887.

[3] Fenn, A.J., "Evaluation of Adaptive Phased Array Antenna Far-Field Nulling Performance in the Near-Field Region," *IEEE Trans. Antennas and Propagat.*, Vol. 38, No. 2, February 1990, pp. 173-185.

[4] Fenn, A.J., H.M. Aumann, F.G. Willwerth, and J.R. Johnson, "Focused Near-Field Adaptive Nulling: Experimental Investigation," *1990 IEEE Antennas and Propagat. Society Int. Symposium Digest,* Vol. 1, May 7-11, 1990, pp. 186-189.

[5] Scharfman, W.E., and G. August, "Pattern Measurements of Phased-Arrayed Antennas by Focusing into the Near Zone," in *Phased Array Antennas (Proc. of the 1970 Phased Array Antenna Symposium)*, A.A. Oliner and G.H. Knittel, (eds.), Dedham, MA: Artech House, 1972, pp. 344-350.

[6] Henderson, R.I., "Crosspolarisation Due to Radial Field Components in Reflector Antennas," *Electronics Letters*, Vol. 21, No. 14, 1985, pp. 617-618.

[7] Eid, D.A.M, and L. Shafai, "Focal Region Field of Conical Reflectors," *1986 IEEE Antennas and Propagat. Society Int. Symposium Digest,* June 8-13, 1986, pp. 511-514.

[8] Fenn, A.J., "On the Radial Component of the Electric Field for a Monopole Phased Array Antenna Focused in the Near Zone," *IEEE Trans. Antennas and Propagat.*, Vol. 40, No. 6, 1992, pp. 723-727.

[9] Fenn, A.J., H.M. Aumann, and F.G. Willwerth, "Linear Array Characteristics with One-Dimensional Reactive-Region Near-Field Scanning: Simulations and Measurements," *IEEE Trans. Antennas and Propagation*, Vol. 39, No. 9, 1991, pp. 1305-1311.

[10] Herper, J.C., and A. Hessel, "Performance of $\lambda/4$ Monopole in a Phased Array," *1975 IEEE Antennas and Propagation Society International Symposium Digest*, pp. 301-304.

[11] Fenn, A.J., "Theoretical and Experimental Study of Monopole Phased Array Antennas," *IEEE Trans. Antennas Propagat.*, Vol. 33, No. 10, 1985, pp. 1118-1126.

[12] Schelkunoff, S.A., *Electromagnetic Waves*, New York: D. Van Nostrand Co., 1943, pp. 370-371.

[13] Stutzman, W.L., and G. A. Thiele, *Antenna Theory and Design*, New York: Wiley, 1981.

[14] Richmond, J.H., "Radiation and Scattering by Thin-Wire Structures in a Homogeneous Conducting Medium (Computer Program Desc.)," *IEEE Trans. Antennas Propagat.*, Vol. 22, No. 2, 1974, p. 365.

[15] Sinnott, D.H., "An Improved Algorithm for Matrix Analysis of Linear Antenna Arrays," Australian Defense Scientific Service, Weapons Research Establishment, Adelaide, South Australia, WRE-TECH. NOTE-1066(AP), 1974.

11

Displaced Phase Center Antenna Measurements Using Near-Field Scanning

11.1 INTRODUCTION

As discussed in Chapter 2, the displaced phase center antenna (DPCA) concept can be considered for use in airborne or space deployable phased array radar systems, for purposes of cancelling ground clutter (refer to Figure 2.1). Figure 2.2 showed examples of overlapped and split aperture distributions for two-phase center DPCA. Chapter 2 investigated, by simulations, the size of DPCA arrays and showed that array mutual coupling has a significant impact on the clutter-cancellation performance. In the case of a planar array, the ability to perform DPCA is limited, additionally, by the amplitude and phase errors produced in the transmit/receive array modules as well as in the array beamformer (refer to Figure 2.3). The amount of clutter cancellation that can be achieved by a DPCA array depends strongly on the displaced phase center radiation pattern similarity or match between phase centers. An example of the radiation pattern similarity (amplitude and phase) between two displaced phase centers was shown in Figure 2.13. To characterize such antennas requires accurate measurements of the far-field radiation patterns generated by two or more independent aperture illuminations having physically separated phase centers. Due to long-range requirements and mechanical considerations, direct far-field measurements may not be practical for large airborne systems or fragile space-deployable DPCA systems, and so an alternative measurement approach is addressed in this chapter. In general, near-field measurement techniques (planar, planar-polar, cylindrical, and spherical) as depicted in Figure 11.1 can be used to obtain far-field performance of antennas. Of the techniques shown in Figure 11.1, planar near-field measurements are the most appropriate for measuring the radiation patterns for displaced phase centers of a planar phased array. The mathematical formulation to obtain far-field radiation patterns from planar near-field scanning data is discussed in detail in Chapter 12.

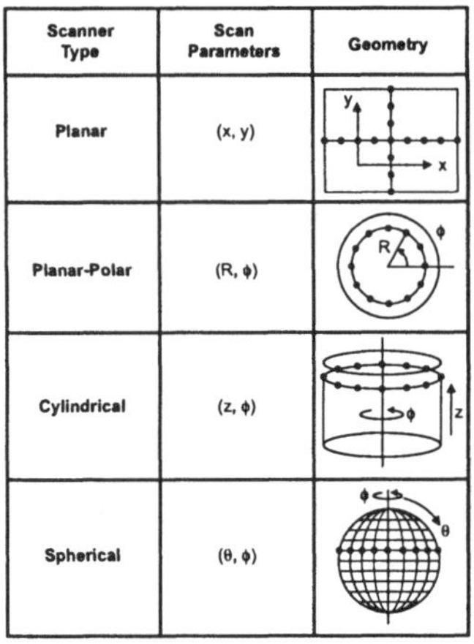

Scanner Type	Scan Parameters	Geometry
Planar	(x, y)	
Planar-Polar	(R, φ)	
Cylindrical	(z, φ)	
Spherical	(θ, φ)	

Figure 11.1 Depiction of different types of near-field scanning systems.

An investigation of the use of planar near-field measurements to characterize the performance of displaced phase center antennas is described in this chapter. The design and implementation of a 96-element subscale DPCA corporate-fed phased array is discussed. DPCA results are quantified experimentally, using near-field measurements, under a number of test conditions; scan angle, frequency, phase center displacement, and simulated module outages.

11.2 DISPLACED PHASE CENTER ANTENNA CLUTTER CANCELLATION

Both airborne and space-based radar (SBR) systems must cope with detecting and tracking targets against the strong background clutter of the Earth. With a conventional pulse doppler radar system, the doppler spectrum of the clutter is controlled by utilizing apertures with narrow-beam radiation patterns. In contrast, smaller phased array apertures can be realized by considering the DPCA concept. With "simultaneous beam" DPCA, clutter is cancelled rather than avoided by employing two independent receive phase centers to effectively form two colocated monostatic radars [1]. This concept was depicted in Chapter 2 in Figure 2.1, where a moving target and a moving SBR DPCA platform were shown. In a DPCA array, the full aperture is used for two successive pulse transmissions and on receive two overlapping portions of the aperture are used. The phase-center displacement between the receive apertures (denoted A and B) is adjusted to compensate for the platform velocity. Thus, for two pulses separated in time by one pulse repetition interval (PRI), the first reception occurs at the forward phase center and the second reception is made at the trailing phase center. During a PRI the clutter is assumed to be stationary; however, during this interval the target moves. Due to this motion, the target has a relative phase shift. There is no such phase shift from the clutter during this time. The result is that when the signals received by the two phase centers are subtracted, the clutter is significantly cancelled, leaving a signal return that depends on the amount of target phase shift in the PRI. Optimizing the target return is accomplished by varying the PRI, which requires variable phase center separations. The amount of clutter cancellation achieved is limited by how well the two phase center radiation patterns are matched in amplitude and phase, primarily over the main beam. To compute the upper-bound clutter cancellation capability of two DPCA radiation patterns, as discussed in Chapter 2, it is necessary to form the pattern correlation matrix

$$M = \begin{bmatrix} M_{11} & M_{12} \\ M_{21} & M_{22} \end{bmatrix} \qquad (11.1)$$

where the correlation between two channels is expressed as

$$M_{11} = \int\int |E_o(\theta,\phi)|^2 |E_1(\theta,\phi)|^2 A(\theta,\phi) d\theta d\phi \qquad (11.2)$$

$$M_{22} = \int\int |E_o(\theta,\phi)|^2 |E_2(\theta,\phi)|^2 A(\theta,\phi) d\theta d\phi \qquad (11.3)$$

$$M_{12} = \int\int |E_o(\theta,\phi)|^2 E_1(\theta,\phi) E_2^*(\theta,\phi) A(\theta,\phi) d\theta d\phi \qquad (11.4)$$

$$M_{21} = \int\int |E_o(\theta,\phi)|^2 E_2(\theta,\phi) E_1^*(\theta,\phi) A(\theta,\phi) d\theta d\phi \qquad (11.5)$$

where (θ,ϕ) are standard spherical coordinates, $E_o(\theta,\phi)$ is the electric field pattern of the transmitting antenna, $E_1(\theta,\phi)$ and $E_2(\theta,\phi)$ are the electric field patterns of the two receiving antennas (* denotes conjugate), and $A(\theta,\phi)$ is a weighting function that depends on the radar waveform, the clutter model, and the geometry of the problem. These integrals can be evaluated by numerical integration. In this chapter, it is assumed that $A(\theta,\phi) = 1$, so the clutter cancellation is dependent on the antenna pattern match only. The DPCA array electric field patterns $E_1(\theta, \phi)$ and $E_2(\theta, \phi)$ are measured with respect to their assumed phase center positions, which are taken to be the geometric centers of the excited portions of the respective apertures. The range of integration will be $0 < \phi < 2\pi$ and $0 < \theta < \theta_{max}$ where θ_{max} is the maximum angle for which background clutter is taken into account. The clutter-cancellation measure was derived in Chapter 2 and is given in terms of the correlation matrix elements by

$$C = 1 - \frac{|M_{12}|^2}{M_{11} M_{22}} \qquad (11.6)$$

In the measured results that follow, clutter cancellation is expressed as a positive decibel value by taking $10 \log_{10}(1/C)$.

11.3 DISPLACED PHASE CENTER ANTENNA NEAR-FIELD MEASUREMENT TECHNIQUE

The radiation patterns of a DPCA array can be measured directly on a conventional far-field antenna range. Such measurements were performed on the Multiple Antenna Surveillance Radar (MASR) [2, 3]. These measurements required a linear positioner for establishing a common center of rotation and identical multipath for the different phase centers. Consider Figure 11.2 which shows a DPCA phased array antenna having displaced phase centers A and B. The small circle superimposed at the center of the array represents a common, desired center of rotation for each phase center. For far-field measurements, a horn antenna can be located at a distance $2D^2/\lambda$ or greater, each phase center would be linearly translated to the circle position, and then the array would be rotated in angle to record the far-field radiation pattern. For small durable arrays it is straightforward to provide one-dimensional movement (translation of the array), but for relatively large and delicate space deployable antennas or airborne antennas this may be difficult and impractical due to both mechanical and range length considerations. As an example of the large far-field test distance that would be needed, for a large space-deployable aperture of, say, 30m at 1.3 GHz, the far-field criterion $2D^2/\lambda = 7.5$ km. An alternate approach is to use planar near-field antenna measurements to compute the far-field radiation pattern, as discussed in Chapter 12. This concept is illustrated in Figure 11.3 for two phase centers displaced by the distance p. Array physical translation can be avoided by shifting two independent near-field probe scan planes by the same distance p. (Alternately, the near-field scan plane can be fixed and the array can be physically translated, if desired.) The near-field test distance is typically only a few wavelengths from the antenna under test. Near-field measurements have become a conventional means for evaluating far-field antenna performance. The technique has been applied to multiple phase center arrays. This chapter shows experimentally that near-field DPCA measurements are practical. The chapter is organized as follows: The details of a subscale SBR phased array are described in Section 11.3.1. A description of the near-field scanning system is given in Section 11.3.2. In Section 11.4, measured DPCA results are shown.

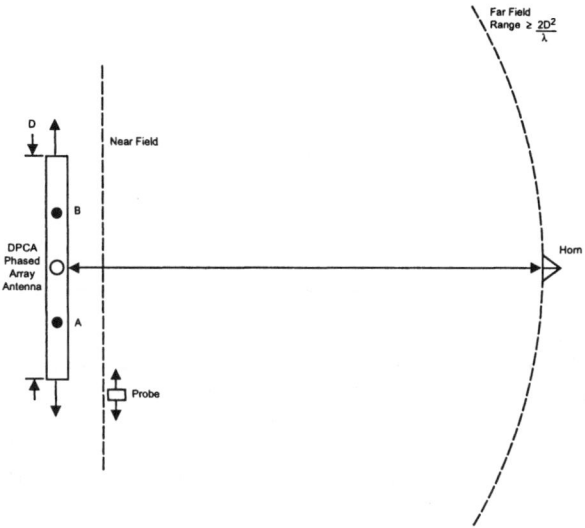

Figure 11.2 DPCA array testing in the near field and in the far field.

11.3.1 DPCA TEST ARRAY DESCRIPTION

The pertinent geometry for an SBR in low altitude orbit was shown in Figure 9.1. For a downward pointed SBR platform, the maximum and minimum scan angles are approximately 60° and 30°, respectively. A 30° cone has been excluded from the scan sector, primarily due to clutter considerations at high grazing angles. In this case the phased array radiating elements are not required to have maximum gain at broadside, but rather a pattern minimum or null at broadside is desirable. For uniform coverage between the 30° and 60° cones, the choice of an omnidirectional array radiating element becomes apparent. Monopole [4, 5] and loop-fed slotted cylinder [6] array elements have been designed and tested, the simplest being a vertically polarized monopole as shown previously in Figure 9.2. The subscale SBR test array, as shown in Figure 11.4, was chosen to have 96 radiating elements arranged in 8 rows and 12 columns with a hexagonal lattice using 12.7 cm element spacing.

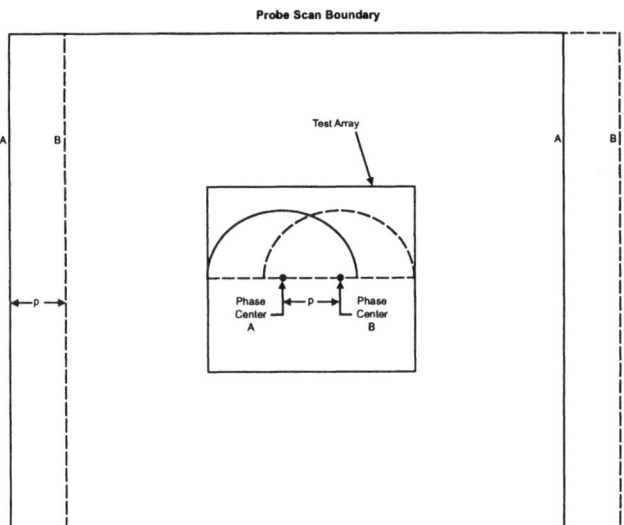

Figure 11.3 DPCA near-field measurement concept. The probe scan boundary is shifted by the amount of phase center displacement.

Two rows of passively terminated (50-ohm resistively loaded) elements are used to reduce array and ground plane edge effects. A photograph of the prototype SBR test array is shown in Figure 11.5. The array is shown mounted on a linear slide positioner that was used to align the displaced phase centers over the center of rotation for far-field pattern measurements.

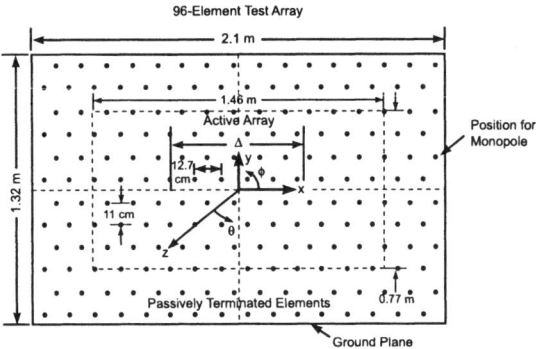

Figure 11.4 DPCA test array layout.

A representative DPCA beamforming architecture for airborne or spaceborne radar is shown in Figure 11.6. An actual DPCA radar transmit/receive

Figure 11.5 Photograph of DPCA test array.

module would contain an RF power amplifier and a low noise amplifier. However, as indicated in Figure 11.6, in the subscale array described in this chapter, only the necessary phase shifters and variable attenuators to control the beam steering and phase center displacement were included. The DPCA architecture described here is implemented by two independent beam forming networks. The DPCA test array modules consist of two channels, each containing 6-bit digital phase shifters (64 phase states with 5.625° phase steps) and 7-step (coarse) attenuators (0 dB, 1.3 dB, 2.8 dB, 4.6 dB, 6.8 dB, 10 dB, and >55 dB (*off*) attenuation states). The phasers provide beam agility, while the attenuators provide the function of phase center movement. To achieve, say, 40 dB cancellation, maximum rms errors of 3.0° and 0.5 dB at the element level were required.

11.3.2 NEAR-FIELD SCANNER

A vertically oriented 1.5m by 3.0m planar scanner was constructed for measuring the near field of the previous test array. The primary purpose of the experiment was a proof of concept for DPCA near-field measurements. Near-field measurements sometimes truncate the near-field scan at a specified

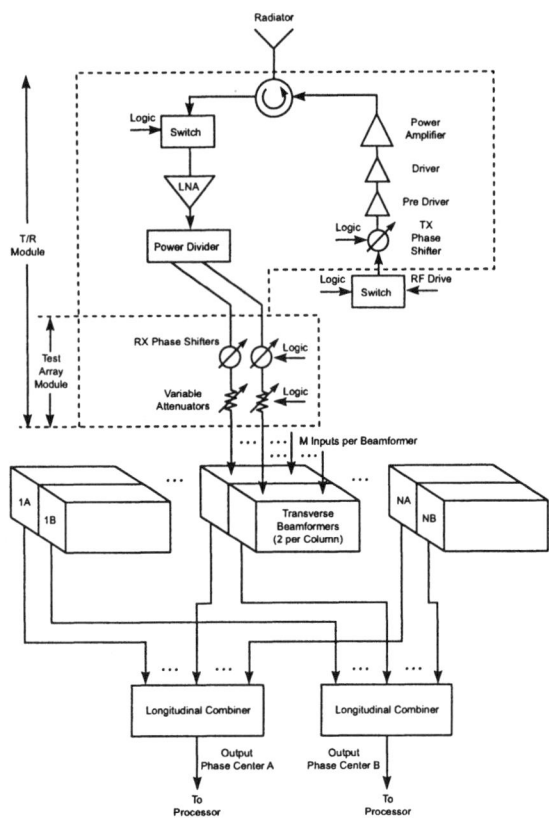

Figure 11.6 Block diagram showing a representative displaced phase center antenna beamforming architecture.

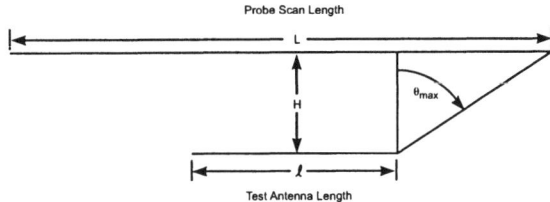

Figure 11.7 Diagram depicting the near-field scanning geometry for a test antenna.

angle beyond the edge of the test antenna to accurately quantify the radiation pattern sidelobe performance out to a specified maximum angle. This near-field truncation geometry is depicted in Figure 11.7. Since clutter cancellation depends primarily on the main beams being matched, no attempt was made to accurately characterize wide-angle sidelobes. Most of the near-field scans

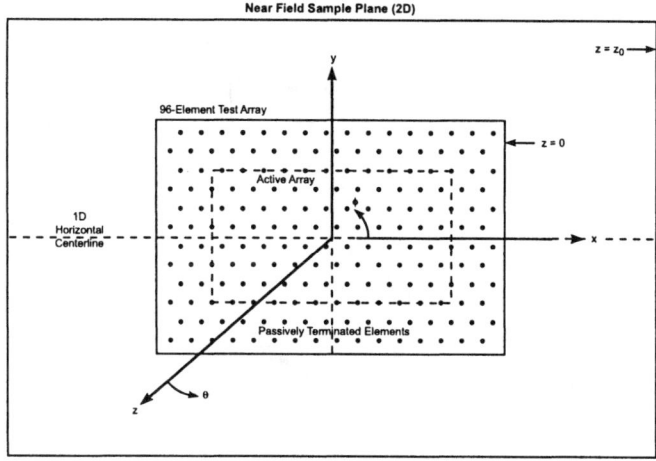

Figure 11.8 Geometry for general two-dimensional sample plane and one-dimensional horizontal centerline cut for near-field measurements of the 96-element DPCA test array.

were truncated at a level 15 to 20 dB down from the peak amplitude of the near field. Center line near-field cuts, shown in Figure 11.8, were used to measure the principal plane azimuth patterns in the direction of the phase center displacement of the array. A block diagram of the near-field scanning system is shown in Figure 11.9. The probe x, y position is measured accurately with a pair of linear interferometer scales. The system is controlled with a desktop computer. The near-field probe is a dual-polarized circular waveguide, which is surrounded with absorber. A photograph of the DPCA test array mounted in front of the near-field scanner is shown in Figure 11.10.

11.4 RESULTS

Prior to array pattern measurements, the test array was calibrated by positioning the near-field probe close to each monopole element (offset from the null) and stepping through all phase and amplitude states of the modules. The amplitude and phase states were calibrated by linear approximation and table lookup. After calibration, the overall module rms amplitude and phase errors were less than 0.5 dB and 2.7°, respectively. To compute DPCA performance, two receive patterns and one transmit pattern are required. For transmit and receive patterns, a two-dimensional cosine taper with −10 dB edge illumination was used. As a large number of array test conditions were of interest, it was desired to reduce the number of near-field data samples. Since the array illumination used here is separable, a single centerline

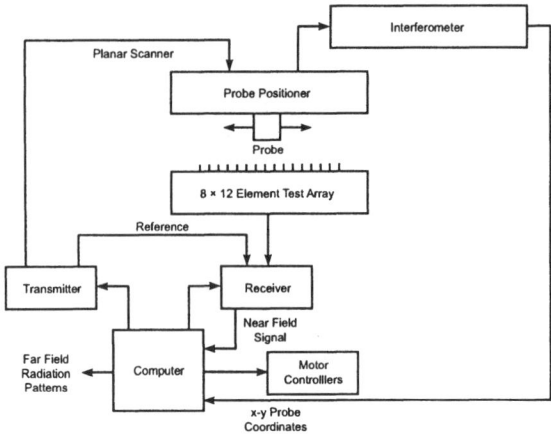

Figure 11.9 Block diagram for near-field scanning system.

Figure 11.10 Photograph of the displaced phase center antenna under test with the near-field scanning system.

cut produces nearly the same principal plane far-field pattern as would be computed from a two-dimensional set of near-field data. Centerline cuts have been demonstrated by Newell and Crawford as being useful in obtaining approximate far-field patterns [7]. All of the data which follow are derived from centerline near-field measurements at a distance approximately 1.5λ from the test array. The sample spacing used was 0.2λ, which was selected primarily to reduce errors that may occur due to multipath. The conventional near-field to far-field transformation with probe compensation discussed by Joy and Paris and described in Chapter 12 is used [8, 9].

Measured near-field amplitude and phase data for phase center A with the main beam steered to $-40°$ are shown in Figure 11.11. The corresponding near-field to far-field transformed radiation pattern (transmit or receive) is shown in Figure 11.12.

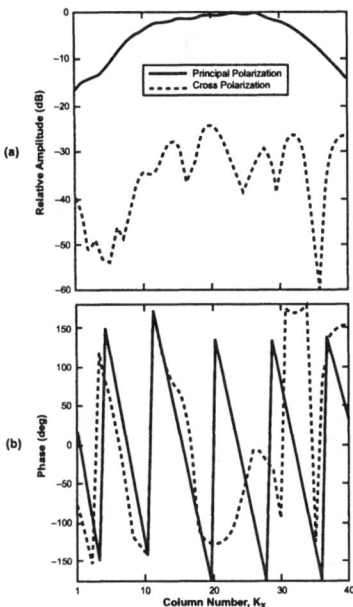

Figure 11.11 Measured one-dimensional near-field data for the DPCA test array steered to $\theta = -40°$. The data were measured in the horizontal centerline, as depicted in Figure 11.8.

According to (11.1), the DPCA clutter cancellation [10] depends on the product of the transmit pattern (for the full array) and receive patterns for the two receive phase centers. Hence, a two-way (transmit times receive) amplitude pattern cut for the two phase centers displaced by 12.7 cm (1 column of the array) is shown in Figure 11.13. The two main beams (denoted A and B) appear to be well matched, and this can be quantified by computing the beam correlation matrix from which the beam cancellation can be calculated by (11.6). The cancellation as a function of pattern threshold is shown in Figure 11.14. It is seen that the cancellation is relatively converged once a pattern threshold of about -20 dB is reached. In other words, main beam match is the primary factor in achieving the desired goal of pattern cancellation. The beam cancellation as a function of frequency is shown in

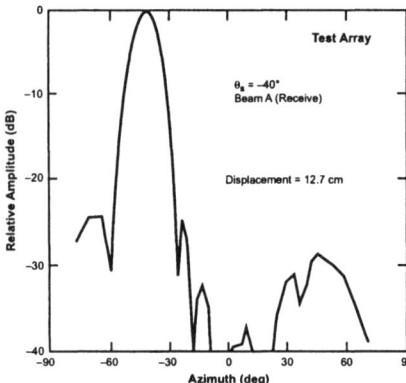

Figure 11.12 Far-field radiation pattern computed from measured one-dimensional near-field data (see Figure 11.11) for the DPCA test array steered to $\theta = -40°$.

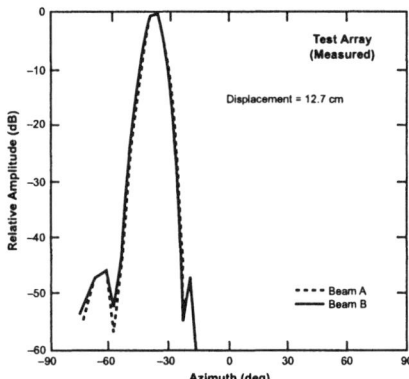

Figure 11.13 Comparison of the radiation patterns of phase centers A and B based on near-field measured data for the DPCA test array steered to $\theta = -40°$. The phase center displacement for this example is 12.7 cm.

Figure 11.15. The cancellation is greater than 40 dB over nearly a 15 percent bandwidth. The degradation at 1.2 GHz is attributed to excessive phase errors in the module. Next, the DPCA cancellation as a function of scan angle is shown in Figure 11.16. A cancellation of greater than 40 dB is achieved over much of the 30–60° scan sector. The cancellation degrades rapidly as the beam is steered toward the null of the monopole element pattern (refer to Figure 9.16). Finally, the DPCA cancellation as a function of module failures is shown in Figure 11.17 for uncompensated and compensated conditions. Here, uncompensated means that a failure in one phase center does not affect

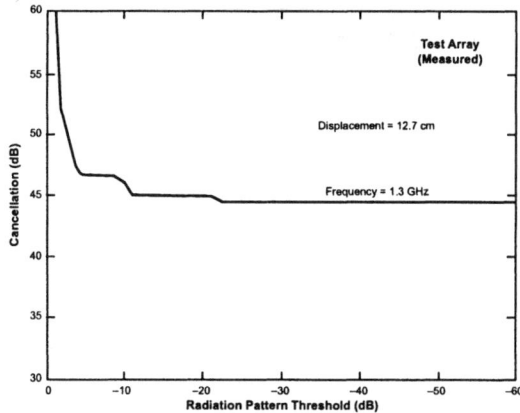

Figure 11.14 DPCA cancellation as a function of pattern threshold for phase centers A and B based on near-field measured data for the DPCA test array steered to $\theta = -40°$. The phase center displacement for this example is 12.7 cm.

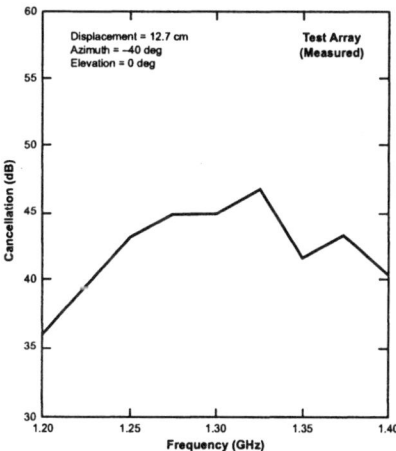

Figure 11.15 DPCA cancellation as a function of frequency for phase centers A and B based on near-field measured data for the DPCA test array steered to $\theta = -40°$. The phase center displacement for this example is 12.7 cm.

the other phase center. Compensated refers to a situation where a module fails in one phase center and, to maintain good match between the phase centers, the corresponding element in the second phase center is purposely turned off. The result is that the compensated array has a slow degradation, as opposed to the uncompensated array that degrades rapidly with increase in module failures. Compensation is clearly an effective means for achieving

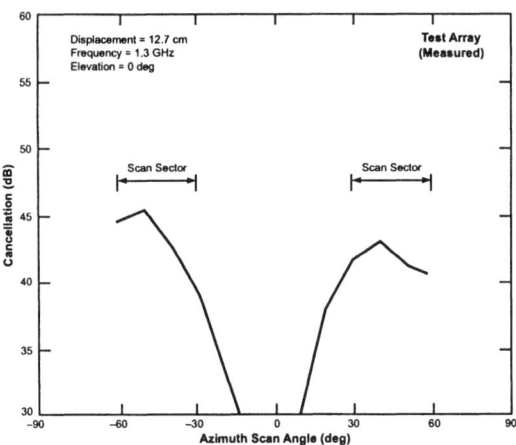

Figure 11.16 DPCA cancellation as a function of scan angle for phase centers A and B based on near-field measured data for the DPCA test array steered to $\theta = -40°$. The phase center displacement for this example is 12.7 cm.

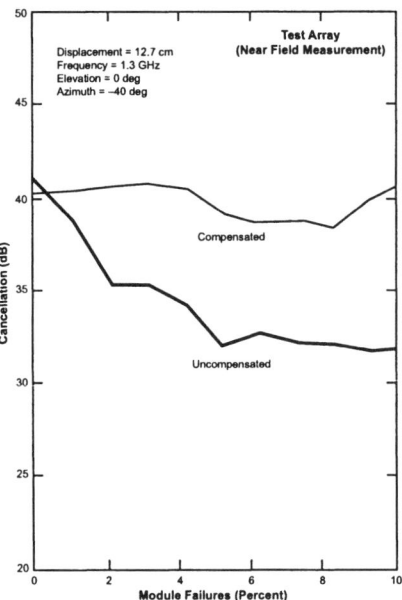

Figure 11.17 DPCA cancellation as a function of simulated T/R module failures for phase centers A and B based on near-field measured data for the DPCA test array steered to $\theta = -40°$. The phase center displacement for this example is 12.7 cm.

good pattern match.

11.5 SUMMARY

This chapter has described a method for evaluating the cancellation performance of large-aperture displaced phase center antennas with the planar near-field scanning method. A subscale 96-element phased array designed for displaced phase center antenna operation has been tested using conventional planar near-field measurements. The 96-element corporate-fed phased array used monopole radiating elements to achieve wide-angle scan coverage. Displaced phase center antenna beam cancellation has been measured using the planar near-field scanning technique under a number of test conditions: scan angle, frequency, phase center displacement, and simulated module failures. The measurements indicate that 40 dB cancellation is achieved for this small test array for scan angles 30° to 60° from broadside, 15 percent bandwidth, and up to 10 percent module failures with compensation. Larger arrays would be capable of greater cancellation based on the analysis presented in Chapter 2. The next chapter describes the planar near-field scanning technique for evaluating the performance of low sidelobe antennas.

References

[1] Kelly, E.J., and G.N. Tsandoulas, "A Displaced Phase Center Antenna Concept for Space Based Radar Applications," *IEEE Eascon*, September 1983, pp. 141-148.

[2] Stone, M.L., and W.J. Ince, "Air-to-Ground MTI Radar Using a Displaced Phase Center Phased Array," *IEEE International Radar Conference*, April 1980, pp. 225-230.

[3] Tsandoulas, G.N., "Unidimensionally Scanned Phased Arrays," *IEEE Trans. Antennas Propagat.*, Vol. 28, No. 1, 1980, pp. 86-99.

[4] Herper, J.C., and A. Hessel, "Performance of λ/4 Monopole in a Phased Array," *IEEE Antennas and Propagat. Society, 1975 Symposium Digest*, pp. 301-304.

[5] Fenn, A.J., "Theoretical and Experimental Study of Monopole Phased Array Antennas," *IEEE Trans. Antennas Propagat.*, Vol. 33, No. 10, 1985, pp. 1118-1126.

[6] Fenn, A.J., "Arrays of Horizontally Polarized Loop-Fed Slotted Cylinder Antennas," *IEEE Trans. Antennas Propagat.*, Vol. 33, No. 4, 1985, pp. 375-382.

[7] Newell, A.C., and M.L. Crawford, "Planar Near-Field Measurements on High Performance Array Antennas," Electromagnetics Division, Institute for Basic Standards, National Bureau of Standards, NBSIR 74-380, July 1974.

[8] Paris, D.T., W.M. Leach, and E.B. Joy, "Basic Theory of Probe-Compensated Near-Field Measurements," *IEEE Trans. Antennas Propagat.*, Vol. 26, No. 3, 1978, pp. 373-379.

[9] Joy, E.B, et al. "Applications of Probe-Compensated Near-Field Measurements," *IEEE Trans. Antennas Propagat.*, Vol. 26, No. 3, 1978, pp. 379-389.

[10] Fenn, A.J., F.G. Willwerth, and H.M. Aumann, "Displaced Phase Center Antenna Near Field Measurements for Space Based Radar Applications," *Phased Arrays 1985 Symposium Proceedings,* October 15-18, 1985, RADC-TR-85-171, pp. 303-318.

12

Low-Sidelobe Phased Array Antenna Measurements Using Near-Field Scanning

12.1 INTRODUCTION

The planar near-field scanning technique is commonly used in determining the far-field radiation patterns of antennas [1-4]. The technique is particularly well suited for evaluating the performance of low-sidelobe planar phased array antennas [5]. In a phased array antenna, low sidelobes are achieved using appropriate aperture amplitude taper together with precision element weighting and array calibration to achieve the desired accuracy in the array illumination. To evaluate the performance of these low-sidelobe array antennas, high-quality near-field measurements are required. The near field of the antenna is measured typically using a low-gain antenna probe that is moved to a regular grid of points on a planar surface as depicted in Figure 12.1.

The probe can be used either to transmit or receive the desired RF signal (usually a CW tone). After collecting the near-field data, a general purpose computer is used to mathematically transform the data, by means of the antenna plane wave spectrum, to a far-field pattern. The plane wave spectrum is obtained by a mathematical transformation of the antenna aperture illumination and consists of radiating and evanescent plane waves. The radiating plane waves reach the antenna far-field region and, thus, contribute to the far-field pattern. For a phased array antenna, an example of an evanescent plane wave is a non-propagating grating lobe.

Planar near-field measurements are commonly performed in the radiating near-field region [3]. In this region (typically 2 to 10 wavelengths from

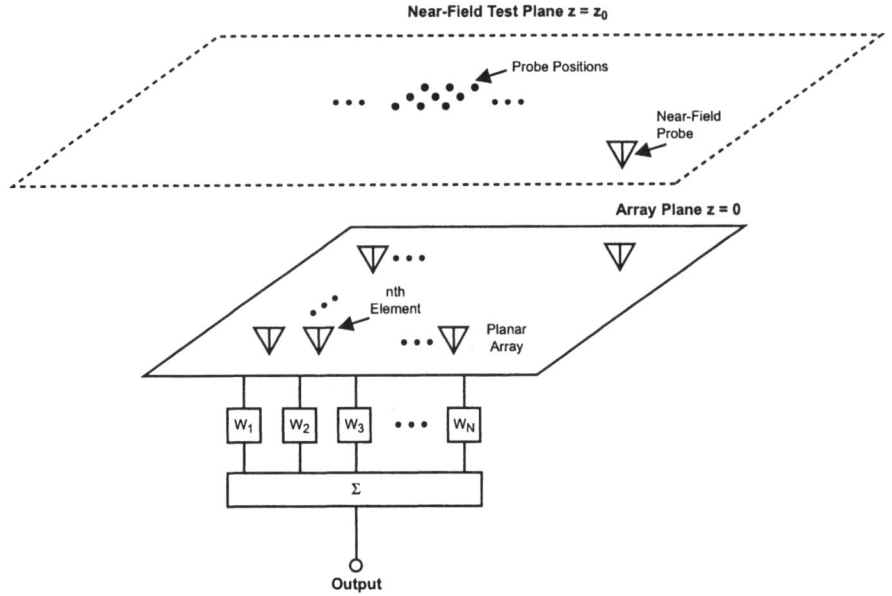

Figure 12.1 Phased array antenna and near-field test (scan) plane.

the test antenna aperture), the evanescent contributions of the near field are sufficiently attenuated so that conventional one-half wavelength data sample spacing can be used. It is well known that, as the near-field test distance decreases, the near-field data sample spacing needs to decrease in order to take into account the increase that occurs in the evanescent component [6]. Data collection at a distance of approximately one wavelength or less is implemented in what has been referred to as the reactive or evanescent near-field region [3]. The effect of the evanescent near field on the required sample spacing is a function of the aperture illumination and aperture diameter, as has been addressed theoretically by Wang [7]. For a low-sidelobe phased array antenna, effects of the near-field probe on the near-field data, plane wave spectrum, and transformed far-field pattern have been theoretically quantified [8].

To better understand phased array characteristics as determined by the planar near-field scanning technique, two theoretical models are investigated. Both theories are intended to apply in the evanescent and radiating near-field region. The first theory is referred to in this chapter as the field point theory, which is the same as near-field scanning without probe compensation. Here, the tangential electric field component radiated by an antenna is computed at a series of near-field points. The second theory is referred to as the dipole probe theory, where the received voltage at a V-dipole near-field antenna probe is computed at a series of probe positions, and this is an application of the probe-compensated near-field scanning technique. Both theoretical models include the effects of array polarization and mutual coupling through the use of the method of moments, as described in Section 12.2. The theory is applied to the case of a monopole phased array antenna. The design of a low-sidelobe corporate-fed linear phased array of monopole antenna elements is discussed in Section 12.3. In Section 12.4, a brief discussion of the near-field measurements system is given. Theoretical and experimental results are presented in Section 12.5. Centerline scanning at less than one wavelength distance from the antenna under test is used to obtain the plane wave spectrum and far-field pattern. The effects of the probe on the near-field theoretical data, plane wave spectrum, and transformed far-field pattern are demonstrated. It is shown that a thin-wire V-dipole theoretical probe antenna can accurately model an experimental near-field measurement probe that consists of a rectangular waveguide surrounded with anechoic material.

12.2 THEORY

12.2.1 PLANAR NEAR-FIELD SCANNING FORMULATION

The purpose of this section is to briefly review the theory for determining far-field radiation patterns from measured or theoretical near-field data. The formulation differs somewhat from that presented in [1] with respect to the the notation and the test antenna orientation in the reference rectangular coordinate system. To begin, let the time-harmonic electric field radiated from an aperture be denoted by $E(x, y, z)$ and let the aperture be located on the xy plane as shown in Figure 12.2(a).

At any distance z in front of the aperture the electric field can be represented as a superposition of plane waves [1, 8-10]. In general, these plane waves can be evanescent or propagating depending on the distance from the aperture.

From Maxwell's equations in free space (described by permittivity ϵ_o and permeability μ_o) and a vector identity, the vector wave equation is expressed as

$$\nabla^2 E + k^2 E = 0 \qquad (12.1)$$

where $k = \omega\sqrt{\mu_o\epsilon_o} = 2\pi/\lambda$ is the wavenumber, with λ the wavelength, and ∇^2 is the Laplacian operator. A solution to the wave equation, for outgoing waves, with the $e^{j\omega t}$ time variation suppressed is of the form

$$E(x, y, z) = A(k)e^{-j\boldsymbol{k}\cdot\boldsymbol{r}} \qquad (12.2)$$

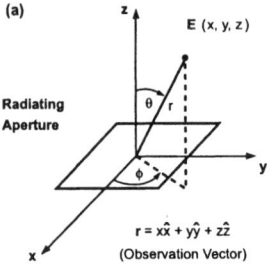

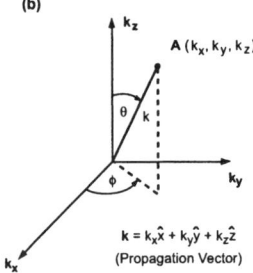

Figure 12.2 Coordinate systems for describing (a) electric field and (b) plane wave spectrum.

where
$$r = x\hat{x} + y\hat{y} + z\hat{z} \tag{12.3}$$
is the observation postion vector,
$$k = k_x\hat{x} + k_y\hat{y} + k_z\hat{z} \tag{12.4}$$
is the propagation vector, and as is depicted in Figure 12.2(b)
$$A = A_x\hat{x} + A_y\hat{y} + A_z\hat{z} \tag{12.5}$$
is the commonly referred to complex plane wave amplitude function or plane-wave spectrum. The wavenumber components in terms of spherical coordinates are
$$k_x = k\sin\theta\cos\phi \tag{12.6}$$
$$k_y = k\sin\theta\sin\phi \tag{12.7}$$
$$k_z = k\cos\theta \tag{12.8}$$

From (12.6)-(12.8) observe that

$$k^2 = k_x^2 + k_y^2 + k_z^2 \qquad (12.9)$$

In (12.9), k_x and k_y are chosen to be the independent variables with k_z being the dependent variable, such that

$$k_z = \begin{cases} \sqrt{k^2 - k_x^2 - k_y^2} & k_x^2 + k_y^2 \leq k^2 \\ -j\sqrt{k_x^2 + k_y^2 - k^2} & k_x^2 + k_y^2 > k^2 \end{cases} \qquad (12.10)$$

Only two components of the plane wave spectrum A are independent, A_x and A_y, with A_z chosen to be the dependent variable. Now since $\nabla \cdot E = 0$, from (12.2) and (12.5) it is readily shown that

$$A_z(k_x, k_y) = -\frac{1}{k_z}(k_x A_x + k_y A_y) \qquad (12.11)$$

The general solution for E can be expressed as a linear combination of A over all values of k_x and k_y as

$$E(x, y, z) = \int_{-\infty}^{\infty} \int_{-\infty}^{\infty} A(k_x, k_y) e^{-j\mathbf{k} \cdot \mathbf{r}} dk_x dk_y \qquad (12.12)$$

12.2.1.1 Test Antenna Has Known Electric Field

If the tangential electric field E_t is known over a planar surface at $z = z_o$, the plane wave spectrum can be determined from the inverse Fourier transform relationship

$$A(k_x, k_y) = \frac{1}{4\pi^2} \int_{-\infty}^{\infty} \int_{-\infty}^{\infty} E_t(x, y, z_o) e^{j\mathbf{k} \cdot \mathbf{r}} dx dy \qquad (12.13)$$

Specifically, for $E_t = E_x \hat{x} + E_y \hat{y}$ the independent components of A are

$$A_x(k_x, k_y) = e^{jk_z z_o} A_x(k_x, k_y, z_o) \qquad (12.14)$$

$$A_y(k_x, k_y) = e^{jk_z z_o} A_y(k_x, k_y, z_o) \qquad (12.15)$$

where

$$A_x(k_x, k_y, z_o) = \frac{1}{4\pi^2} \int_{-\infty}^{\infty} \int_{-\infty}^{\infty} E_x(x, y, z_o) e^{j(k_x x + k_y y)} dx dy \qquad (12.16)$$

$$A_y(k_x, k_y, z_o) = \frac{1}{4\pi^2} \int_{-\infty}^{\infty} \int_{-\infty}^{\infty} E_y(x, y, z_o) e^{j(k_x x + k_y y)} dx dy \qquad (12.17)$$

are the plane wave spectrum components computed from the electric field on the measurement plane. As suggested in Figure 12.1, the infinite integrals in (12.16) and (12.17) are truncated according to the near-field scan length L (refer to Figure 11.7). For uniformly spaced near-field data, the resulting finite integrals are evaluated efficiently using the fast Fourier transform (FFT) algorithm.

From Rhodes [11], in the far field of the antenna the electric field can be written in terms of the plane wave spectrum as

$$E(r, \theta, \phi) = j2\pi k \cos\theta \frac{e^{-jkr}}{r} A(k_x, k_y) \qquad (12.18)$$

To obtain E_θ and E_ϕ components from (12.18) a standard conversion from rectangular components to spherical components is made.

12.2.1.2 Test Antenna Has Known Received Voltage: Near-Field Probe Pattern Compensation

In practice the near-field measured data are collected by utilizing a moveable antenna probe. The probe is assumed here to be linearly polarized and is used as a transmitting source. In this case, the test antenna received voltage is known, rather than its electric field. At each near-field position, two probe orientations are in general required – one where the probe is $\hat{x}$ polarized and the other where the probe is $\hat{y}$ polarized. Let the far-zone electric field vector pattern of the near-field probe with x,y orientations in free space be expressed as

$$e^x(\theta, \phi) = \hat{\theta} e^x_\theta(\theta, \phi) + \hat{\phi} e^x_\phi(\theta, \phi) \qquad (12.19)$$

$$e^y(\theta, \phi) = \hat{\theta} e^y_\theta(\theta, \phi) + \hat{\phi} e^y_\phi(\theta, \phi) \qquad (12.20)$$

It is further assumed that the probe pattern is not affected by the presence of the test antenna and that the probe far-field pattern can be obtained by conventional far-field measurements.

Next, let the test antenna far-field pattern be given by

$$E(\theta, \phi) = \hat{\theta} E_\theta(\theta, \phi) + \hat{\phi} E_\phi(\theta, \phi) \qquad (12.21)$$

In (12.21), E_θ and E_ϕ represent the two desired components of the antenna under test. These components will be determined in what follows by solving two simultaneous equations.

Let a^x, a^y denote the probe plane wave spectrum for $\hat{x}$-, $\hat{y}$-polarized probe orientations, respectively. The received voltage at the array output, for a near-field transmitting probe at position (x, y, z_o) is given as [1]

$$v^x(x, y, z_o) = \frac{8\pi^2}{\omega\mu} \int_{-\infty}^{\infty} \int_{-\infty}^{\infty} k_z A(k_x, k_y) \cdot a^x(k_x, k_y) e^{-j\mathbf{k}\cdot\mathbf{r}_o} dk_x dk_y \tag{12.22}$$

$$v^y(x, y, z_o) = \frac{8\pi^2}{\omega\mu} \int_{-\infty}^{\infty} \int_{-\infty}^{\infty} k_z A(k_x, k_y) \cdot a^y(k_x, k_y) e^{-j\mathbf{k}\cdot\mathbf{r}_o} dk_x dk_y \tag{12.23}$$

In words, the received voltage $v(x, y, z_o)$ is equal to the superposition over all k_x, k_y of the dot product of the test antenna plane wave spectrum and probe plane wave spectrum. Taking the inverse Fourier transform of (12.22) and (12.23) and defining an apparent or probe-distorted plane wave spectrum as

$$A'^x(k_x, k_y, z_0) = \frac{1}{4\pi^2} \int_{-\infty}^{\infty} \int_{-\infty}^{\infty} v^x(x, y, z_o) e^{j(k_x x + k_y y)} dx dy \tag{12.24}$$

$$A'^y(k_x, k_y, z_0) = \frac{1}{4\pi^2} \int_{-\infty}^{\infty} \int_{-\infty}^{\infty} v^y(x, y, z_o) e^{j(k_x x + k_y y)} dx dy \tag{12.25}$$

yields the simplified expressions

$$k_z A(k_x, k_y) \cdot a^x = \frac{\omega\mu}{8\pi^2} e^{jk_z z_0} A'^x(k_x, k_y, z_0) \tag{12.26}$$

$$k_z A(k_x, k_y) \cdot a^y = \frac{\omega\mu}{8\pi^2} e^{jk_z z_0} A'^y(k_x, k_y, z_0) \tag{12.27}$$

As in (12.16) and (12.17), the infinite integrals in (12.24) and (12.25) are replaced by finite integrals according to the actual near-field scan lengths. Next, analogous to (12.18) the probe far-zone electric field is written in terms of the probe plane wave spectrum as

$$e^x(r, \theta, \phi) = j2\pi k \cos\theta \frac{e^{-jkr}}{r} a^x(k_x, k_y) \tag{12.28}$$

$$e^y(r, \theta, \phi) = j2\pi k \cos\theta \frac{e^{-jkr}}{r} a^y(k_x, k_y) \tag{12.29}$$

where e^x, e^y denote the far-field vector patterns for the x-, y-directed probe, respectively. Substituting (12.18), (12.28), and (12.29) into (12.26) and (12.27) and dropping the e^{-jkr}/r dependence as usual in the far field yields

$$\mathbf{E}(\theta, \phi) \cdot \mathbf{e}^x(\theta, \phi) = C \cos\theta e^{jk_z z_0} A'^x \tag{12.30}$$

$$\mathbf{E}(\theta, \phi) \cdot \mathbf{e}^y(\theta, \phi) = C \cos\theta e^{jk_z z_0} A'^y \tag{12.31}$$

where the constant C is given as

$$C = -\frac{k\omega\mu_0}{2} \tag{12.32}$$

Equations (12.30) and (12.31) are solved simultaneously for the components E_θ and E_ϕ by utilizing (12.19), (12.20), and (12.21) with the result

$$E_\theta = C\cos\theta e^{jk_z z_0} \frac{e_\phi^y A'^x - e_\phi^x A'^y}{\Delta} \tag{12.33}$$

$$E_\phi = C\cos\theta e^{jk_z z_0} \frac{-e_\theta^y A'^x + e_\theta^x A'^y}{\Delta} \tag{12.34}$$

where

$$\Delta = e_\theta^x e_\phi^y - e_\phi^x e_\theta^y \tag{12.35}$$

For a $\phi = 0°$ pattern cut of a predominantly E_θ-polarized linear array antenna $A'^x \gg A'^y$. Assuming a probe with low cross-polarization characteristics (that is, $e_\phi^x \ll e_\theta^x$, $e_\phi^y \ll e_\theta^x$) then (12.33) reduces to

$$E_\theta \approx C\cos\theta e^{jk_z z_0} \frac{A'^x}{e_\theta^x} \tag{12.36}$$

In evaluating the plane wave spectrum function A'^x, given by (12.24), an N_x-point (including zero filling) FFT is used. Assuming the near-field sample spacing is denoted by d_x and that the unfolded FFT transform index is denoted by K_x, then the transform wavenumber is given by

$$k_x = \frac{2\pi(K_x - N_x/2)}{d_x N_x} \tag{12.37}$$

The far-field angle (θ) is computed from the wavenumber k_x by the use of (12.6) and (12.37).

12.2.2 MONOPOLE PHASED ARRAY NEAR-FIELD MODELING USING THE METHOD OF MOMENTS

In the method of moments [12] (also referred to as the method of weighted residuals), boundary conditions are used to find the antenna response to a given excitation. The excitation here is the amplitude and phase incident at each element of the phased array. Due to mutual coupling or the mutual impedance between array elements, the actual illumination achieved will be different from that which is theoretically desired.

It is assumed here for convenience that there is one unknown complex current function per element of the array. A piecewise-sinusoidal current distribution is used as the moment method basis and testing functions. Since the basis functions and testing functions are the same, this is known as a Galerkin's formulation. For a piecewise-sinusoidal Galerkin's moment method formulation, the mutual impedance between array elements is readily computed [13]. In this chapter, the array elements are assumed to be resonant monopoles and, referring to Chapter 9, it is known that one unknown per element is adequate for pattern computation based on comparison with measurements and simulations [14].

12.2.2.1 Field Point Approach

The geometry for a finite array of monopoles over infinite ground plane is shown in Figure 12.3(a). Standard spherical coordinate angles (θ, ϕ) are used to describe the observation position for far-field pattern computation. The ground plane is located in the $z = 0$ plane and the monopoles are $\hat{z}$ polarized. Using image theory, the ground plane can be removed from the analysis. For a monopole array, an equivalent dipole array results (Figure 12.3(b)). From far-field theory, the monopole radiates (or receives) only the E_θ electric field component. The nth array element is located at the position (x_n, y_n) and the electric current that flows along the wire is assumed to be of the form

$$i_n(z) = i_n \frac{\sin(k(l - |z|))}{\sin(kl)} \qquad (12.38)$$

where i_n is the complex terminal current, l is the dipole half-length (monopole length), and $k = 2\pi/\lambda$ is the wavenumber. For the nth array element, the tangential near-field components on the plane $z = z_o$ are expressed as

$$E_{nx}(x, y, z_o) = E_{n\rho'} \cos \phi' \qquad (12.39)$$

$$E_{ny}(x, y, z_o) = E_{n\rho'} \sin \phi' \qquad (12.40)$$

where [15]

$$E_{n\rho'} = \frac{30j i_n}{\rho' \sin(kl)} (e^{-jkr_1} \cos \theta_1 + e^{-jkr_2} \cos \theta_2 - 2\cos(kl) e^{-jkr_o} \cos \theta_o) \qquad (12.41)$$

is the radial component of the electric field in cylindrical coordinates and where r_o, r_1, r_2, ρ', θ_o, θ_1, and θ_2 are defined in Figure 12.3(b) and $\phi' = \tan^{-1}((y - y_n)/(x - x_n))$. To compute the antenna near field including array mutual-coupling effects it is necessary to determine the array terminal currents i_n defined earlier using the method of moments.

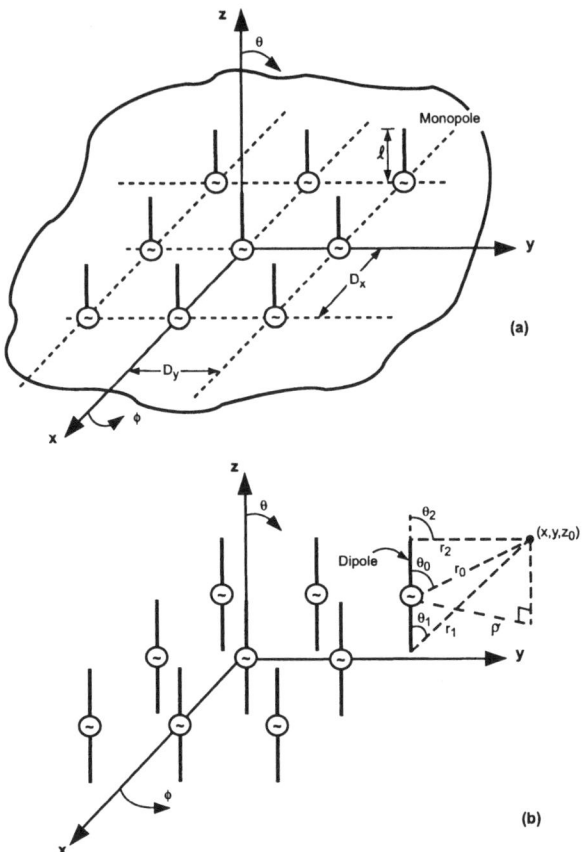

Figure 12.3 Geometry for finite array antennas: (a) monopole array over ground plane, and (b) equivalent dipole array with ground plane removed and monopole images included.

Let Z represent the mutual impedance matrix for the equivalent dipole array. Referring to Figure 12.4(a), Z is expressed as

$$Z = Z^{o.c.} + Z_L I \qquad (12.42)$$

where $Z^{o.c.}$ is the open-circuit mutual impedance matrix for the array, I is the identity matrix, and Z_L is the load impedance.

The mutual impedance Z_{mn} between two identical parallel thin-wire elements is a function only of the length of the elements and the element spacing. The self-impedance Z_{mm} of the array elements is taken to be equal to the mutual impedance between two thin elements separated by one wire radius.

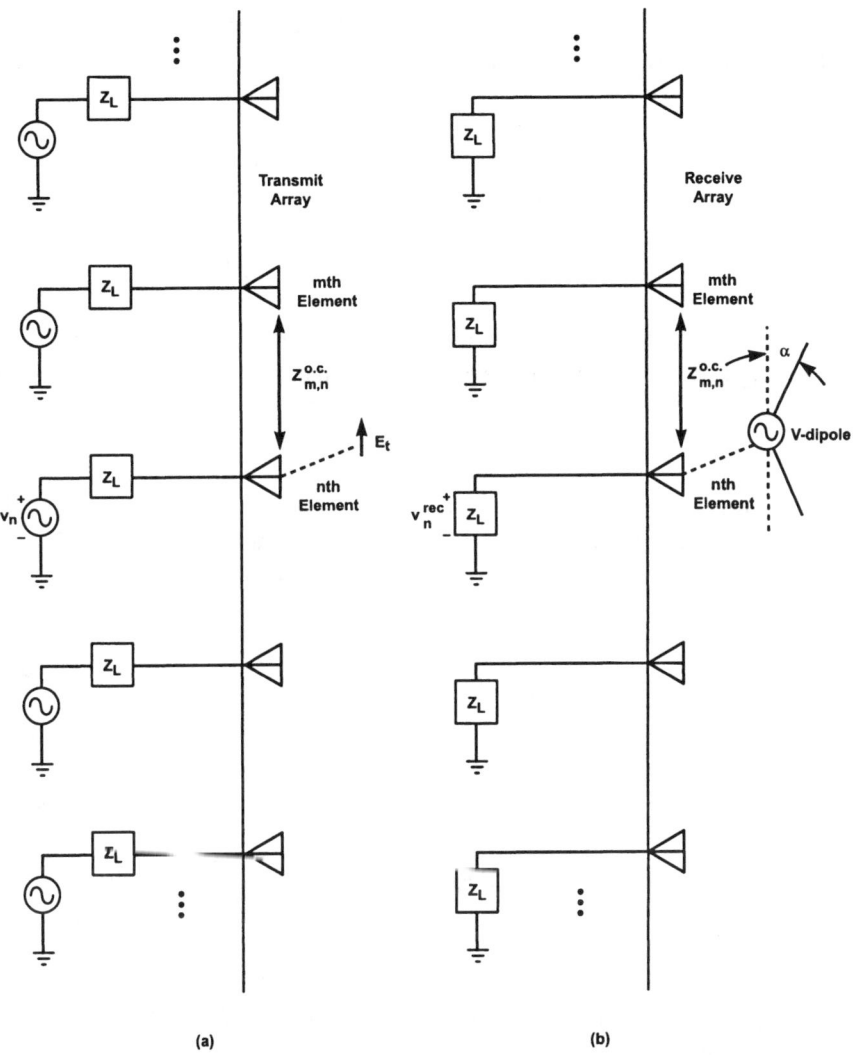

Figure 12.4 Near-field modeling for phased array antenna: (a) field point theory approach where array is transmitting and the near-zone electric field is computed, and (b) dipole probe theory where a V-dipole antenna is transmitting and the voltage received by the array is computed.

Define v as the voltage excitation matrix of the array. Then the array element terminal currents, denoted i, are found by solving the system of equations written in matrix form as

$$v = Z \cdot i \qquad (12.43)$$

The nth element of the voltage excitation matrix for a phased array antenna is given by (refer to (8.3) and (8.4))

$$v_n = A_n e^{j\Psi_{ns}} \tag{12.44}$$

where A_n is the amplitude illumination and

$$\Psi_{ns} = -k\sin\theta_s(x_n \cos\phi_s + y_n \sin\phi_s) \tag{12.45}$$

is the phase progression which scans the main beam in the direction (θ_s, ϕ_s).

Using (12.42) in (12.43), the array terminal currents are found according to

$$i = [Z^{o.c.} + Z_L I]^{-1} v \tag{12.46}$$

where $^{-1}$ means matrix inverse. Finally, using the array terminal currents in (12.41), together with (12.39) and (12.40), the array near-zone field including mutual-coupling effects is expressed by superposition as

$$E_x(x, y, z_o) = \sum_{n=1}^{N} E_{nx} \tag{12.47}$$

$$E_y(x, y, z_o) = \sum_{n=1}^{N} E_{ny} \tag{12.48}$$

Upon evaluating (12.47) and (12.48) numerically, the field point theory plane wave spectrum, given by (12.16) and (12.17), can be computed, from which the far-field pattern is determined from (12.18).

Having determined the array terminal currents, the *direct* far-field pattern can, of course, be computed directly using the product of the isolated element far-field pattern and the array factor as,

$$P_\theta(\theta, \phi) = p_\theta(\theta) \sum_{n=1}^{N} i_n e^{jk \sin\theta(x_n \cos\phi + y_n \sin\phi)} \tag{12.49}$$

where

$$p_\theta(\theta) = \frac{\cos(kl \sin\theta) - \cos(kl)}{\cos\theta} \tag{12.50}$$

is the pattern of a vertical dipole. Equation (12.49) will be used later to show that a centerline near-field scan that is transformed to the far field is a close approximation to the direct far-field pattern.

12.2.2.2 Dipole Probe Approach

In this section, an expression for the voltage received by array antenna elements, due to a near-field radiating probe antenna, is derived. A similar formulation for the case of a far-field source is addressed in [16]. The near-field scanning formulation described here has also found use in the application of near-field adaptive nulling [17] as discussed in Chapter 4.

A V-dipole probe is selected here because its pattern shape is readily adjusted by varying the tilt angle of the dipole arms (refer to Chapter 14). Thus, the pattern of the V-dipole can be made approximately equal to that of a practical near-field probe, which is typically an open-ended waveguide that can be surrounded by microwave absorber. The current distribution on the V-dipole is assumed to be piecewise-sinusoidal. Consider Figure 12.4(b), which depicts the circuit model for a receive array and a source antenna. Let $v_n^{rec}(x, y, z_o)$ be the voltage received in the nth array element due to a near-field source antenna (in our case a transmitting V-dipole antenna) at position (x, y, z_o). The array elements are assumed to be terminated in a load impedance denoted Z_L, which in general is complex. The open-circuit mutual impedance between the mth and nth array elements is denoted by $Z_{m,n}^{o.c.}$. Similarly, the open-circuit mutual impedance between the nth array element and the near-field V-dipole source antenna is denoted $Z_{n,probe}^{o.c.}(x, y, z_o)$, which is evaluated for thin-wire antennas using computer subroutines discussed in [18]. Now, $i_1, i_2, \cdots, i_n, \cdots, i_N$ are the received terminal currents for the N array elements. The received voltages are related to the terminal currents and load impedances using

$$v_n^{rec}(x, y, z_o) = -i_n^{rec}(x, y, z_o) Z_L \quad n = 1, 2, \cdots, N \quad (12.51)$$

Let i_t be the terminal current of the near-field source antenna. It is assumed that the array antenna does not affect the terminal current of the source antenna. This means that multiple interaction between the source antenna and array antenna is ignored. The array received voltages can be written as

$$v_n^{rec} = i_1^{rec} Z_{n,1}^{o.c.} + i_{2,j}^{rec} Z_{n,2}^{o.c.} + \cdots + i_N^{rec} Z_{n,N}^{o.c.} + i_t Z_{n,probe}^{o.c.}(x, y, z_o)$$
$$(12.52)$$

where $n = 1, 2, \cdots, N$. In (12.52), the term $i_t Z_{n,probe}^{o.c.}(x, y, z_o)$ is the open-circuit voltage at the nth array element due to a near-field probe at position (x, y, z_o).

Now, define

$$v_n^{o.c.}(x, y, z_o) = i_t Z_{n,probe}^{o.c.}(x, y, z_o), \quad (12.53)$$

and using (12.51) and (12.53) in (12.52) and rearranging terms yields

$$-v_n^{o.c.} = i_1^{rec} Z_{n,1}^{o.c.} + \cdots + i_n^{rec}(Z_{n,n}^{o.c.} + Z_L) + \cdots + i_N^{rec} Z_{n,N}^{o.c.} \quad (12.54)$$

where $n = 1, 2, \cdots, N$. Equation (12.54) can be written in matrix form as

$$-v^{o.c.} = [Z^{o.c.} + Z_L I] \, i^{rec} \tag{12.55}$$

where $v^{o.c.}$ is the array open-circuit voltage matrix, $Z^{o.c.}$ is the array open-circuit mutual impedance matrix, I denotes the identity matrix, and i^{rec} is the array received terminal current matrix. From (12.51) it is clear that

$$i^{rec} = -\frac{v^{rec}}{Z_L} \tag{12.56}$$

Substituting (12.56) in (12.55) and solving for v^{rec} yields

$$v^{rec}(x, y, z_o) = Z_L \left[Z^{o.c.} + Z_L I\right]^{-1} v^{o.c.}(x, y, z_o) \tag{12.57}$$

which gives the element received voltages as in Chapter 4. To compute the array beamformer (coherent power combiner) output, define an array weight vector, w, where the nth element is given by

$$w_n = A_n e^{j\Psi_{ns}} \tag{12.58}$$

with A_n being the amplitude illumination and Ψ_{ns} being the phase illumination given by (12.45). Thus, the complex received voltage at the array output port, due to the near-field radiating probe, is expressed as

$$v_{output}^{rec}(x, y, z_o) = w^\dagger \cdot v^{rec}(x, y, z_o) \tag{12.59}$$

where $\dagger$ means complex conjugate transpose.

To implement probe compensation for a V-dipole probe it is convenient to use a closed-form expression for the far-field radiation pattern. Using the notation of (12.36), the far-field pattern c_θ of a V-dipole antenna, with arm length l_v and having a sinusoidal current distribution of the form of (12.38), is readily obtained (see Chapter 14) by adding the contributions of two monopoles forming a V shape [19]. The result is

$$e_\theta^x = \frac{j30 i_t}{\sin(kl_v)} \Big(\frac{e^{jkl_v \sin(\theta+\alpha)} - j\sin(\theta+\alpha)\sin(kl_v) - \cos(kl_v)}{\cos(\theta+\alpha)}$$
$$+ \frac{e^{-jkl_v \sin(\theta-\alpha)} + j\sin(\theta-\alpha)\sin(kl_v) - \cos(kl_v)}{\cos(\theta-\alpha)} \Big) \tag{12.60}$$

To demonstrate the accuracy of this closed-form expression (which assumes one piecewise-sinusoidal current function across the length of the dipole), J.H. Richmond's moment method computer code [20], which utilizes overlapping piecewise-sinusoidal current functions, was used for comparison. Consider the case where the dipole arms are tilted by 45° and the arm-length is one-quarter wavelength, that is, $\alpha = 45°$ and $l_v = 0.25\lambda$ in (12.40). (Note: These parameters will be used in the theoretical probe model discussed in the next section.) The results are presented in Figure 12.5 for one, three, and five piecewise-sinusoidal current functions or unknowns. The indication is that one unknown current function produces a far-field pattern that agrees to within about 0.3 dB of the pattern obtained with five unknowns. This is consistent with other simulations of V-dipole antenna elements [21]. Thus, (12.60) is sufficiently accurate for purposes of performing probe compensation in this chapter.

12.3 LOW-SIDELOBE PHASED ARRAY ANTENNA PROTOTYPE

A prototype precision 32-element linear phased array antenna [22] was constructed, in part, for developing and evaluating near-field measurement techniques for low-sidelobe antennas. A photograph of the array is shown in Figure 12.6.

The array operates at L-band (1.25–1.35 GHz) and consists of coaxially fed monopole antenna elements arranged in a square lattice having 5 rows and 36 columns with spacing equal to 10.922 cm. To reduce edge effects, two

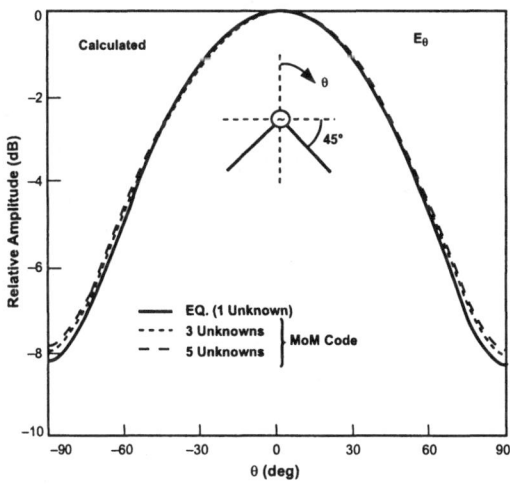

Figure 12.5 Comparison of far-field radiation patterns for $\lambda/2$ center-fed V-dipole antenna in free space using (12.40) and Richmond's thin-wire moment method code.

Figure 12.6 Photograph of 32-element linear phased array antenna. Two guard bands of elements surround the 32 driven monopole elements.

guard bands of passively terminated elements completely surround the center row of 32 active elements. The passive terminations are 50-ohm resistive loads. The length and diameter of the monopole elements are 6.096 cm and 0.3175 cm, respectively. Each monopole consists of a brass rod mounted on a type-N connector. The ground plane dimensions are 0.61m by 4.37m with a flatness that was measured to be within 0.05 cm peak. A simplified block diagram of the receive portion of the transmit/receive (T/R) module is shown in Figure 12.7.

The phase shifter is implemented in the 1100-MHz first local oscillator line. Beginning at a 275-MHz carrier frequency, $0°$ to $90°$ phase shift with 12 bits is effected. Next, frequency multiplication by four and filtering is used to generate the $0°$ to $360°$ phase controlled 1100-MHz LO. The RF signal is mixed initially down to the band 150–250 MHz, and then it is mixed with a 120–220 MHz tone down to a fixed 30 MHz. It is here that amplitude control is effected, also with 12 bits. A 40-dB attenuation range is implemented using two cascaded voltage controlled attenuators. Measured amplitude and phase tolerances of these modules were typically less than 0.02 dB and $0.2°$, respectively. The slightly nonlinear characteristics of the voltage controlled attenuators and phase shifters require calibration using a microprocessor that was built into each module. All modules were connected to a serial data bus. This bus permitted all modules to simultaneously receive commands from a general purpose desktop computer. The power combiner (beamformer) is implemented at 30 MHz with an isolation greater than 40 dB between ports.

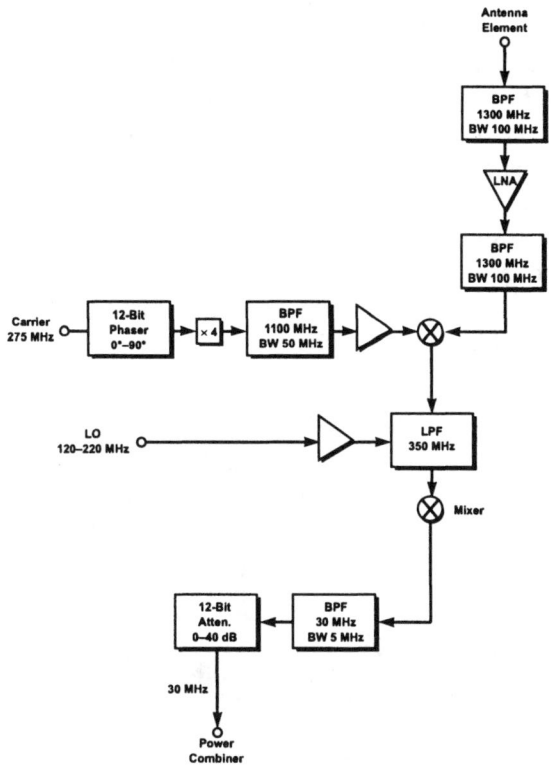

Figure 12.7 Block diagram for receive portion of T/R module. © 1991 IEEE [8].

12.4 NEAR-FIELD MEASUREMENTS SYSTEM

A linear scanner was built [22] to evaluate the performance of this phased array antenna. The near-field probe position in the x-direction is determined by a laser range finder and adjusted by means of a stepper motor driven cogged belt. The probe position accuracy is to within 0.005 cm of the desired position. For the test antenna, amplitude and phase measurements are made easier and more accurately by the fact that the output of the array is at a 30 MHz intermediate frequency. Thus, a low frequency network analyzer can be used to make these measurements with the desired degree of accuracy.

The near-field probe consists of a standard L-band coaxial to rectangular waveguide transition with the flange removed and surrounded with a 0.6m by 0.6m square sheet of pyramidal anechoic material. The rectangular waveguide (WR-650) inside dimensions are 16.51 cm by 8.255 cm. A photograph of the near-field probe is shown in Figure 12.8. The measured VSWR of this near-field probe antenna, over the frequency range 1.2-1.4 GHz, is less than 1.8:1 as depicted in Figure 12.9. In particular, at the center frequency of 1.3 GHz, the VSWR is equal to 1.7:1. Far-field E-plane radiation patterns of the rectangular waveguide near-field probe (with and without the surrounding absorber sheet). were measured on a conventional far-field antenna range. The measured far-field pattern data are shown in Figure 12.10. It is clear that the absorber modifies the free-space pattern shape. All near-field data presented in this chapter were collected with the absorber/waveguide probe combination. The purpose of the absorber is to reduce reflections that might occur between the test antenna and the near-field probe support arm.

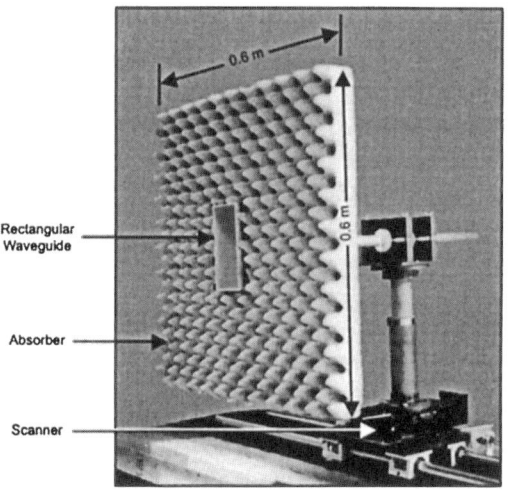

Figure 12.8 Photograph of rectangular waveguide near-field probe.

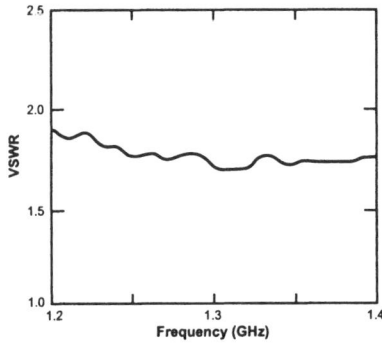

Figure 12.9 Measured VSWR for rectangular waveguide near-field probe.

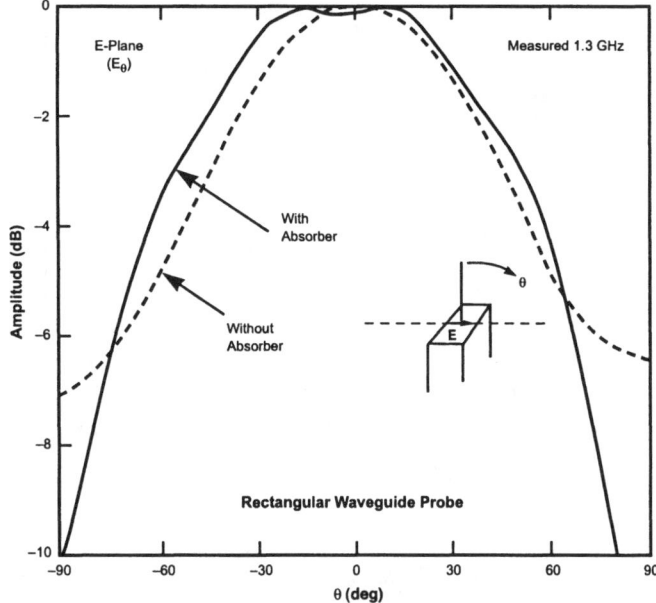

Figure 12.10 Measured far-field patterns for rectangular waveguide near-field probe with and without absorber.

12.5 RESULTS

12.5.1 SELECTION OF THEORETICAL PROBE MODEL

As mentioned earlier, the purpose of using a V-dipole probe (with arbitrary arm tilt angle) in the theoretical model is to have flexibility in designing or selecting the probe radiation pattern shape. This flexibility allows the theoretical probe model to closely match the pattern characteristics of the measurement probe. The closed-form far-field pattern for a V-dipole was computed, using (12.60), for various arm tilt angles. Based on the results shown in Figure 12.11, a V-dipole with $\lambda/4$ arm length and 45-degree arm tilt produces a theoretical pattern that is within about 1.0 dB of the measurements (over a $\pm 70°$ field of view) at 1.3 GHz for the open-ended rectangular waveguide probe with absorber. Notice that the $\alpha = 0°, 30°$ cases have a substantially larger variation from the measured probe data. Thus, the $\alpha = 45°$ V-dipole is selected here as the theoretical probe model.

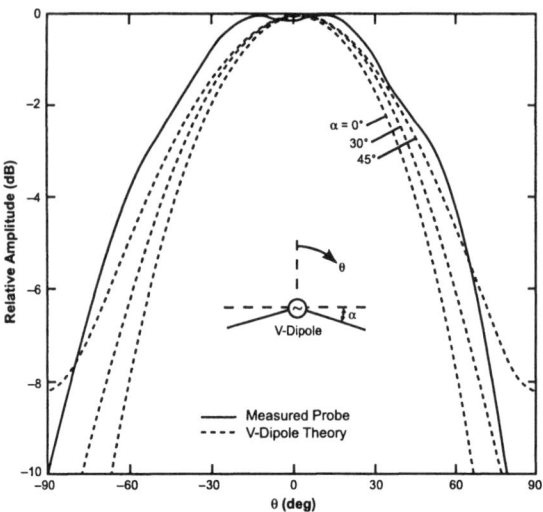

Figure 12.11 Comparison of far-field radiation patterns for theoretical V-dipole antenna with various arm tilt angles and experimental rectangular waveguide near-field probe with absorber. © 1991 IEEE [8].

12.5.1.1 Grating Lobe Positions for a Phased Array

The grating lobe positions, in terms of wavenumber, for a linear array with element spacing D_x and scanned to the angle θ_s are given by [23]

$$k_{x_m} = k \sin \theta_s + \frac{2\pi}{D_x} m; \quad m = \pm 1, \pm 2, \cdots \quad (12.61)$$

These, grating lobes are nonpropagating, provided that they fall within the invisible region where $|k_x| > 2\pi$. For an element spacing $D_x = 0.473\lambda$ (corresponding to the experimental array) and scan angle $\theta_s = -30°$, the first grating lobes ($m = \pm 1$ in (12.61)) are located at $k_{x_{-1}} = -16.44$ and $k_{x_1} = 10.16$. Similarly, the second grating lobes are located at $k_{x_{-2}} = -29.71$ and $k_{x_2} = 23.43$. The array factor, denoted $F(k_x)$, for an isotropic linear array can be expressed as

$$F(k_x) = \sum_{n=1}^{N} w_n e^{j k_x x_n} \quad (12.62)$$

Assuming a Taylor 40-dB array taper ($\overline{n} = 10$) and 32 isotropic antenna elements, the calculated array factor versus k_x is shown in Figure 12.12. Clearly, the first and second grating lobes fall outside the visible region and do not propagate to the far field. Depending on the near-field distance, these lobes may or may not be present in the resulting plane wave spectrum.

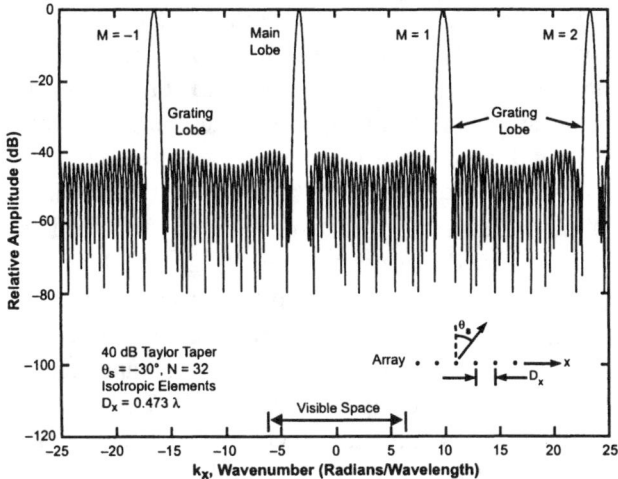

Figure 12.12 Calculated array factor versus the wave number k_x for a 32-element linear array with 0.473λ spacing.

The largest wavenumber that is produced in the FFT process is found by substituting $K_x = N_x$ in (12.37) with the result

$$k_x^{max} = \frac{\pi}{d_x} \tag{12.63}$$

For the theoretical simulations and experimental measurements, a near-field sample spacing of $d_x = 0.141\lambda$, which produces $k_x^{max} = 22.24$, was chosen. This means that, in computing the FFT, any large plane wave spectrum amplitude components above $|k_x| = 22.24$ will be aliased [24]. For example, the second grating lobe ($m = 2$) at $k_x = 23.43$ will alias to the position $k_x = -21.05$. This plane wave spectrum grating lobe and aliasing effect is demonstrated in what follows next.

12.5.1.2 Field Point Theory Compared to V-Dipole Probe Theory

The experimental array discussed in Section 12.3 was simulated using the moment method formulation presented in Section 12.2. The array illumination was assumed to be a 40-dB Taylor taper ($\bar{n} = 10$) with the main beam steered to $\theta_s = -30°$.

A sketch of the 32-element linear test array and near-field scan geometry is given in Figure 12.13. Notice that the shaded elements along the center row form the active (driven) linear array while the unshaded elements form two guard bands with passive terminations (50-ohm resistive). The length of the

32-element driven array is 3.39m. A centerline scan (x variable, $y = 0$) of length 4.14m was taken at the distance $z_0 = 12.7$ cm. This centerline scan sampled the near field down to about the -40 dB (or less) level. The number of near-field data samples is 128 and, to evaluate the plane wave spectrum, a 2048-point FFT (with zero-filling) is used. Note: The effect of zero filling is to interpolate the far-field pattern to a finer grid. At 1.3 GHz, the near-field distance is 0.55λ, which implies that the scan is performed in the reactive region. Here it is expected that nonpropagating grating lobes will be present in the invisible region ($|k_x| > 2\pi$) of the plane wave spectrum.

The principal polarization for this array along the centerline scan is the $\hat{x}$ component. For the field point theory, the tangential electric field component $E_x(x, 0, z_0)$ is computed using (12.47) and the plane wave spectrum component A_x is computed according to (12.16). The far-zone E_θ component is computed based on (12.18). In the case of the V-dipole probe theory, (12.59) is used to compute directly the received voltage $v^x(x, 0, z_0)$ from which (12.24) is implemented to obtain the probe-distorted plane wave spectrum component A'^x. The probe-compensated E_θ component is then computed according to (12.36).

A comparison of the two theories is now made. The near-field amplitude and phase are shown in Figure 12.14, where the field point and dipole theories yield similar results. However, the field point theory has a substantially higher peak-to-peak amplitude ripple (1.0 dB) compared to the V-dipole theory (0.2 dB). The number of ripples, 32, is equal to the number of driven array elements. Over the driven portion of the phased array, the root mean square (rms) differences between the two theories are 0.74 dB and 8.0°. The ripple amplitude is directly related to the grating lobe amplitude observed in the antenna plane wave spectrum. Figure 12.15 shows the plane wave spectrum for both theories using the data in Figure 12.14. Notice the presence of the expected first grating lobes at $k_x = 10.16$ and $k_x = -16.44$. The field point theory grating lobes are higher than the corresponding V-dipole probe theory

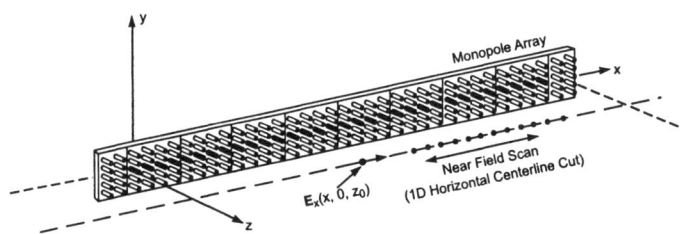

Figure 12.13 Geometry for monopole array and near-field centerline scan. © 1991 IEEE [8].

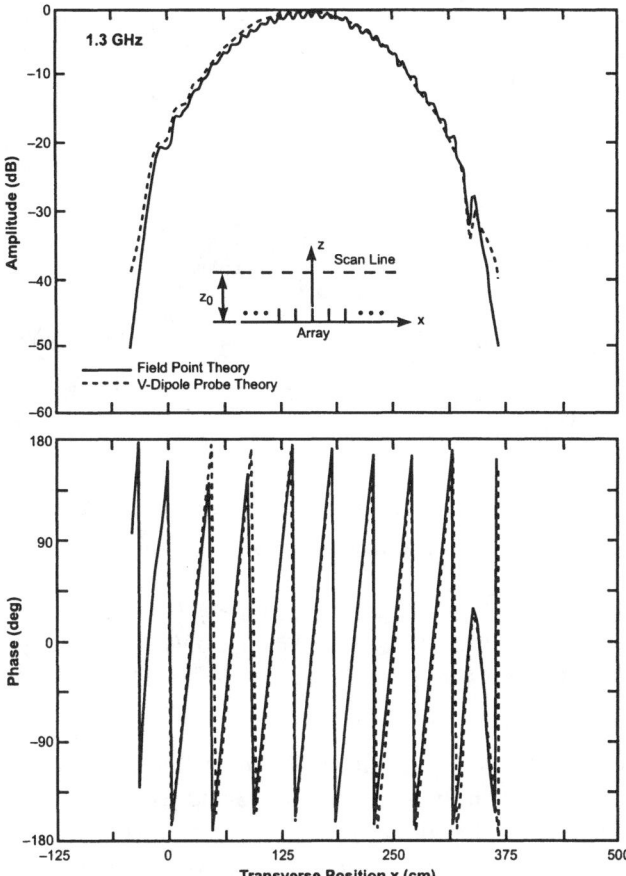

Figure 12.14 Comparison of 32-element monopole array near-zone electric field $E_x(x)$ using the field point theory and the array received voltage $v^x(x)$ due to a V-dipole transmitting antenna using the V-dipole theory. The near-zone data are computed at the distance $z_0 = 0.55\lambda$.

lobes. However, the probe theory lobes should be distorted, in amplitude, because of the effect of the V-dipole probe. Presumably, probe compensation of the plane wave spectrum would produce grating lobes of the correct amplitude; however, this was not of interest here. What is important is the presence of an aliased grating lobe for the field point theory. This is located at the wavenumber $k_x = -21.05$. No such alias occurs for the V-dipole theory. This seems to indicate that evanescent lobes may not be as significant a problem, in terms of creating aliases or errors in the evanescent and radiating portion of the plane wave spectrum, for a V-dipole probe when compared

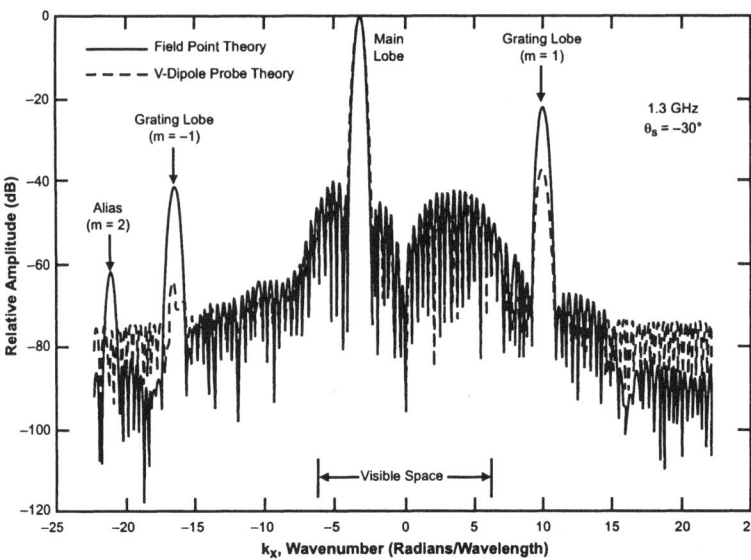

Figure 12.15 Comparison of 32-element monopole array field point theory plane wave spectrum $A_x(k_x, z_0 = 0.55\lambda)$ and V-dipole theory probe-distorted plane wave spectrum $A'^x(k_x, z_0 = 0.55\lambda)$. © 1991 IEEE [8].

to the field point theory. As will be shown in the next section, the same conclusion can be drawn for a practical waveguide probe.

Next, the far-field patterns (transformed from the field point and V-dipole near-field data), E_θ component, are shown in Figure 12.16. Note that probe compensation has been used in the case of the V-dipole theory, and that the agreement between the two theories is very good. To show that a centerline scan provides adequate information for the near-field to far-field transformation, Figure 12.17 compares the field point theory transformed pattern with the direct far-field pattern computed using (12.49). Very little differences between the two patterns can be seen.

12.5.1.3 V-Dipole Probe Theory Compared to Near-Field Measurements

The purpose of developing the V-dipole probe near-field theory was to be able to make accurate predictions of the behavior of a phased array antenna. In this section, comparisons of the results obtained using theory and experiments are made and good agreement is achieved.

The experimental 32-element monopole array was calibrated at 1.3 GHz and commanded to steer the main beam to $\theta_s = -30°$ with a 40-dB Taylor taper. A full-scan comparison of the near-field measurements and V-dipole

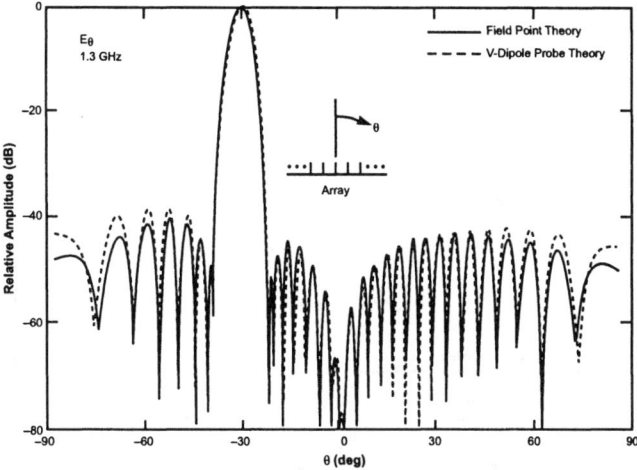

Figure 12.16 Comparison of 32-element monopole array far-field patterns computed using the field point theory and the V-dipole probe theory.

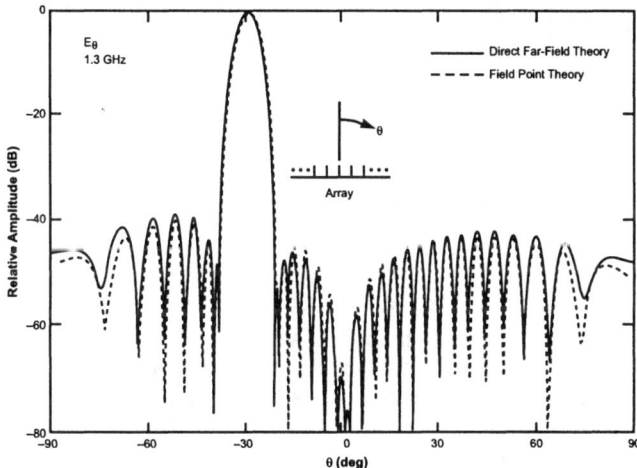

Figure 12.17 Comparison of 32-element monopole array far-field patterns based on the field point theory (12.47) and the direct method (12.49). © 1991 IEEE [8].

probe theory (this is the same theoretical data that was compared against the field point theory) is shown in Figure 12.18. An expanded scale plot of the same near-field data covering 0 to −3 dB is shown in Figure 12.19. Clearly, the agreement between the measured and simulated near-field data is good.

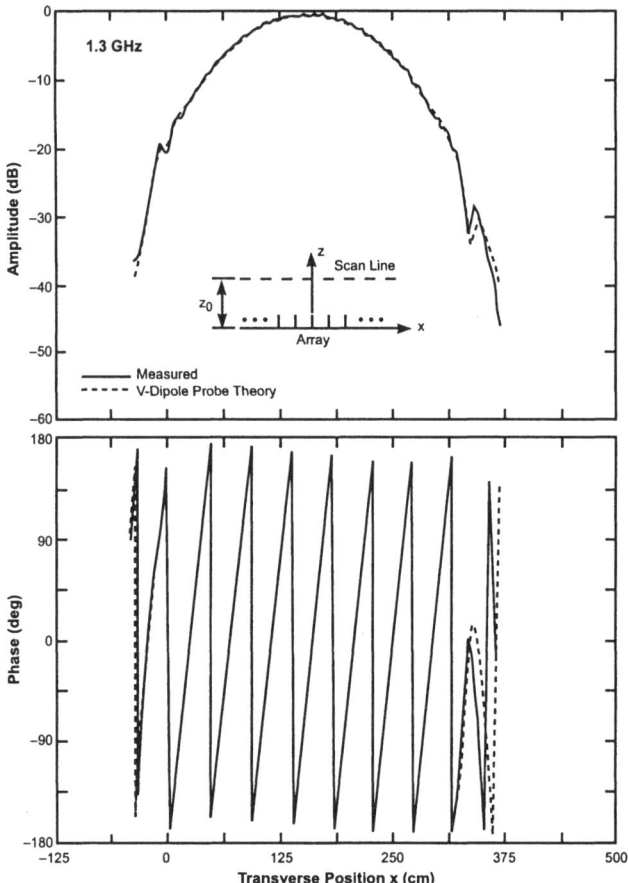

Figure 12.18 Comparison of measured and theoretical near-zone received voltage for the 32-element monopole array. The measurement probe is a open-ended rectangular waveguide surrounded with anechoic material. The theoretical probe is a V-dipole. The near-zone distance is $z_0 = 0.55\lambda$. © 1991 IEEE [8].

The peak-to-peak ripple is 0.5 dB for the measurements compared to 0.2 dB for the theory. Thus, the measured grating lobes should be higher than the theoretical grating lobes, and this is the case as the probe-distorted plane wave spectrum shows in Figure 12.20. The first grating lobes are evident in this figure, and there is no indication of any large aliased grating lobes. The probe-compensated far-field patterns are shown in Figure 12.21 and are generally in good agreement. The average sidelobes are −46.7 dB measured and −48.6 dB theoretical. There is a slight filling-in of the monopole null near

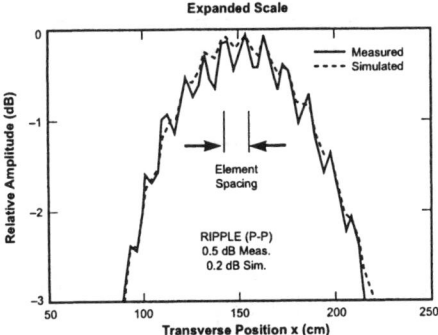

Figure 12.19 Expanded scale comparison of measured and theoretical near-zone received voltage for the 32-element monopole array. The measurement probe is a open-ended rectangular waveguide surrounded with anechoic material. The theoretical probe is a V-dipole. The near-zone distance is $z_0 = 0.55\lambda$. © 1991 IEEE [8].

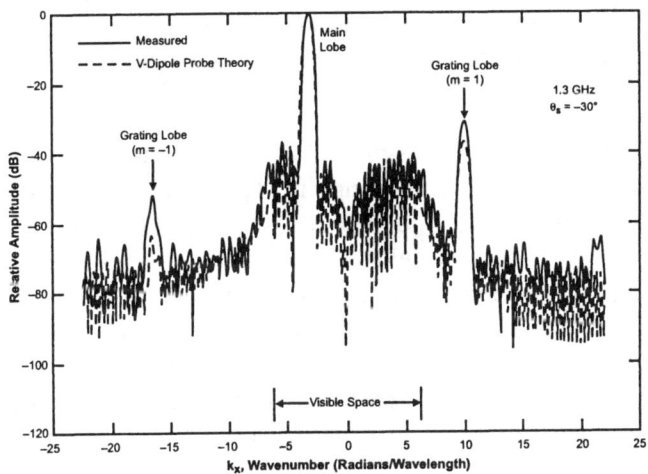

Figure 12.20 Comparison of 32-element monopole array probe-distorted plane wave spectrum based on near-zone measurements and V-dipole probe theory. The near-zone distance is $z_0 = 0.55\lambda$. © 1991 IEEE [8].

$\theta = 0°$ for the measurements, but this occurs at the -60 dB level.

12.6 SUMMARY

In this chapter, planar near-field scanning, with and without probe compensation, has been implemented theoretically using the method of moments to

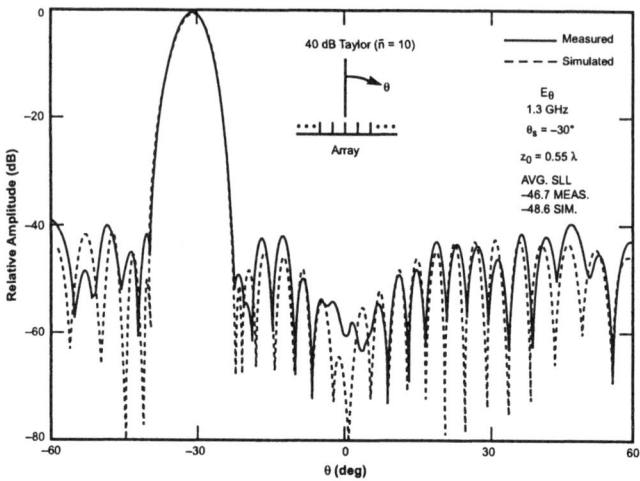

Figure 12.21 Comparison of 32-element monopole array probe-compensated far-field patterns based on near-zone measurements and V-dipole probe theory. The near-zone distance is $z_0 = 0.55\lambda$. © 1991 IEEE [8].

determine the characteristics of phased array antennas. A monopole phased array antenna has been analyzed using a field point theory where the near-zone electric field of the antenna is computed. The near field is transformed, via the plane wave spectrum, to a far-field pattern. Probe compensation is not required in this case. A probe theory, where the array near-zone received voltage due to a V-shaped dipole radiating antenna is computed, has also been presented. For the V-dipole theory, probe compensation was implemented using a closed-form expression for the far-field pattern of the V-dipole. For further discussion of the theory and practice of near-field measurements, the interested reader can refer to Slater [25].

The design of an experimental low-sidelobe monopole phased array antenna has been described here. Near-field simulations of this antenna using both theories have been made. The simulations indicate that evanescent (non-propagating) grating lobes, in the plane wave spectrum, have a lower amplitude when observed by a V-dipole probe compared to those observed by the field point theory. Adjusting the tilt angle of the V-dipole probe arms allowed the V-dipole far-field pattern to match approximately the far-field pattern of an experimental rectangular waveguide probe surrounded with absorber. It was shown that the V-dipole probe theory can accurately model the observed experimental performance of a monopole array. The theory presented here can be applied readily to other types of array antenna elements.

12.7 PROBLEM SET

12.1 Derive (12.1). Demonstrate that (12.2) is a solution of (12.1).
12.2 Derive (12.11) from (12.2) and (12.5).
12.3 Derive (12.30) and (12.31).
12.4 Derive (12.33) and (12.34) from (12.30) and (12.31).
12.5 Using (12.6) and (12.37), solve for the far-field angle θ.
12.6 Write software that will compute the far-field pattern of a V-shaped dipole as given by (12.60) and plot the amplitude in dB versus angle as in Figure 12.11 for tilt angles of $\alpha = 0°, 30°, 45°$.
12.7 Using (12.62) compute the array factor versus wavenumber k_x from -25 radians/wavelength to 25 radians/wavelength for the case of a 32-element array of isotropic elements with uniform illumination ($w_n = 1$), scan angle $\theta_s = -30°$, and array element spacing $D_x = 0.473\lambda$. Refer to Figure 12.12.
12.8 Repeat Problem 12.7, but now with $D_x = 0.55\lambda$, and identify the grating lobe positions in terms of radians/wavelength.

References

[1] Paris, D.T., W.M. Leach, and E.B. Joy, "Basic Theory of Probe-Compensated Near-Field Measurements," *IEEE Trans. Antennas Propagat.*, Vol. 26, No. 3, 1978, pp. 373-379.

[2] Joy, E.B., W.M. Leach, G.P. Rodrigue, and D.T. Paris, "Applications of Probe-Compensated Near-Field Measurements," *IEEE Trans. Antennas Propagat.*, Vol. 26, No. 3, 1978, pp. 379-389.

[3] Yaghjian, A.D., "An Overview of Near-Field Antenna Measurements," *IEEE Trans. Antennas Propagat.*, Vol. 34, No. 1, 1986, pp. 30-45.

[4] Gillespie, E.S., Guest Editor, "Special Issue on Near-Field Scanning Techniques," *IEEE Trans. Antennas Propagat.*, Vol. 36, No. 6, 1988.

[5] Newell, A.C., and M.L. Crawford, "Planar Near-Field Measurements on High Performance Array Antennas," National Bureau of Standards Report No. NBSIR 74-380, July 1974.

[6] Joy, E.B., and D.T. Paris, "Spatial Sampling and Filtering in Near-Field Measurements," *IEEE Trans. Antennas Propagat.*, Vol. 20, No. 3, 1972, pp. 253-261.

[7] Wang, J.H., "An Examination of the Theory and Practices of Planar Near-Field Measurement," *IEEE Trans. Antennas Propagat.*, Vol. 36, No. 6, 1988, pp. 746-753.

[8] Fenn, A.J., H.M. Aumann, and F.G. Willwerth, "Linear Array Characteristics with One-Dimensional Reactive-Region Near-Field Scanning: Simulations and Measurements," *IEEE Trans. Antennas and Propagat.*, Vol. 39, No. 9, September 1991, pp. 1305-1311.

[9] Hanfling, J.D., G.V. Borgiotti, and L. Kaplan, "The Backward Transform of the Near Field for Reconstruction of Aperture Fields," *1979 IEEE Antennas Propagat. Society Symposium Digest,* Vol. 2, IEEE, New York, pp. 764-767.

[10] Borgiotti, G., "Fourier Transforms Method in Aperture Antennas Problems," *Alta Frequenza*, Vol. 32, No. 11, 1963, pp. 808-816.

[11] Rhodes, D.R., *Synthesis of Planar Antenna Sources*, Oxford: Oxford University Press, 1974.

[12] Stutzman, W.L., and G.A. Thiele, *Antenna Theory and Design*, New York: Wiley, 1981.

[13] Castello, D., and B.A. Munk, "Tables of Mutual Impedance of Identical Dipoles in Echelon," The Ohio State University, ElectroScience Laboratory, Technical Report 2382-1, AD-822013, October 17, 1967.

[14] Fenn, A.J., "Theoretical and Experimental Study of Monopole Phased Array Antennas," *IEEE Trans. Antennas Propagat.*, Vol. 33, No. 10, October 1985, pp. 1118-1126.

[15] Schelkunoff, S.A., *Electromagnetic Waves*, New York: D. Van Nostrand Co., 1943, pp. 370-371.

[16] Gupta, I.J., and A.A. Ksienski, "Effect of Mutual Coupling on the Performance of Adaptive Arrays," *IEEE Trans. Antennas Propagat.*, Vol. 31, No. 5, 1983, pp. 785-791.

[17] Fenn, A.J., "Moment Method Analysis of Near Field Adaptive Nulling," *IEE Sixth International Conf. on Antennas and Propag., ICAP 89 Proceedings*, April 4-7, 1989, pp. 295-301.

[18] Richmond, J.H., and N.H. Geary, "Mutual Impedance of Nonplanar-Skew Sinusoidal Dipoles," *IEEE Trans. Antennas Propagat.*, Vol. 23, No. 3, 1975, pp. 412-414.

[19] Schelkunoff, S.A., and H.T. Friis, *Antennas, Theory and Practice*, New York: Wiley, 1952, pp. 499-501.

[20] Richmond, J.H., "Radiation and Scattering by Thin-Wire Structures in a Homogeneous Conducting Medium (Computer Program Description)," *IEEE Trans. Antennas Propagat.*, Vol. 22, No. 2, 1974, p. 365.

[21] Fenn, A.J., "Element Gain Pattern Prediction for Finite Arrays of V-Dipole Antennas Over Ground Plane," *IEEE Trans. Antennas Propagat.*, Vol. 36, No. 11, 1988, pp. 1629-1633.

[22] Aumann, H.M., and F.G. Willwerth, "Near-Field Testing of a Low-Sidelobe Phased Array Antenna," *Proceedings of the Antenna Measurement Techniques Association 1987 Meeting*, September 28-October 2, 1987, pp. 3-7.

[23] Hansen, R.C., (ed.), *Microwave Scanning Antennas, Volume II, Array Theory and Practice*, Los Altos, CA: Peninsula Publishing, 1985, pp. 199-208.

[24] Newell, A.C., "Error Analysis Techniques for Planar Near-Field Measurements," *IEEE Trans. Antennas Propagat.*, Vol. 36, No. 6, 1988, pp. 754-768.

[25] Slater, D., *Near-Field Antenna Measurements*, Norwood, MA: Artech House, 1991.

13

Arrays of Horizontally Polarized Omnidirectional Elements

13.1 INTRODUCTION

As discussed in Chapter 8, conventional wide-angle scanning planar phased array antennas generally utilize radiating elements that have peak gain at broadside [1]. This choice of radiating element is usually the result of a scanning requirement by the radar or communications system to cover a portion of a hemisphere typically from 0° to 60° from broadside. For certain phased array antenna applications, such as for space-based radar, to reduce radar clutter it is desirable to have maximum element gain away from broadside with minimum gain or a null occurring at broadside [2]. Such an element can be, for example, a monopole array element for vertical polarization [3], as discussed in Chapter 9, or a loop-fed slotted cylinder array element for horizontal polarization [4], as is considered in the present chapter.

In this chapter, measured data are given for small arrays of loop-fed slotted cylinders. First, the return loss and radiation pattern of a single element on a ground plane over a 15 percent bandwidth are given. Next, the center element gain pattern, return loss, and mutual coupling in a passively terminated seven-element hexagonal array are investigated. Finally, linear arrays of up to seven elements are considered, and it is shown that the element gain pattern has only a small variation with array size. Overall, the measured data presented here will show that the loop-fed slotted cylinder has an array element gain pattern that is well suited for wide-angle scanning applications.

13.2 THEORY

13.2.1 ARRAY GAIN AND ELEMENT GAIN

As discussed in Chapter 8, in a large array where each element is subject to essentially the same mutual-coupling effects, the element gain pattern is invariant with position [5]. The array main-beam gain $G_a(\theta, \phi)$ for uniform amplitude illumination is then given by

$$G_a(\theta, \phi) = g_e(\theta, \phi) \cdot N_a \qquad (13.1)$$

where $g_e(\theta, \phi)$ is the element gain pattern, N_a is the number of active array elements, and θ, ϕ are standard spherical coordinates. For an infinite array without grating lobes, the element gain pattern is a function of the unit cell area A_e, the scanned array transmission coefficient $T(\theta, \phi)$, and the projected aperture factor $\cos \theta$ as given by

$$g_e(\theta, \phi) = \frac{4\pi A_e}{\lambda^2} |T(\theta, \phi)|^2 \cos \theta \qquad (13.2)$$

where

$$|T(\theta, \phi)|^2 = 1 - |\Gamma(\theta, \phi)|^2 \qquad (13.3)$$

with $\Gamma(\theta, \phi)$ being the scanned array reflection coefficient. In (13.2), A_e and $\cos \theta$ are factors common to all phased array antenna radiating elements. For common array antenna elements such as dipoles and waveguides, $T(\theta, \phi)$ can be computed [1]. For complicated array elements, such as the loop-fed slotted cylinder discussed in this chapter, direct measurement of the element gain pattern and mutual coupling in a small array is a reliable approach for characterizing array performance.

One way to simulate the environment of a phased array is to surround the element of interest with a number of identical elements that are terminated in resistive loads matched to the feed lines [5]. In a large array, each element has an element gain pattern $g_e(\theta, \phi)$ which, by superposition, produces the array gain given by (13.1). As shown in Figure 13.1, the approach to compute or measure the element gain pattern is to feed only one element and terminate all others with the load Z_L. The mutual coupling (denoted as S_{mn} in Figure 13.1) between array elements can be measured.

13.3 LOOP-FED SLOTTED CYLINDER ANTENNA

A uniform current loop antenna [6-9] over a ground plane is known to have a horizontally polarized omnidirectional pattern with a null at broadside. An

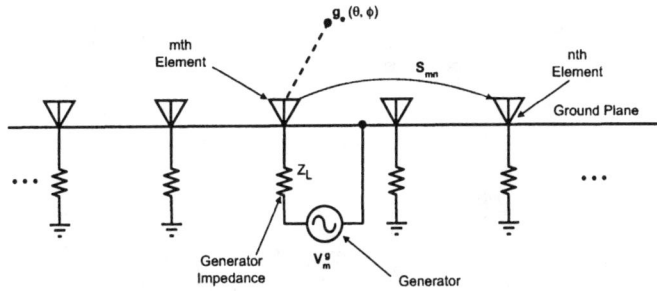

Figure 13.1 Single driven element in an array environment for determination of the array element gain pattern. © 1985 IEEE [4].

example of a double-sided printed circuit version of a uniform current loop antenna is shown in Figure 13.2. The center pin of a semirigid RG-141 coaxial cable is soldered to the top portion of a balanced two-wire transmission line. The lower portion of the two-wire transmission line is connected to the outer conductor of the coaxial line. The two balanced transmission lines feed two one-half wavelength dipoles in parallel, which forms a uniform current loop. One of the difficulties with this antenna is its inherent narrow bandwidth due to a large input reactance. While the input resistance is close to 50 ohms the input reactance of a one-wavelength perimeter uniform-current loop is typically several hundred ohms inductive. The inductive component can be reduced by capacitively loading the loop. One implementation of this capacitive loading is to surround the uniform-current loop by an open-ended metallic cylinder with four vertical slots as depicted in Figure 13.3. The resonance of the antenna is controlled by the slot width and by the diameter and height of the cylinder. The slots can be loaded with a dielectric material to lower the resonant frequency. Assuming the slots are loaded with a dielectric slab material with dielectric constant equal to 2.6, typical dimensions for resonance (determined empirically) are cylinder height $H = 0.425\lambda$, cylinder diameter $D = 0.35\lambda$, and slot width $W = 0.04\lambda$.

13.4 RESULTS

13.4.1 SINGLE ELEMENT

A photograph of a single radome-enclosed loop-fed slotted cylinder antenna that covers the frequency band 1.2 to 1.4 GHz (Model #5241-L1-D, Antenna Corporation of America, Harleysville, Pennsylvania) is shown in Figure 13.4. The antenna was mounted on a square aluminum ground plane with side dimension 1.22m. This configuration and the associated measurement

Figure 13.2 Photograph showing an example of a uniform current loop antenna. © 1985 IEEE [4].

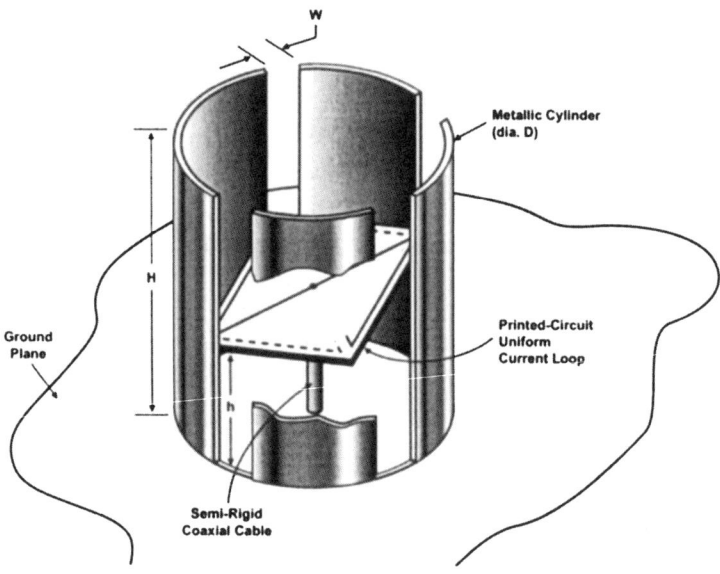

Figure 13.3 Diagram showing a loop-fed slotted cylinder antenna. © 1985 IEEE [4].

coordinate system are shown in Figure 13.5. For this antenna, the dominant polarization is E_ϕ.

The measured swept return loss or VSWR for a single loop-fed slotted cylinder on a 1.22m ground plane is shown in Figure 13.6. Over the 1.2- to 1.4-GHz band the measured VSWR is less than 2.0:1. The measured

Arrays of Horizontally Polarized Omnidirectional Elements 325

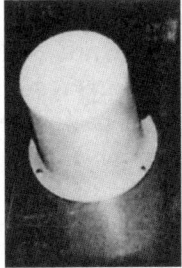

Figure 13.4 Photograph of a radome-enclosed loop-fed slotted cylinder antenna. © 1985 IEEE [4].

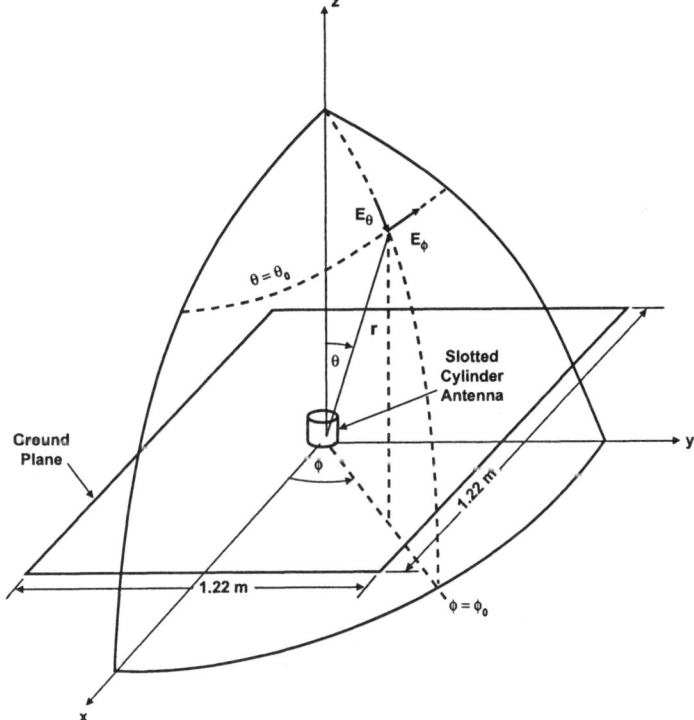

Figure 13.5 Diagram depicting a loop-fed slotted cylinder antenna on a square ground plane. © 1985 IEEE [4].

principal plane ($\phi = 0°$) patterns of the single, isolated, loop-fed slotted cylinder antenna on the 1.22m ground plane at 1.2, 1.3, and 1.4 GHz are shown in Figure 13.7. The measured peak gains are 3.5 dBi, 5.1 dBi, and 4.1 dBi at 1.2, 1.3, and 1.4 GHz, respectively. The peak gain occurs in the vicinity of $\theta = 52°$ as desired for wide-angle scanning. The null depth at $\theta = 0°$ is greater than 30 dB, which would provide suppression of nadir clutter and/or nadir jamming for a space-based radar. Figure 13.8 shows conical pattern cuts for $\theta = 55°$, that is, close to the gain peak. Over the 15 percent bandwidth the amplitude is omnidirectional to within ± 0.25 dB. Also, the phase uniformity for a constant cone angle $\theta = 55°$ was measured at 1.2, 1.3, and 1.4 GHz, as shown in Figure 13.9. The azimuthal phase variation over $-180° < \phi < 180°$ is less than $\pm 4°$.

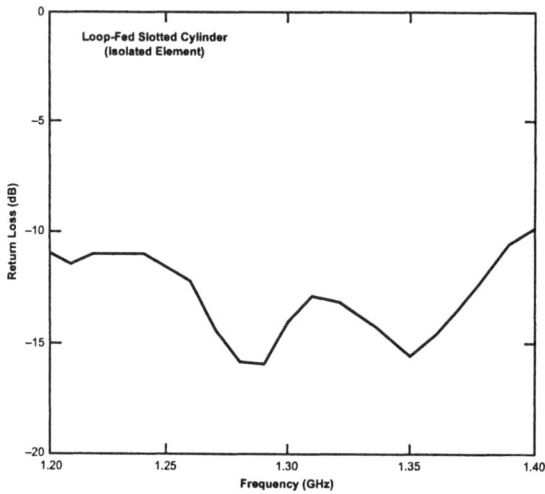

Figure 13.6 Measured return loss for a single (isolated) loop-fed slotted cylinder antenna on a 1.22m ground plane. © 1985 IEEE [4].

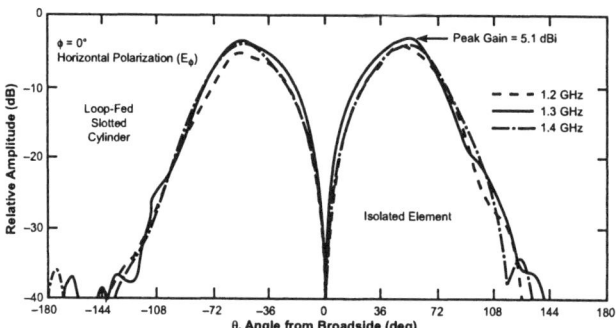

Figure 13.7 Measured principal-plane ($\phi = 0°$) gain radiation patterns (E_ϕ component) for a single (isolated) loop-fed slotted cylinder antenna on a 1.22m ground plane. © 1985 IEEE [4].

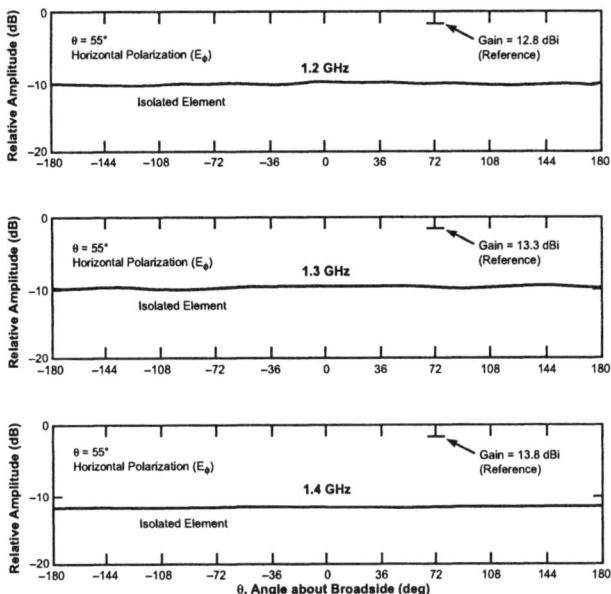

Figure 13.8 Measured conical-cut ($\theta = 55°$) gain radiation patterns (E_ϕ component) for a single (isolated) loop-fed slotted cylinder antenna on a 1.22m ground plane. © 1985 IEEE [4].

13.4.2 SEVEN-ELEMENT HEXAGONAL ARRAY

A seven-element hexagonal array of loop-fed slotted cylinders with 12.7 cm spacing (0.55λ at 1.3 GHz) was assembled as shown in Figure 13.10. A square aluminum ground plane with side dimension equal to 1.22m was used. The six antenna elements surrounding the center element are terminated in 50-ohm resistive loads. Note: There was no attempt to optimize the design of the loop-fed slotted cylinder in the array environment; however, the element match is shown to be good. The measured swept return loss for the center element in the seven-element hexagonal array is shown in Figure 13.11. The maximum VSWR is 1.6:1 over the 15 percent bandwidth. The amplitude of the mutual coupling between the center element and one of the surrounding elements is shown in Figure 13.12. The average coupling over the band is approximately -12.5 dB.

The measured center element gain pattern for both $\phi = 0°$ and $\phi = 30°$ planes is shown in Figure 13.13 at 1.2, 1.3, and 1.4 GHz. Note the $\phi = 0°$ cut is taken through the plane containing three elements. Both the pattern peak location and broadside null depth are very similar to the single element case.

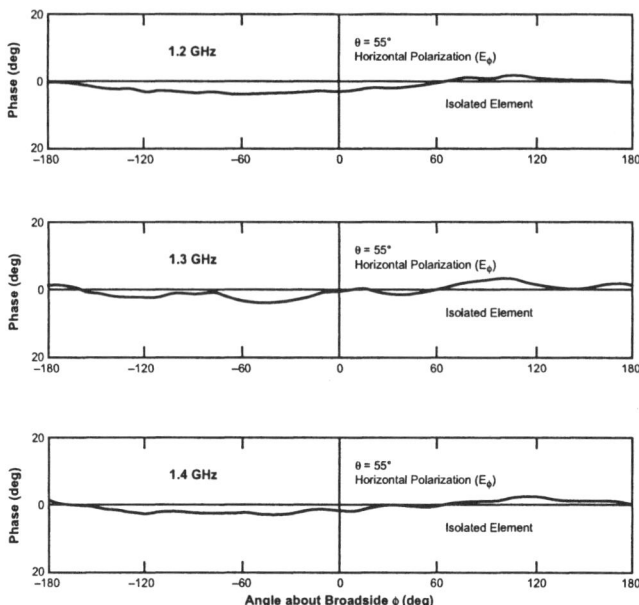

Figure 13.9 Measured conical-cut ($\theta = 55°$) phase radiation patterns (E_ϕ component) for a single (isolated) loop-fed slotted cylinder antenna on a 1.22m ground plane. © 1985 IEEE [4].

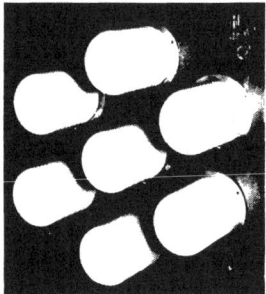

Figure 13.10 Photograph of a seven-element hexagonal array of loop-fed slotted cylinder antennas. © 1985 IEEE [4].

Wide-angle scanning with substantial radiation pattern gain is clearly possible with this element. Conical pattern cuts for $\theta = 55°$ are shown in Figure 13.14. The pattern amplitude differs by no more than ±0.8 dB from omnidirectional over the 15 percent bandwidth. Also shown is the cross-polarized component E_θ, which is down by more than 23 dB from the principal polarization. It should be kept in mind that this planar array is considered small, being only

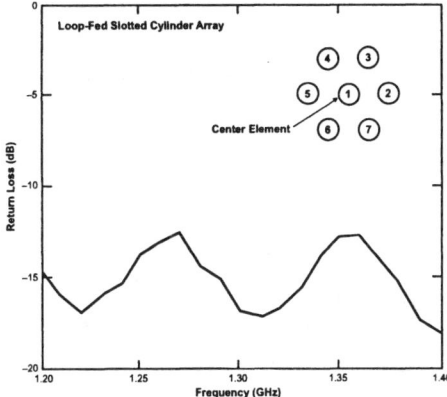

Figure 13.11 Measured return loss for the center element of a seven-element hexagonal array of loop-fed slotted cylinder antennas shown in Figure 13.10. © 1985 IEEE [4].

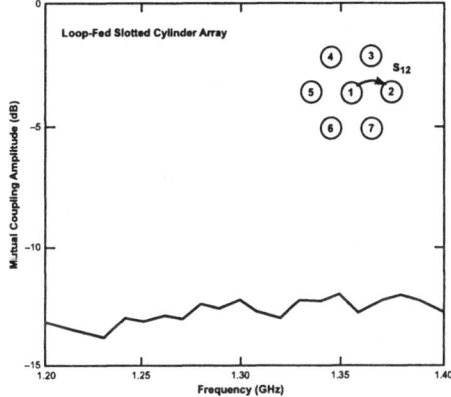

Figure 13.12 Measured mutual coupling amplitude (S_{12}) from the center element to one of the surrounding elements in a seven-element hexagonal array of loop-fed slotted cylinder antennas shown in Figure 13.10. © 1985 IEEE [4].

seven elements in a hexagonal pattern. It would not be possible to infer the performance of the center element in a much larger array based on the results for this small hexagonal array. In the next section, the same seven elements are reconfigured to form a linear array of a variable number of elements. This is done to see if an undesired blind spot (or pattern null) may exist away from broadside. It is shown that there is no evidence of any such blind spot.

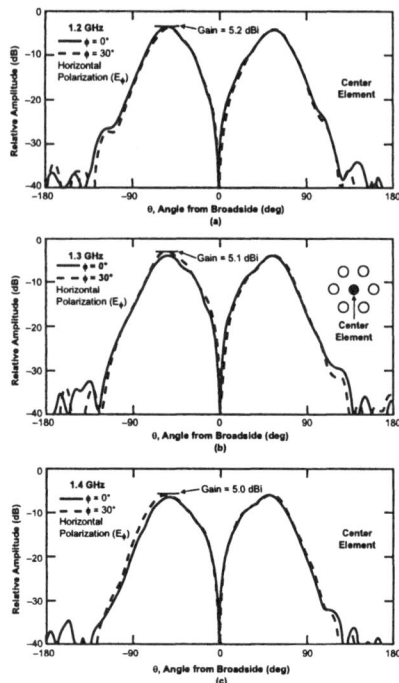

Figure 13.13 Measured principal-plane ($\phi = 0°$) gain radiation patterns (E_ϕ component) for the center element of the seven-element hexagonal array of loop-fed slotted cylinder antennas on a 1.22m ground plane. © 1985 IEEE [4].

13.4.3 LINEAR ARRAYS

To investigate the effect on the element pattern of increasing the array size in one dimension, linear arrays with three, five, and seven elements were investigated. Data are presented only at the center frequency of 1.3 GHz. The element spacing is 12.7 cm, which corresponds to 0.55λ. Figure 13.15 shows a photograph of a linear array of seven elements. The ground plane is 1.22m square. Pattern cuts were taken through the principal plane containing the line of the array. The center element is fed and the surrounding elements are terminated in 50-ohm resistive loads. The measured center element gain patterns for one, three, five, and seven elements are compared in Figure 13.16. Overall, the pattern shape is seen to be relatively insensitive to the number of surrounding elements. Initially, there is a small reduction in peak gain as the array size increases, but the gain seems to be converging to a constant value as seven elements is reached. There is no indication that for any further

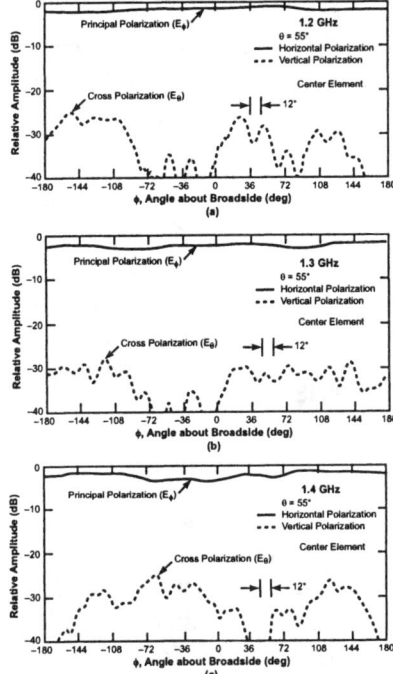

Figure 13.14 Measured conical-cut ($\theta = 55°$) gain radiation patterns (E_ϕ component) for the center element of the seven-element hexagonal array of loop-fed slotted cylinder antennas on a 1.22m ground plane. © 1985 IEEE [4].

increase in array size the pattern gain will deviate substantially from the seven-element linear array case shown here. From these data there appears to be no evidence of a potential blind spot between broadside and endfire for this type of element.

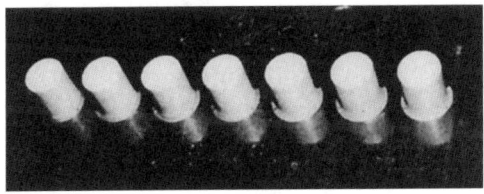

Figure 13.15 Photograph of a seven-element linear array of loop-fed slotted cylinder antennas. © 1985 IEEE [4].

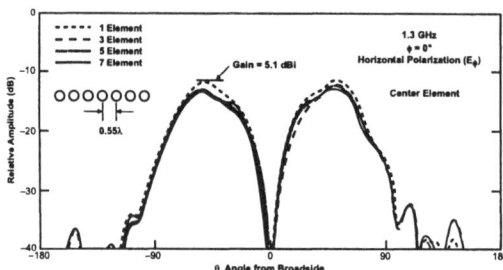

Figure 13.16 Measured principal-plane ($\phi = 0°$) gain radiation patterns (E_ϕ component) for the center element of linear arrays (1, 3, 5, and 7 elements) of loop-fed slotted cylinder antennas on a 1.22m ground plane. © 1985 IEEE [4].

13.5 SUMMARY

A loop-fed slotted cylinder element has been measured in a passively terminated array on a ground plane. For a seven-element hexagonal array, the center element gain pattern is omnidirectional with horizontal polarization and has a null at broadside. Peak gain for this element occurs in the vicinity of $\theta = 52°$ from broadside. The center element VSWR over a 15 percent bandwidth was shown to be less than 1.6:1. Mutual coupling between the center element and one of the surrounding elements was shown. Also, to investigate the element gain pattern as a function of the array size, linear arrays of three, five, and seven elements were considered. Overall, the data indicate that wide-angle phased array scanning from broadside is possible with this element. The size of the permissible scan sector depends largely on the minimum allowed element gain. Since the element gain pattern has a null at broadside, this region is clearly restricted from the scan sector. Future studies could consider the development of electromagnetic simulation models and measurements of large arrays of loop-fed slotted cylinder antenna elements.

References

[1] Mailloux, R.J., "Phased Array Theory and Technology," *Proc. IEEE*, Vol. 70, No. 3, 1982, pp. 246-291.

[2] Kelly, E.J., and G.N. Tsandoulas, "A Displaced Phase Center Antenna Concept for Space Based Radar Applications," *IEEE Eascon*, 1983, pp. 141-148.

[3] Fenn, A.J., "Theoretical and Experimental Study of Monopole Phased Array Antennas," *IEEE Trans. Antennas Propagat.*, Vol. 34, No. 10, 1985, pp. 1118-1126.

[4] Fenn, A.J., "Arrays of Horizontally Polarized Loop-Fed Slotted Cylinder Antennas," *IEEE Trans. Antennas Propagat.*, Vol. 33, No. 4, 1985, pp. 375-382.

[5] R. C. Hansen, (ed.), *Microwave Scanning Antennas, Vol. II, Array Theory and Practice* New York: Academic, 1966, Ch. 3.

[6] Alford, A. and A. G. Kandoian, "Ultrahigh-Frequency Loop Antennas," *Trans. AIEE*, Vol. 59, 1940, pp. 843-848.

[7] Anderson, S.R., "VHF Omnirange Accuracy Improvements," *IEEE Trans. Aerospace and Navigational Elec.*, Vol. 12, No. 1, 1965, pp. 26-35.

[8] Sengupta, D.L., "Theory of V.O.R. Antenna Radiation Patterns," *Electron. Lett.*, Vol. 7, No. 15, 1971, pp. 418-420.

[9] Gupta, R.K., "On Radiation Properties of the Alford Loop and a Dipole Antenna," *Proc. Inst. Elec. Eng. India*, 1971, pp. 145-150.

14

Finite Arrays of Crossed V-Dipole Elements

14.1 INTRODUCTION

Dipole phased array antennas are commonly used in radar and communications applications, and these elements have been researched extensively in the literature [1-8]. Dipoles can be designed to radiate or receive linear polarization, dual polarization, or circular polarization. Dipoles typically have straight arms that are parallel to a conducting ground plane; however, for wide-angle pattern coverage for a single element, or wide-angle scanning for an array, the dipole arms can be swept downward toward the ground plane forming a V-shaped antenna [7-11]. The element gain pattern for finite phased arrays of dual-polarized crossed V-dipole (pronounced *Vee* dipole) antennas above a ground plane is addressed in this chapter, both with computer simulation and by measurements. The method of moments is used to compute the center element gain pattern of a finite array of V-dipoles. An experimental 19-element passively terminated planar array is described, and mutual coupling, center element scanned array reflection coefficient, and center-element gain pattern measurements are presented. The simulated data are shown to be in good agreement with measurements.

14.2 THEORY

As discussed in Chapter 8, it is known that for certain radiating elements blind spots [1] can occur in large planar phased-array antennas. In phased-array antenna applications, a blind spot is identified with an array element pattern null (minimum) or a scanned array reflection coefficient with amplitude equal to unity. For example, a blind spot has been observed in a balun-fed straight-arm dipole phased array [5, 6]. A center-fed straight dipole with balun is shown in Figure 14.1(a). The dipole has a half length denoted l, with strip width w, and is located a distance h above a conducting ground plane. The balun in this case is a balanced two-wire line that is shorted at the ground plane, which for a one-quarter wavelength transmission line, appears as an open-circuit at the feed terminals to restrict current flow on the balun. When this element is situated in a large phased array with the main beam steered away from broadside, the illumination of the balun from the array elements is not symmetrical, and unbalanced currents can flow on the balun. The presence of induced currents flowing on the balun can radiate and in some cases cancel the primary field of the dipole in a certain direction, creating the blind spot. This effect can be reduced or eliminated by tilting the dipole arms toward the ground plane at an angle α forming a V-shaped element [7-9] (see Figure 14.1(b)).

It can be inferred that the tilted arms of the V-dipole act to reduce the amplitude of currents that flow on the balun stubs. If this inference is true, then it can be assumed that the balun and feedline scattering can be ignored in a theoretical analysis of the element gain pattern.

The radiation pattern of a single V-dipole antenna in free space can be computed as follows. Schelkunoff and Friis have given an expression for the radiation pattern of a dipole arm with sinusoidal current distribution [10]. If it is assumed that the dipole arm is oriented in the z' direction, the radiation field pattern is expressed as

$$E_\theta = \frac{j30i_t}{\sin kl} e^{jkl\cos\theta'} \frac{j\cos\theta' \sin kl - \cos kl}{\sin\theta'} \tag{14.1}$$

where $k = 2\pi/\lambda$, and λ is the wavelength. From the diagram shown in Figure 14.2 for a V-dipole in free space, the pertinent angles θ'_L for the left half dipole and θ'_R for the right half of the V-dipole, in terms of the angles θ and α, are given by

$$\theta'_L = \frac{\pi}{2} - (\theta + \alpha) \tag{14.2}$$

and

$$\theta'_R = \frac{\pi}{2} - (\theta - \alpha) \tag{14.3}$$

The total electric field due to the left and right halves of the V-dipole antenna is given by

$$E_\theta = E_{\theta_L} + E_{\theta_R} \tag{14.4}$$

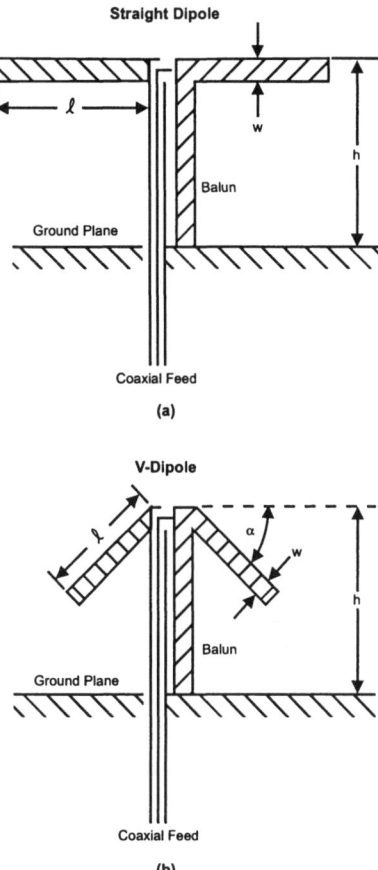

Figure 14.1 Balun-fed dipoles above a ground plane. (a) Straight arm, and (b) V-dipole. © 1988 IEEE [9].

Using (14.2) and (14.3) for the left and right halves of the V-dipole in (14.1) can be combined to yield the total E-field for the V-dipole in free space as

$$E_\theta = \frac{j30i_t}{\sin(kl)} \left(\frac{e^{jkl\sin(\theta+\alpha)} - j\sin(\theta+\alpha)\sin(kl) - \cos(kl)}{\cos(\theta+\alpha)} \right.$$
$$\left. + \frac{e^{-jkl\sin(\theta-\alpha)} + j\sin(\theta-\alpha)\sin(kl) - \cos(kl)}{\cos(\theta-\alpha)} \right) \quad (14.5)$$

Equation (14.5) is applied to near-field probe scanning in Chapter 12 (for example, see Figure 12.4(b)). If the V-dipole is located above a conducting

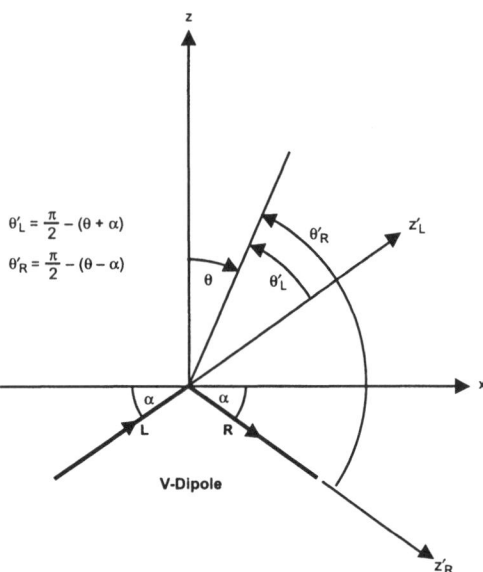

Figure 14.2 V-dipole element in free space.

ground plane, the method of images can be used to include an image V-dipole forming a two-element array in free space, from which the combined primary plus image V-dipole radiation pattern can be determined. The radiation patterns of an array of V-dipole antennas can be computed by using the appropriate array factor.

To include array mutual coupling effects, a general thin-wire moment method computer code [12] can be used to compute the V-dipole array element gain patterns. The moment method analysis assumes overlapping piecewise-sinusoidal expansion and test functions. For convenience, for the analysis in this chapter an infinite ground plane will be assumed. The method of images is used to replace the infinite ground plane with an image array of V-dipoles. To enforce the boundary condition of zero tangential electric field on the ground plane, the image dipoles are illuminated 180° out of phase with respect to the primary dipoles. An array of crossed V-dipoles can be analyzed using two independent arrays, as shown in Figure 14.3. In the moment method simulation model, the strip dipole arm (used in the experimental array) is approximated using a cylindrical dipole with diameter one-half the strip width [13]. As will be shown by comparison of simulations and measurements for the element gain pattern, the assumption of independent operating (orthogonal) arrays is valid. In other words, the orthogonal dipole in the experimental array has little influence on the primary dipole radiation pattern. The principally polarized dipole elements are defined here as being parallel to the xz plane, as depicted in Figure 14.3(a). The center principally polarized dipole (or xz dipole) is used to generate E-plane (E_θ component) patterns. The elements parallel to the yz plane are used to generate the orthogonal polarization, as depicted in Figure 14.3(b). The center orthogonally polarized dipole (or yz dipole) is used to produce H-plane (E_ϕ component) patterns. Both E- and H-plane patterns are computed and measured in the $\phi = 0°$ cut. Thus the dual-polarized coverage over a common scan sector is addressed in this chapter.

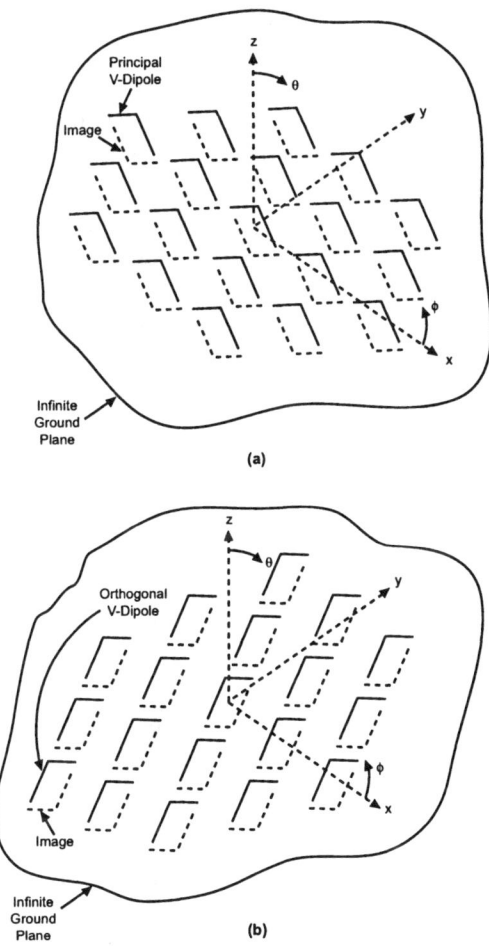

Figure 14.3 Models used to analyze element gain patterns in an array of V-dipoles. (a) E-plane model and (b) H-plane model. © 1988 IEEE [9].

Several infinite array analyses for dipole radiators with feedlines have been developed [6, 10, 11]. In [6], straight-arm dipole arrays with feedlines were analyzed in terms of the transverse magnetic (TM) feed region modes. It was shown that the dominant TM mode can go into cut-off, producing a blindspot, prior to the onset of grating lobes. This mode cut-off effect was presented in terms of the propagation constant of the TM mode. The theoretical results presented in [11] indicate the presence of blind spots in the active scan impedance for straight dipole elements with feedlines. For swept-back dipoles with feedlines it was demonstrated that a blind spot does not tend to occur. It is desirable to be able to predict the V-dipole element gain pattern in a finite array, including the effects of mutual coupling. Many phased-array antennas are designed to scan up to about 60° from broadside. The purpose of this chapter is to show that over this ±60° scan sector these predictions can be made accurately for V-dipoles using a simplified method of moments model. Only the current flowing on the dipole arms is included in the moment method model. Balun or feedline effects are not considered here, and the ground plane is assumed infinite. Crossed dipole elements are useful in generating circular or dual-linear polarization. Dual linearly polarized center-element gain patterns are considered here. Passively terminated arrays are convenient for determining the element gain pattern [5]. In this situation a single element is driven, usually the center element, and the surrounding elements are terminated in resistive loads.

14.3 DIPOLE ELEMENT PROTOTYPES DESCRIPTION

A dipole array element design for straight and V-dipoles is discussed in this section. The desired operating band of interest here is 1.2-1.4 GHz (15.4 percent bandwidth). At the center frequency (1.3 GHz), the dipole terminals are located approximately $\lambda/4$ above the ground plane. Both straight-arm and V-dipole elements were fabricated, for comparing their measured radiation patterns when each element is mounted above a conducting ground plane. A photograph of the crossed straight-arm dipole antenna used in these measurements is shown in Figure 14.4. The dipole arms ($l = 5.461$ cm, $w = 0.356$ cm, $h = 5.588$ cm, as in Figure 14.1(a)) were etched on a double-sided printed circuit (PC) board. Two of these PC boards are notched and interleaved, forming the desired crossed dipole antenna. The crossed dipoles are fed by two separate RG-141 (0.358 cm diameter) coaxial cables (with 50-ohm characteristic impedance), which provide independent orthogonally polarized signals. The balun/feedline is made from a section of RG-141 semi-rigid coaxial cable and a solid brass rod (0.358 cm diameter). Electrical connections at the pair of dipole terminals are facilitated by the use of a

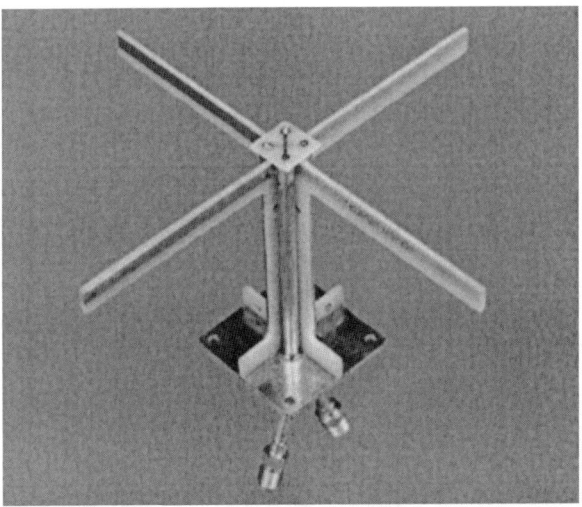

Figure 14.4 Photograph of a crossed straight-arm dipole antenna.

double-sided printed circuit board. The balun feedlines (stubs) form a two-wire balanced line with spacing 0.754 cm. For the given line diameter and spacing, the balun characteristic impedance is 165 ohms [14].

The crossed V-dipole element used in these experiments was fabricated in the same manner, but with the arms swept down toward the ground plane at a 45° angle, as shown in Figure 14.5 ($l = 5.461$ cm, $w = 0.356$ cm, $h = 5.588$ cm, $\alpha = 45°$, as in Figure 14.1(b)). Nineteen such V-dipoles were fabricated for array testing. The ground plane used for the single element and V-dipole array measurements was a 3.175 mm sheet of aluminum with dimensions 1.22m × 1.22m square.

14.4 MEASUREMENTS OF SINGLE STRAIGHT DIPOLE AND V-DIPOLE ELEMENTS ABOVE A GROUND PLANE

To compare dipole designs, straight-arm and crossed V-dipole antennas were mounted at the center of a 1.22m square aluminum ground plane, and linearly polarized radiation patterns were measured in an anechoic chamber. The measured E-plane and H-plane patterns are shown in Figures 14.6 and 14.7, respectively.

In the measured E-plane patterns (E_θ component) shown in Figure 14.6, as expected the V-dipole has a wider angular pattern coverage compared to the straight-arm dipole. For angles greater than about 45° from broadside, the

342 Adaptive Antennas and Phased Arrays for Radar and Communications

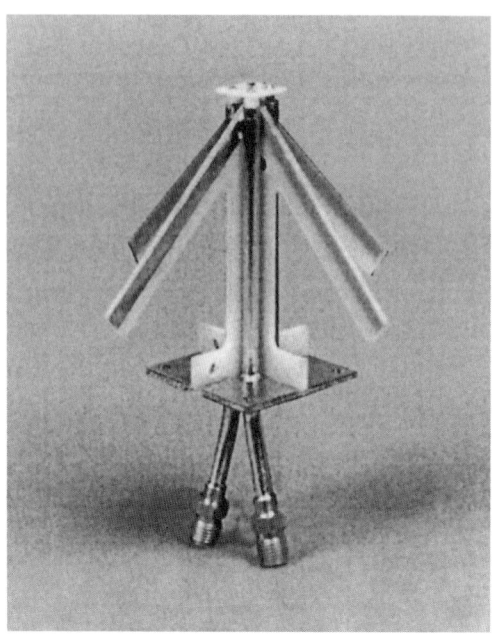

Figure 14.5 Photograph of a crossed V-dipole antenna. © 1988 IEEE [9].

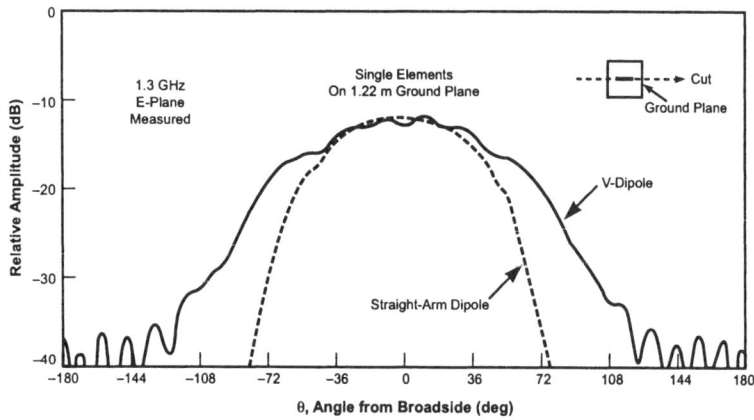

Figure 14.6 Measured E-plane radiation patterns for crossed straight-arm and crossed V-dipole antennas.

E-plane radiation patterns of the straight and V-dipoles differ significantly. For example, at an angle 60° from broadside, the V-dipole gain is about 5 dB higher than the straight dipole gain as observed in Figure 14.6. This wider pattern coverage should allow wider scanning coverage compared to

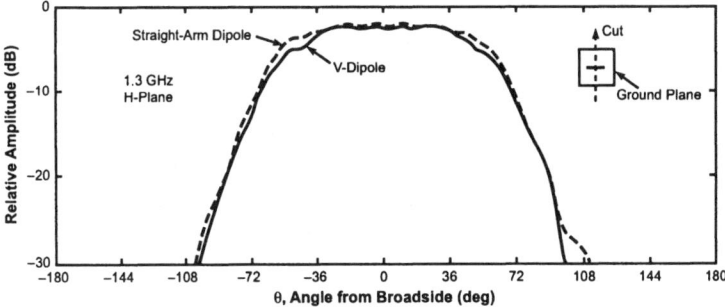

Figure 14.7 Measured H-plane radiation patterns for crossed straight-arm and crossed V-dipole antennas.

the straight-arm dipole, which is a motivating factor for tilting the dipole arms. In the measured H-plane patterns (E_ϕ component) shown in Figure 14.7, the radiation patterns of the straight-arm and V-dipole single elements are approximately the same.

14.5 V-DIPOLE ARRAY DESCRIPTION

A layout for the finite array of 19 V-dipole antenna elements on a 1.22m square ground plane is shown in Figure 14.8. For wide-angle scanning the V-dipole element spacing in the array is chosen as 12.7 cm (corresponding to 0.55λ at the center frequency of 1.3 GHz) on a hexagonal lattice. This lattice spacing allows scanning without the formation of grating lobes in any plane out to 60° from broadside. A photograph of the array of 19 V-dipoles is shown in Figure 14.9. For center element gain pattern measurements and moment method simulations, the center element is driven and 50-ohm resistive loads are assumed at the terminals of each surrounding passively terminated dipole element. For array mutual coupling (S parameter) measurements, the center element is driven and all elements except the receive dipole element are terminated in 50-ohm resistive loads.

14.6 MEASURED ARRAY MUTUAL COUPLING RESULTS

Center element return loss and mutual coupling data were measured for the 19-element V-dipole array, and these data are given in Figure 14.10. At 1.3 GHz (center frequency) the measured return loss ($20\log_{10}|S_{11}|$) was -8.5 dB. In the E-plane ($\phi = 0°$), the measured coupling to the nearest

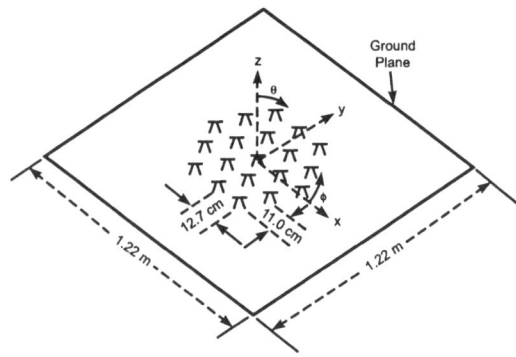

Figure 14.8 Layout for 19-element crossed V-dipole array. © 1988 IEEE [9].

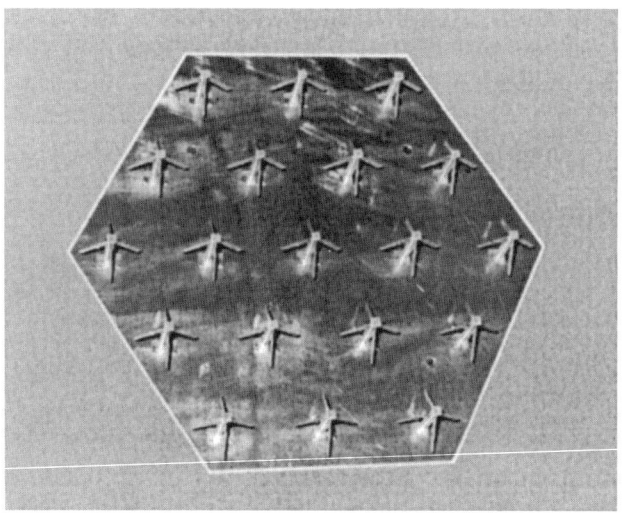

Figure 14.9 Photograph of 19-element crossed V-dipole array.

neighbor was −25.5 dB. In the H-plane, the measured coupling to the nearest neighbor was −19.7 dB. These measured complex mutual coupling data can be summed according to (8.16) to compute the array scan reflection coefficient for the center element versus scan angle in both E- and H-planes, as depicted in Figure 14.11. Based on these calculations, the array scan mismatch loss ranges from about 0.5 dB to 1 dB over scan angles from 0° to 60°, respectively.

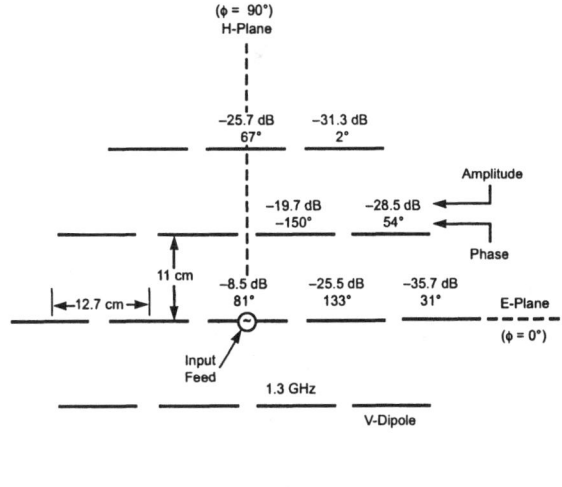

Figure 14.10 Measured return loss and mutual coupling for 19-element crossed V-dipole array.

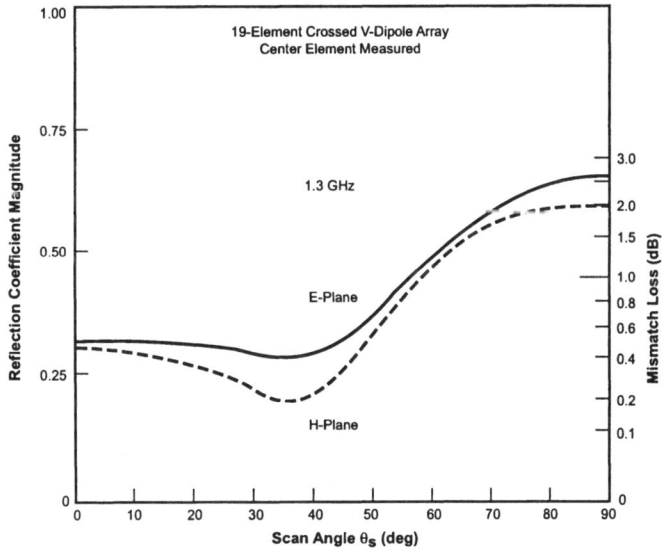

Figure 14.11 Calculated scan reflection coefficient versus scan angle as computed from measured mutual coupling data (Figure 14.10) for the 19-element crossed V-dipole array.

14.7 CENTER ELEMENT GAIN PATTERN MEASUREMENTS AND SIMULATIONS

A study was made, initially, to determine in the moment method model the required number of piecewise-sinusoidal expansion functions per dipole to obtain converged element gain patterns in the array environment. It was found that three unknowns per dipole are sufficient for pattern convergence. This demonstration of convergence is shown in Figure 14.12, where the E-plane center element gain pattern is computed at 1.3 GHz for one and three unknowns per dipole. In comparing these two curves over ±60° from broadside, the maximum amplitude difference is 0.25 dB. For the 19-element array with three unknowns per element and including the image array, there are 114 unknowns for the moment method array currents.

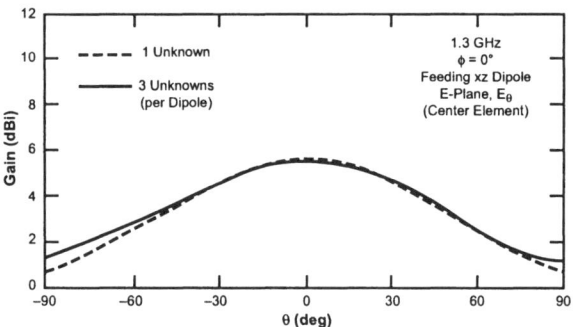

Figure 14.12 Moment method calculated center element gain patterns in the E-plane ($\phi = 0°$) with one and three piecewise-sinusoidal current expansion functions per dipole element for the 19-element crossed V-dipole array. © 1988 IEEE [9].

Measured and theoretical center element gain patterns, in the passively terminated array, at 1.2, 1.3, and 1.4 GHz are presented in Figures 14.13 (E-plane) and 14.14 (H-plane). These patterns are taken in the $\phi = 0°$ cut for the E-plane and $\phi = 90°$ for the H-plane. The measured and theoretical data are represented by solid and dashed curves, respectively. The patterns indicate peak gain (approximately 5 to 6 dBi) occurring in the vicinity of broadside. For any particular pattern cut the measurements show no more than 4 dB amplitude taper at 60° from broadside. Also, there is no evidence of a blind spot. As was mentioned earlier, many phased arrays are designed to scan at most 60° from broadside, and for this sector in both E- and H-planes the theoretical model is generally in good agreement with the measurements. Over the 15.4% bandwidth shown, the theoretical peak gains agree to within 1 dB of the experimental data. The maximum amplitude variation between

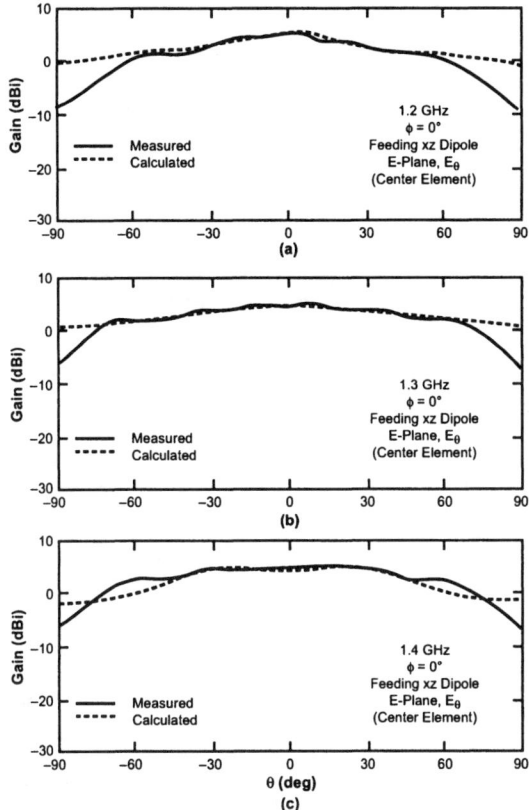

Figure 14.13 Measured and computed center element gain patterns for 19-element V-dipole array. Feeding xz dipole, E-plane, $\phi = 0°$ cut. (a) 1.2 GHz, (b) 1.3 GHz, and (c) 1.4 GHz. © 1988 IEEE [9].

theory and experiment is 1.5 dB over ±50° in the E-plane. In the vicinity of 60° from broadside in the E-plane, at 1.4 GHz, the difference between theory and experiment is 2.5 dB. The maximum amplitude difference in the H-plane at 1.4 GHz is 2.5 dB and occurs near $\theta = 50°$. Increased accuracy of the calculated pattern could possibly be achieved by incorporating finite ground plane edge diffraction effects into the moment method formulation [15].

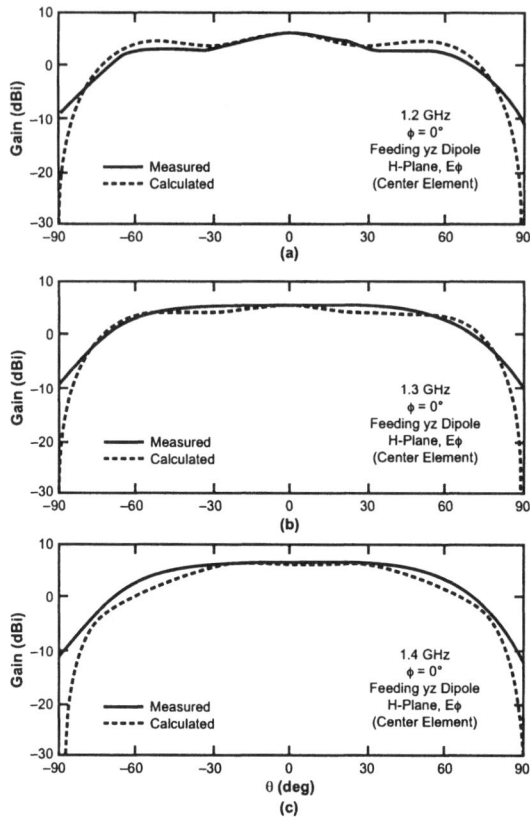

Figure 14.14 Measured and computed center element gain patterns for 19-element V-dipole array. Feeding yz dipole, H-plane, $\phi = 0°$ cut. (a) 1.2 GHz, (b) 1.3 GHz, and (c) 1.4 GHz. © 1988 IEEE [9].

14.8 SUMMARY

This chapter has addressed the element gain patterns and scan reflection coefficient for V-dipole array elements. An approach for calculating the element gain pattern for finite arrays of dual-polarized crossed V-dipoles above an infinite ground plane was presented. The theoretical model assumes that currents flow only on the dipole arms and not on the feedlines. A general thin-wire method of moments code was used to analyze the theoretical model. To verify the theory, a prototype array was built and tested. Good agreement between the theory and measurements was demonstrated. Based on the measured data, there is no evidence of a blind spot for this phased-array radiator design. The assumption that the balun feed can be ignored in the theoretical model was confirmed. Such an assumption is valid only when a blind spot associated with the balun is not present.

References

[1] Allen, J., "Gain and Impedance Variation in Scanned Dipole Arrays," *IEEE Trans. Antennas and Propagat.*, Vol. 10, No. 5, 1962, pp. 566-572.

[2] Hansen, R.C., (ed.), *Microwave Scanning Antennas, Vol. 2*, Los Altos, CA: Peninsula Publishing, 1985, pp. 162-170, 301-308.

[3] Hansen, R.C., *Phased Array Antennas*, New York: Wiley, 1998, pp. 273-303.

[4] Kraus, J.D., *Antennas,* 2nd ed., New York: McGraw-Hill, 1988, pp. 435-496.

[5] Mailloux, R.J., "Phased Array Theory and Technology," *Proc. IEEE*, Vol. 70, No. 3, 1982, pp. 246-291.

[6] Mayer, E.D., and A. Hessel, "Feed Region Modes in Dipole Phased Arrays," *IEEE Trans. Antennas Propagat.*, Vol. 30, No. 1, 1982, pp. 66-75.

[7] Stark, L., "Comparison of Array Element Types," in *Phased Array Antennas (Proc. 1970 Phased Array Antenna Symp.)*, A.A. Oliner and G.H. Knittel, (eds.), Dedham, MA: Artech House, 1972, pp. 51-66.

[8] Knittel, G.H., (Brookner, E., ed.), Chapter 21 in *Radar Technology*, Dedham, MA: Artech House, 1977, pp. 289-301.

[9] Fenn, A.J., "Element Gain Pattern Prediction for Finite Arrays of V-Dipole Antennas over Ground Plane," *IEEE Trans. Antennas Propagat.*, Vol. 36, No. 11, 1988, pp. 1629-1633.

[10] Schelkunoff, S.A., and H.T. Friis, *Antennas: Theory and Practice*, New York: Wiley, 1952, pp. 499-502.

[11] Schuman, H.K, D. R. Pflug, and L. D. Thompson, "Infinite Planar Arrays of Arbitrarily Bent Thin Wire Radiators," *IEEE Trans. Antennas Propagat.*, Vol. 32, No. 4, 1984, pp. 364-377.

[12] Richmond, J.H., "Radiation and Scattering by Thin-Wire Structures in a Homogeneous Conducting Medium, Computer Program Description," *IEEE Trans. Antennas Propagat.*, Vol. 22, No. 2, 1974, p. 365.

[13] Hallen, E., *Electromagnetic Theory*, New York: Wiley, 1962, p. 62.

[14] Jasik, H., *Antenna Engineering Handbook,* New York: McGraw-Hill, 1961, p. 30-3.

[15] Thiele, G.A., and T. H. Newhouse, "A Hybrid Technique for Combining Moment Methods with the Geometrical Theory of Diffraction," *IEEE Trans. Antennas Propagat.*, Vol. 23, No. 1, 1975, pp. 62-69.

15

Experimental Ultrawideband Dipole Antenna Array

15.1 INTRODUCTION

Ultrawideband phased-array antennas are of interest for a number of radar and communications applications [1]. An article by Hansen [2] has presented a theoretical analysis of short, thin dipole elements in a 5:1 bandwidth two-dimensional phased array. In this chapter, a design for a ultra-high frequency band (UHF) broadside-scanned linearly polarized, ultrawideband dipole array antenna for receive and/or transmit applications is briefly investigated. The ultrawideband array design presented here makes use of multiple large-diameter cylindrical tubular dipole elements that are closely spaced. Section 15.2 describes the design and fabrication for a linearly polarized prototype 3×12 dipole array antenna (0.61m $\times$ 3.05m) that consists of three columns of dipoles, with 12 dipoles per column, mounted above a reflector formed by three conducting tubes. Section 15.3 discusses measured results for this array.

15.2 ULTRAWIDEBAND DIPOLE ARRAY DESIGN

The diagram in Figure 15.1 depicts the conceptual development of an ultrawideband dipole array that is investigated here. As depicted in this figure, a single thin dipole antenna in free space tends to be narrowband and radiates to the left and right. A single thick dipole antenna in free space tends to have wideband performance [3] and radiates to the left and right. Adding a ground plane or reflector a suitable distance to the left of the thick wideband dipole decreases radiation to the left and increases the radiation to the right. Adding dipole elements in a closely spaced collinear fashion can improve the tuning of the dipoles [2]. Multiple columns of collinear arrays can then be included to increase the aperture size and gain as desired. A single collinear array could be used as a focal line feed for a parabolic cylinder reflector to produce a large aperture size.

Ultrawideband array performance can be achieved by providing dipole arms with a large-diameter conducting tube and central feed arrangement as shown conceptually for one dipole element in Figure 15.2. Close spacing between the tips of the dipoles in the array can be used to extend the bandwidth of a single isolated dipole element.

At the upper end of the antenna frequency band, the maximum center-to-center electrical spacing d_{max} between the dipoles in an array is chosen to avoid grating lobes according to the following condition (see (8.49)):

$$\frac{d_{max}}{\lambda_h} \leq \frac{1}{1 + \sin \theta_s} \qquad (15.1)$$

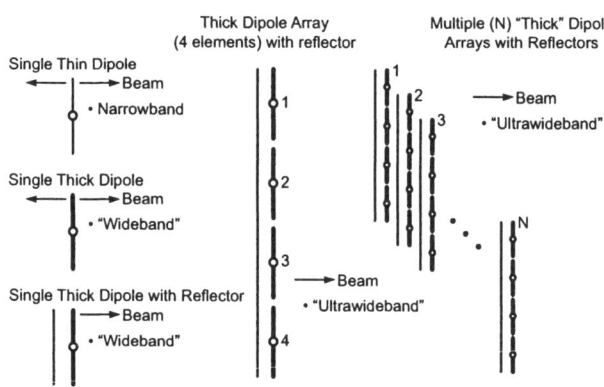

Figure 15.1 Conceptual development of an ultrawideband dipole array.

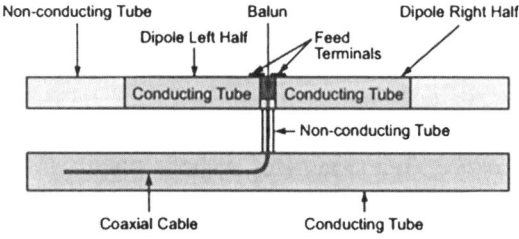

Figure 15.2 Conceptual ultrawideband dipole array element details.

where λ_h is the wavelength at the highest frequency and θ_s is the scan angle from broadside. For the array discussed in this chapter, for broadside scan $\theta_s = 0°$, so from (15.1), $d_{max} \leq \lambda_h$. The frequency is computed from the wavelength, λ, as

$$f = c/\lambda \qquad (15.2)$$

where the speed of light is $c = 3 \times 10^8$ meters per second. Thus, if the dipole spacing is chosen to be 25.4 cm, the upper frequency at which a grating lobe will occur is computed from (15.1) to be 1.18 GHz.

At the lower end of the operating band for an ultrawideband antenna, the return loss tends to degrade leading to a reduced antenna gain. If the array elements become electrically short, the resistive loss of the dipole antenna elements can be significant (compared to the antenna radiation resistance) and the antenna gain is degraded [3, pp. 43-48].

A dipole array antenna can be fabricated conveniently by supporting all of the dipole arms in the array with dielectric (nonconducting) tubes. At UHF frequencies, the dielectric support tube (such as fiberglass, phenolic, or PVC) has little effect on the dipole radiation. Each dipole half can be formed by positioning a conducting tube over the nonconducting support tube or by attaching, and overlapping, copper tape with conductive adhesive to the outside of the nonconducting tube. An array can be fabricated readily by using this approach. A coaxial feedline at right angles to the dipole can penetrate the dielectric support tube for the dipole and can then connect to a transformer balun to feed the dipole terminals in a balanced manner.

A conducting tube mounted parallel to each column of the dipole array can be used as a reflector to provide mechanical support for the array and some amount of front-to-back ratio. A flat conducting ground plane could be used to provide an increased front-to-back ratio. The conducting tube (reflector) can contain multiple dipole coaxial cable feedlines from other collinear array elements and, in this manner, scattering from feedlines is reduced or avoided completely. A series of nonconducting tubes (perpendicular to the dipoles) can be used to provide mechanical support and maintain a desired electrical spacing between the dipole arms and the conducting ground plane tube. A detailed antenna design of a 4-element subarray (empirically determined) for UHF operation is shown in Figure 15.3. This subarray can be considered a building block for fabricating larger apertures. The 25.4 cm element spacing was chosen to allow good performance at the low end of the band while avoiding grating lobes and high sidelobes at the high end of the band.

The 3×12 prototype dipole array was fabricated by attaching copper tape (dipole half-length 10.795 cm with a 1.27 cm feed gap and 2.54 cm dipole tip-to-tip spacing) on dielectric tubing (4.128 cm OD). The reflector,

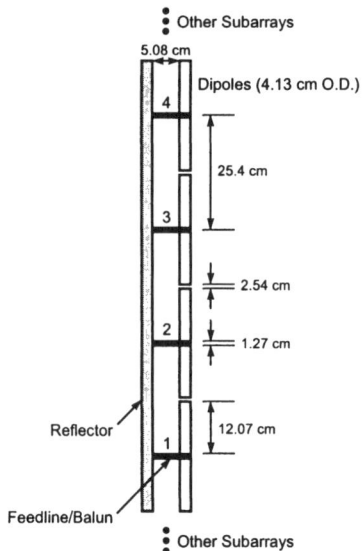

Figure 15.3 Design details for an ultrawideband dipole subarray for UHF operation. (Not to scale.)

which was mounted parallel to the dipoles was fabricated using 3.81 cm square aluminum tubing. Wideband transformer baluns (50 ohm unbalanced to 50 ohm balanced) were used to provide a balanced feed for the terminals of the dipole antenna elements. The typical insertion loss of the transformer balun is about 0.5 dB in the 400- to 500-MHz band, and increasing to about 1.25 dB insertion loss at 1000 MHz. Flexible coaxial cable (0.358 cm outside diameter) with nominal attenuation per foot of 0.085 dB at 500 MHz and 0.122 dB at 1000 MHz, was used to interconnect the dipole baluns to the beamforming circuit. Analog beamforming was accomplished by using four-way in-phase power combiners (insertion loss 0.35 dB at 500 MHz and 0.5 dB at 1000 MHz) to coherently combine 4 dipole elements in a subarray, and then three-way in-phase power combiners (insertion loss 0.6 dB at 500 MHz, 1.2 dB at 1000 MHz) are used to combine the three 4-element subarrays into a 12-element collinear array. Three parallel collinear arrays, with 25.4 cm center-to-center (parallel) spacing, are combined using three-way power combiners. The overall aperture size is 0.61m by 3.05m.

A photograph of one ultrawideband 12-element dipole linear array prototype (one column) is shown in Figure 15.4. For this prototype, the in-phase power combiners were mounted on plates located on the backside of the reflector pole. The dipole array was mechanically supported by six nylon tubes secured by fiberglass threaded rods. A more detailed view of the array prototype is shown in Figure 15.5.

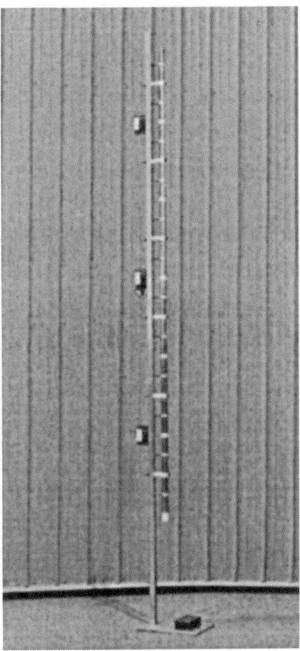

Figure 15.4 Photograph of a single ultrawideband dipole linear array. The array is 3.05 cm long and is composed of three subarrays with four dipoles each.

Figure 15.5 Close-up photograph of a four-element subarray section of the ultrawideband dipole linear array shown in Figure 15.4.

15.3 MEASURED RESULTS FOR PROTOTYPE ULTRAWIDEBAND ARRAY

Return loss measurements of two elements – a central element (element number 6) in the center column, and an edge element (element number 12) in the outer column) in the 3 × 12 dipole array environment – are shown in Figure 15.6. For this measurement, the surrounding dipole array elements were terminated in 50 ohm resistive loads. Both the central dipole element (solid curve) and the edge dipole element (dashed curve) have broadband coverage. The central element return loss has an improved broadband return loss compared to the edge dipole element. This improvement in return loss is due to the tuning effect of array mutual coupling.

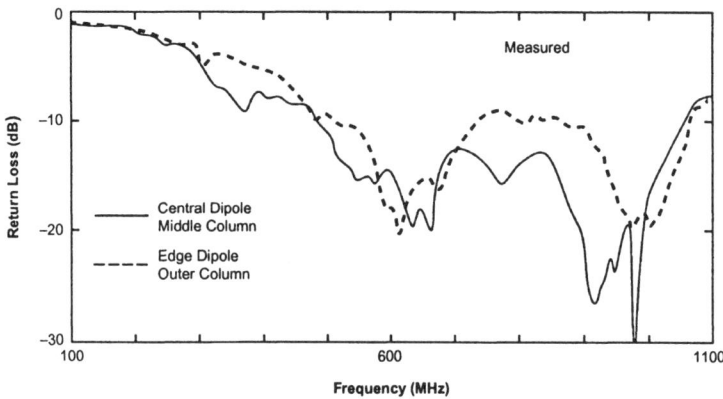

Figure 15.6 Measured return loss for a central dipole element in the center column and an edge element in the outer column of the 3 × 12 UHF ultrawideband dipole array (see Figure 15.7).

The prototype ultrawideband UHF dipole array was mounted on an antenna positioner (large foam column on an azimuth turntable) in the compact range of the MIT Lincoln Laboratory RF System Test Facility [4, 5], as shown in Figure 15.7. This compact range facility utilizes a 7.3m × 7.3m rolled edge reflector to generate a 3.7m × 3.7m plane-wave test zone covering the frequency range from 400 MHz to 100 GHz.

The horizontally polarized antenna radiation patterns of the 3 × 12 dipole array were measured, and the results at 400 MHz, 650 MHz, and 900 MHz are shown in Figure 15.8. At all three frequencies shown, a broadside beam is formed at 0° and the first sidelobes are on the order of −13 dB as expected for uniform illumination. The half-power beamwidth of the array radiation pattern narrows as the frequency increases, as expected.

Figure 15.7 Photograph of the 3 × 12 UHF ultrawideband dipole array under test in a compact range anechoic chamber [5].

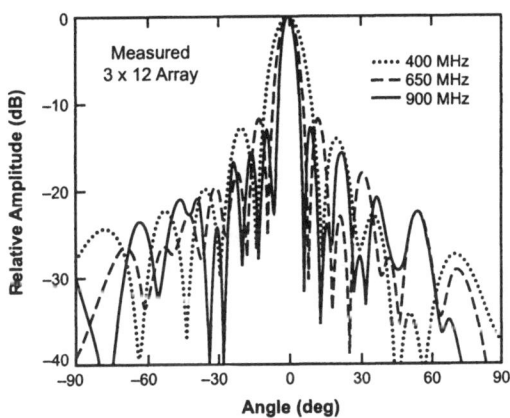

Figure 15.8 Measured radiation patterns for the 3 × 12 UHF ultrawideband dipole array.

15.4 SUMMARY

An ultrawideband dipole array antenna for broadside scan has been described in this chapter. The ultrawideband array elements are large diameter tubular dipoles. Measured data for this prototype array indicate good performance. Large arrays can be fabricated using this ultrawideband antenna design approach.

References

[1] Taylor, J.D., (ed.), *Ultra-Wideband Radar Technology,* Boca Raton, FL: CRC Press, 2001.

[2] Hansen, R.C., "Dipole Array Scan Performance over a Wide-Band," *IEEE Trans. on Antennas and Propag.*, Vol. 47, No. 5, 1999, pp. 956-957.

[3] Stutzman W.L., and G.A. Thiele, *Antenna Theory and Design,* 2nd ed., New York: Wiley, 1998, pp. 172-173.

[4] Fenn, A.J., M.W. Shields, and G.A. Somers, "Introduction to the New MIT Lincoln Laboratory Suite of Ranges," *26th Annual Antenna Measurements Techniques Association Meeting and Symposium*, Atlanta, GA, 2004, pp. 37-41.

[5] Shields, M.W., and A.J. Fenn, "A New Compact Range Facility for Antenna and Radar Target Measurements," *Lincoln Laboratory Journal*, Vol. 16, No. 2, 2007, pp. 381-391.

16

Finite Rectangular Waveguide Phased Arrays

16.1 INTRODUCTION

In large planar arrays, the majority of the elements behave similar to one another in terms of their reflection coefficients and element gain patterns. However, the element characteristics at the array edges or up to approximately five elements from the edge can differ substantially from the inner core elements. It is known that important characteristics of large planar arrays (on the order of 1000 elements or greater) can be determined by using an infinite array model whose elements exhibit uniform behavior throughout the array as described by Stark [1]. However, as the large finite array becomes smaller (on the order of 100 elements or less), the infinite array model tends to be less accurate. This is due to the mutual coupling characteristics of elements in the array being strongly dependent on their location. In small finite arrays, most of the elements are only a few elements away from the array edge, and the radiation characteristics will vary significantly from element to element.

The behavior of small finite planar arrays of waveguides has been investigated by a number of authors. Borgiotti [2] obtained a mutual admittance expression between two identical radiating apertures in the form of a Fourier transform of a function related to the power radiation pattern of the element. Mailloux [3] used the method of moments and a single-mode approximation to the aperture field to analyze the coupling between two closely spaced open-ended waveguide slots. Mailloux [4] also analyzed the coupling between collinear open-ended waveguide slots by expanding the aperture field in a Fourier series. Bird [5] has analyzed the coupling between rectangular waveguides of different sizes. An approximation to the behavior of a finite array of rectangular waveguides was obtained by Amitay, Galindo, and Wu [6] by using infinite array techniques with finite array excitation. Wu also analyzed a finite array of parallel plate waveguides [7]. Cha and Hsiao [8] and Hidayet [9] investigated the reflection coefficients of planar waveguide arrays of size up to 13×13 for a rectangular grid. Hidayet also analyzed a 7×7 planar array arranged in triangular grid. Luzwick and Harrington [10] investigated the coupled power between elements of 7×7 rectangular and triangular grid arrays. Wang [11] investigated the measured E-plane radiation patterns for the center element of a 3×41 array of rectangular waveguides and performed an infinite array analysis of both rectangular and ridged waveguide arrays.

In ideal array theory [12], the radiation patterns of finite arrays can be analyzed by neglecting the mutual coupling between array elements. When mutual-coupling effects are included, a more accurate calculation of the array radiation patterns can be accomplished. To compute the array scan impedance or array scan reflection coefficient, mutual coupling between array elements must be taken into account.

This chapter considers the effects of mutual coupling on the scan reflection coefficient in finite planar arrays of rectangular waveguide-fed apertures. The elements are arranged in a rectangular grid and radiate through an infinite ground plane. The aperture distribution of each array element is found by using the equivalence principle and the method of moments in a manner similar in part to that discussed by Harrington and Mautz [13]. The aperture distribution is used to obtain the scattered field in each waveguide from which the aperture reflection coefficients are computed as investigated by this author [14, 15]. The formulation is valid for arrays of arbitrary size ($P \times Q$ elements). The element reflection coefficients are computed for an 11×11 phased array and compared against published data. The results are shown to be a good agreement for rectangular waveguide elements in both E-plane and H-plane scans - these results are obtained by using singly polarized magnetic current expansion functions. Yavuz [16] analyzed the reflection coefficient and radiation pattern of 5×5 finite arrays of rectangular waveguides. Grassi et al. [17] analyzed 11×11 finite rectangular waveguide arrays using a new generalized scattering matrix approach, and their results for the active reflection coefficient in E- and H-plane scans were in good agreement with the results in [14, 15]. For square waveguide elements, the present technique, as implemented with singly polarized current expansion functions, is only practical for H-plane scans. For E-plane scans with square or large rectangular apertures, higher order modes are present as discussed by Diamond [18] and Stark [19], which requires the addition of orthogonally polarized expansion functions.

16.2 ARRAY FORMULATION

A planar phased array of rectangular waveguide-fed apertures in an infinite ground plane is shown in Figure 16.1. The elements are spaced in a rectangular grid with P rows and Q columns. The array elements have H-plane length L and E-plane width W.

The waveguides are assumed to be excited by sources that are located far from the array aperture and generate the dominant transverse electric (TE_{10}) mode. To achieve beam steering, constant progressive phases ψ_x and ψ_y are applied to the sources in the x and y directions, respectively. The steering phases are related to the scan direction (θ, ϕ) by

$$\psi_x = \frac{2\pi}{\lambda} d_x \sin\theta \cos\phi \qquad (16.1)$$

$$\psi_y = \frac{2\pi}{\lambda} d_y \sin\theta \cos\phi \qquad (16.2)$$

where λ is the wavelength, d_x is the element spacing in the x direction, and d_y is the element spacing in the y direction. When $\phi = 0°, 180°$, scanning is done in the H-plane. E-plane scanning occurs when $\phi = \pm 90°$.

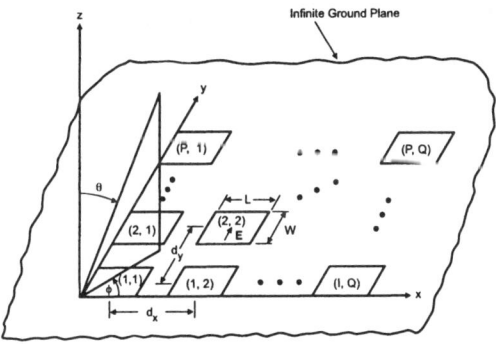

Figure 16.1 Finite phased array of rectangular waveguides with P rows and Q columns. © 1982 IEEE [15].

The formulation of the waveguide array problem utilizes the equivalence principle to remove the interaction between sources in the waveguide and half-space regions. That is, the waveguide and half-space regions are each characterized by an admittance matrix independent of one another. Boundary conditions and the method of moments are used to obtain a set of simultaneous equations involving the unknown aperture distributions [13]. The first step in the analysis is to cover each array element aperture with a perfect electric conductor (the array plane is now an infinite ground plane). Thus, coupling cannot occur between the waveguide and half-space regions. Next, to satisfy the boundary condition that the tangential component of the electric field be continuous across each element aperture, equivalent magnetic surface currents M_{sj} and $-M_{sj}$ are placed on the waveguide and half-space sides of the jth aperture cover, respectively. The equivalent magnetic surface current M_{sj} is related to the electric field E_j in the jth aperture by the relation

$$M_{sj} = \hat{n} \times E_j, \quad j = 1, 2, \cdots, N_a \qquad (16.3)$$

where $\hat{n}$ is the unit normal to the array aperture, and N is the number of array elements. The tangential component of the magnetic field must also be continuous across each element aperture. The magnetic field in the ith waveguide of the array has two sources. The first source, denoted by $H_{ti}^{wg}(M_{si})$, is the tangential component of the magnetic field radiated by the source M_{si} on the waveguide side of the ith aperture cover. The second source, denoted by H_{ti}^{inc}, is the tangential component of the incident magnetic field on the waveguide side of the ith aperture cover. On the half-space side of the ith aperture cover, there is a contribution to the magnetic field from each of the array element sources $-M_{sj}, j = 1, 2, \cdots, N_a$, which radiate in the presence of a perfect electric conductor covering the entire xy plane. Let this half-space contribution be denoted by $H_{tij}^{hs}(-M_{sj})$. Setting the contributions to the tangential magnetic field (of the ith element of the array) on the waveguide side equal to that of the half-space side results in the following set of equations:

$$H_{ti}^{wg}(M_{si}) + H_{ti}^{inc} = \sum_{j=1}^{N_a} H_{tij}^{hs}(-M_{sj}) \qquad (16.4)$$

where $i = 1, 2, \cdots, N_a$.

Rearranging (16.4) and using the linearity of the H_t^{hs} operator yields

$$H_{ti}^{wg}(M_{si}) + \sum_{j=1}^{N_a} H_{tij}^{hs}(M_{sj}) = -H_{ti}^{inc} \qquad (16.5)$$

Equation (16.5) is the operator equation that is used to find the unknown magnetic currents M_s, in each aperture of the array and can be solved by the method of moments. An approximate solution for M_s can be found by using the basis function expansion:

$$M_{sj} = \sum_{n=1}^{N_e} V_{jn} \frac{M_n}{K(n)} \qquad (16.6)$$

where M_n is the basis function (with units of volts/meter), $K^{(n)}$ is a normalization factor for M_n (in volts), V_{jn} is a complex coefficient (with units of volts) to be determined, and N_e is the number of expansions per aperture. One possible basis function expansion for a square waveguide element is shown in Figure 16.2. Two orthogonally polarized sets of overlapping piecewise sinusoidal-uniform surface patches are used to model the various possible waveguide modes that can exist during general scans of large waveguide elements. A typical piecewise sinusoidal-uniform basis function, with length $2l$ and width w, as shown in Figure 16.3, can be expressed in arbitrary (u, v) coordinates as

$$M_n = \hat{u} K_n \frac{\sin \beta(l - |u|)}{\sin \beta l} \quad (16.7)$$

where $-l < u < l$ and $0 < v < w$, $\hat{u}$ is the vector orientation of the basis function, $\beta = 2\pi/\lambda$, and K_n is a constant. This choice of basis function (with the same weighting function) leads to mutual coupling calculations involving only a single integration, as will be discussed later. The normalization factor in (16.6) for the piecewise sinusoidal uniform basis function is found by integrating (16.7) over the width of the patch at $u = 0$, which yields $K^{(n)} = w K_n$.

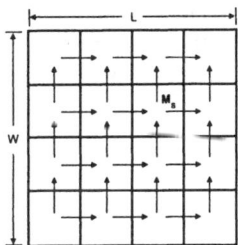

Figure 16.2 Electromagnetic modeling of large waveguide elements ($L > 0.6\lambda$) with two orthogonally polarized sets of overlapping piecewise sinusoidal-uniform magnetic surface sources. © 1982 IEEE [15].

The particular example shown in Figure 16.2 has 24 unknowns per array element, which would allow an analysis of an array of hundreds of waveguide elements based on current personal computer (PC) memory capability. For the case of an E-plane scan of square waveguide elements it has been shown that higher order modes are excited (e.g., TE_{11}, TM_{11}, TE_{20}, TE_{12}, TM_{12}) [18], in which case the earlier expansion is essential. However, for phased arrays with small rectangular waveguide elements ($L < 0.6\lambda$), a singly polarized basis function should be all that is necessary to model the problem where the TE_{10} mode is dominant. Figure 16.4 shows a set of adjacent piecewise

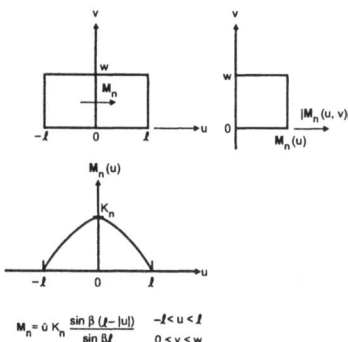

Figure 16.3 A typical piecewise sinusoidal-uniform basis function for the waveguide aperture expansion shown in Figure 16.2. © 1982 IEEE [15].

sinusoidal-uniform basis functions that cover the waveguide aperture. The piecewise sinusoidal function is used to closely approximate the cosine shape of the TE$_{10}$ mode. In the E-plane of the waveguide aperture, the adjacent uniform distributions are a reasonable choice because they can model a smooth distribution in the center portion of the aperture and provide a stepwise approximation to the singularity at the aperture edge. The edge singularity tends to be prominent for elements near the array edge and diminishes for closely spaced elements near the center of the array, as will be demonstrated later in this chapter. Noting in Figure 16.4 that the desired polarization vector is x-directed, (16.7) becomes

$$M_n = \hat{x} K_n \frac{\sin \beta(l - |x|)}{\sin \beta l} \qquad (16.8)$$

where $-l < x < l$ and $0 < y < w$.

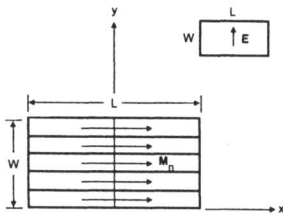

Figure 16.4 Electromagnetic modeling of small waveguide elements ($0.5\lambda < L < 0.6\lambda$) with a set of adjacent piecewise sinusoidal-uniform magnetic surface sources. © 1982 IEEE [15].

Having defined the basis function expansion, (16.6) is now substituted into (16.5) and using linearity of the H_{ti}^{wg} and H_{tij}^{hs} operators yields

$$\sum_{n=1}^{N_e} V_{in} \frac{H_{ti}^{wg}(M_n)}{K^{(n)}} + \sum_{n=1}^{N_e} \sum_{j=1}^{N_a} V_{jn} \frac{H_{tij}^{hs}(M_n)}{K^{(n)}} = -H_{ti}^{inc} \quad (16.9)$$

where $i = 1, 2, \cdots, N_a$.

A Galerkin's method [20] solution requires taking the inner product of (16.9) with normalized testing (or weighting) functions that are equal to the normalized basis functions. Define the inner product to be the integral over the aperture as

$$< M, H > = -\int\int M \cdot H \, ds \quad (16.10)$$

Denoting the normalized testing function as

$$W_m = \frac{M_m}{K^{(m)}} \quad m = 1, 2, \cdots, N_e \quad (16.11)$$

and performing the earlier inner product between (16.9) and (16.11) results in the following set of simultaneous equations:

$$\sum_{n=1}^{N_e} V_{in} \frac{<M_m, H_{ti}^{wg}(M_n)>}{K^{(m)} K^{(n)}} + \sum_{n=1}^{N_e} \sum_{j=1}^{N_a} V_{jn} \frac{<M_m, H_{tij}^{hs}(M_n)>}{K^{(m)} K^{(n)}}$$

$$= -\frac{<M_m, H_{ti}^{inc}>}{K^{(m)}} \quad (16.12)$$

$i = 1, 2, \cdots, N_a, m = 1, 2, \cdots, N_e$.

The first summation on the left side of (16.12) contains a normalized inner product denoted as

$$Y_{mn,i}^{wg} = \frac{<M_m, H_{ti}^{wg}(M_n)>}{K^{(m)} K^{(n)}} \quad (16.13)$$

which has units of Siemens (amperes/volt, sometimes referred to as mhos). This quantity will be referred to as the waveguide admittance (in the ith aperture) between basis function M_n and testing function M_m. The calculation of $Y_{mn,i}^{wg}$ is done in the presence of the waveguide walls and a perfect electric conductor covering the ith aperture. The second summation contains another normalized inner product denoted as

$$Y_{mn,ij}^{hs} = \frac{<M_m, H_{tij}^{hs}(M_n)>}{K^{(m)} K^{(n)}} \quad (16.14)$$

which also has units of Siemens. This quantity represents the mutual admittance in the half-space region (in the presence of a perfect electric conductor) between basis function M_n and testing function M_m. Note that in the half-space region, coupling occurs between the ith and jth apertures, $i = 1, 2, \cdots, N_a$ and $j = 1, 2, \cdots, N_a$.

However, for the waveguide region, coupling only occurs between basis and testing functions sharing the ith aperture – this is due to the effective isolation between waveguides due to the perfect electric conducting aperture covers. On the right side of (16.11) the normalized inner product denoted

$$I_{m,i} = -\frac{<M_m, H_{ti}^{inc}>}{K^{(m)}} \qquad (16.15)$$

has units of amperes. The quantity $I_{m,i}$ represents the current excitation of the mth testing function in the ith aperture.

Equation (16.12) is now expressed in simplified form as

$$\sum_{n=1}^{N_e} V_{in} Y_{mn,i}^{wg} + \sum_{n=1}^{N_e} \sum_{j=1}^{N_a} V_{jn} Y_{mn,ij}^{hs} = I_{m,i} \qquad (16.16)$$

$i = 1, 2, \cdots, N_a$, $m = 1, 2, \cdots, N_e$. The total number of simultaneous equations in (16.16) is given by $N_t = N_a N_e$.

Some properties that are useful in the solution of the previous simultaneous equations can be recognized by writing (16.16) in matrix form as

$$[\boldsymbol{Y}^{wg} + \boldsymbol{Y}^{hs}]\boldsymbol{V} = \boldsymbol{I} \qquad (16.17)$$

where $\boldsymbol{Y}^{wg}$ is a diagonal block matrix (block sizes are $N_e \times N_e$), $\boldsymbol{Y}^{hs}$ is a block-Toeplitz matrix for rectangular grids, $\boldsymbol{I}$ and $\boldsymbol{V}$ are column vectors of length N, $\boldsymbol{I}$ is referred to here as the current excitation matrix, and $\boldsymbol{V}$ is referred to as the voltage response matrix. The sum of Y^{wg} and Y^{hs} is a block Toeplitz matrix that has the form (for an example 3 × 3 array):

$$\boldsymbol{Y} = \begin{bmatrix} \boldsymbol{Y}_0 & \boldsymbol{Y}_1 & \boldsymbol{Y}_2 \\ \boldsymbol{Y}_1 & \boldsymbol{Y}_0 & \boldsymbol{Y}_1 \\ \boldsymbol{Y}_2 & \boldsymbol{Y}_1 & \boldsymbol{Y}_0 \end{bmatrix}$$

where $\boldsymbol{Y}_0$, $\boldsymbol{Y}_l$, and $\boldsymbol{Y}_2$ are symmetric matrices. Sinnott [21] has developed a computer code that solves a block-Toeplitz system of equations. The code uses a recursive relation that avoids matrix inversion of $\boldsymbol{Y}$. Sinnott's code was implemented for the solution of (16.16), which requires storage of just the first block row of the admittance matrix. Both storage and processing requirements are considerably reduced when the previous computer code is used (as compared to other codes not using the block-Toeplitz symmetry such as Crout [22] and Stutzman and Thiele [23]). It is important to note that the admittance matrix in (16.17) is independent of scan angle. Thus, the block-Toeplitz recursion relation is used only once for a series of scan angles (current excitation vectors) with a fixed array geometry. It is recognized that the inclusion of an edge mode would improve convergence. Richmond [24] has used an edge mode to model the edge singularity of strip elements. The edge singularity of an aperture has been treated by Rahmat-Samii and Mittra [25]. The reason for avoiding the addition of an edge mode in the present analysis was to maintain block-Toeplitz symmetry of the admittance matrix.

A substantial savings in computation time is realized when the singly polarized piecewise sinusoidal-uniform basis function expansion (Figure 16.4) is used. Only a single Simpson's rule numerical integration is required to evaluate the half-space mutual admittances $Y^{hs}_{mn,ij}$ in (16.16). This is demonstrated in the following discussion: The half-space mutual admittance expression

$$Y^{hs}_{mn,ij} = \frac{< M_m, H^{hs}_{tij}(M_n) >}{K^{(m)} K^{(n)}} \qquad (16.18)$$

is valid with a perfect electric conductor covering the entire xy plane. Equation (16.18) involves calculating the mutual coupling between two parallel-staggered magnetic surface sources. Recall that the piecewise sinusoidal-uniform magnetic surface current was described by (16.8). Castello and Munk [26] have developed a computer code for calculating the mutual impedance in closed-form between two parallel-staggered electric current thin-wire dipoles in free space. In this analysis, the dipoles have a piecewise-sinusoidal current distribution that results in sine and cosine integrals for their mutual impedance. The thin-wire dipole mutual impedance will be denoted by $Z^{\text{thin wire}}_{mn}$. By relating the magnetic field in the half-space region to the magnetic field that would exist in free space (that is, without a ground plane) and applying duality, an expression involving electric sources in free space can be obtained. Thus, Castello and Munk's work would apply directly to the evaluation of (16.18). The first step is to use the method of images [27] in (16.18) on the source M_n. Both the magnetic source M_n, and the infinite ground plane are located at $z = 0$. The image of M_n is equal to M_n, and thus $2M_n$ radiates in free space. Equation (16.18) can now be written in terms of the free space magnetic field H^{fs}_{tij} as

$$Y^{hs}_{mn,ij} = \frac{2 < M_m, H^{fs}_{tij}(M_n) >}{K^{(m)} K^{(n)}} \qquad (16.19)$$

The normalized inner product in (16.19) is the free space mutual admittance and will be denoted by $Y^{fs}_{mn,ij}$. Duality [27] can now be applied which yields

$$Y^{hs}_{mn,ij} = \frac{2Z^{fs}_{mn,ij}}{\eta_o^2} \qquad (16.20)$$

where $Z^{fs}_{mn,ij}$ is the mutual impedance between two electric surface sources in free space and $\eta_o = 120\pi$ ohms is the impedance of free space.

The geometry involving two parallel-staggered piecewise sinusoidal-uniform electric surface sources is shown in Figure 16.5. A detailed derivation of the mutual impedance between parallel-staggered electric surface sources J_m and J_n (with length a and width b, and with spacings S_x and S_y) has been given elsewhere [14]. A brief discussion of the mutual impedance calculation is as follows: the basic equation for the mutual impedance $Z^{fs}_{mn,ij}$ is a four-fold integration. Two of the integrations (which involve the piecewise sinusoidal currents) are eliminated by using sine and cosine integrals. One of the integrations in the uniform current direction is avoided by using a coordinate transformation [28]. The remaining integral is evaluated numerically by Simpson's rule. The mutual impedance is then expressed as

$$Z^{fs}_{mn,ij} = \frac{2}{3(N-1)^2} \sum_{k=1}^{N} (N-k) C_k Z^{\text{thin wire}}_{mn}(S_x, \sqrt{2}|v_k|) \qquad (16.21)$$

where C_k is Simpson's weighting coefficient, N is the number of sample points in the numerical integration, $v_k = S_y/\sqrt{2} + \Delta v(k-1)$, and $\Delta v = b/\sqrt{2}/(N-1)$.

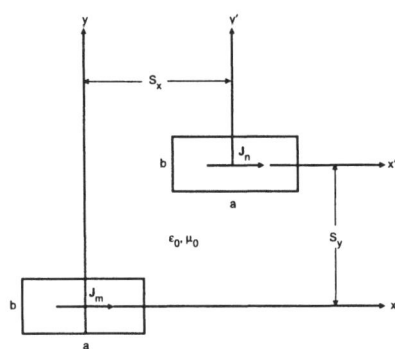

Figure 16.5 The geometry involving two electric surface dipoles in free space. © 1982 IEEE [15].

Substituting (16.21) into (16.20) yields the desired expression for the half-space mutual admittance,

$$Y^{hs}_{mn,ij} = \frac{4}{3(N-1)^2\eta_o^2} \sum_{k=1}^{N} (N-k) C_k Z^{\text{thin wire}}_{mn}(S_x, \sqrt{2}|v_k|) \quad (16.22)$$

Equation (16.22) is the desired expression for determining the half-space mutual admittances. It should be noted that a double numerical integration must be performed if a cross-polarized basis function is included.

The waveguide mutual admittance

$$Y^{wg}_{mn,i} = \frac{<M_m, H^{wg}_{ti}(M_n)>}{K(m)K(n)} \quad (16.23)$$

can be evaluated by applying the method of images to the magnetic source M_n in the presence of the waveguide walls and using a plane wave expansion to find the magnetic field H^{wg}_{ti} radiated by the resulting infinite array set. The number of image arrays can be one, two, or four, depending on the location of source M_n. In general, four image arrays exist when M_n is not centered in the waveguide aperture. A noncentrally located magnetic source example is shown in Figure 16.6, where Roman numerals are used to denote the four image arrays. The array elements have width w and length $2l$ and have interelement spacings D_x and D_y. If M_n is centered in the x-dimension of the waveguide aperture, two infinite arrays with equal interelement spacings exist. When M_n is centered in both the x- and y-dimensions of the aperture, a single infinite array of magnetic surface sources exists. For the aperture expansion given in Figure 16.4 (centered in the x-dimension), two image arrays occur except when M_n is also centered in the y-dimension (single array case). The mutual admittance between the mth magnetic current M_m and any of the image arrays of M_n can be derived by using a plane wave expansion, in a manner similar to that given in Chapter 9, and is given by the following expression [14, 29, 30]:

$$Y^{\text{image}}_{mn} = \frac{1}{2\eta\beta^4 D_x D_y w^2} \sum_{n_2=-\infty}^{\infty} \sum_{n_1=-\infty}^{\infty} \frac{(1-(r_x)^2)}{r_z} |P(n_1,n_2)|^2 e^{j\beta(x'r_x+y'r_y)} \quad (16.24)$$

where D_x and D_y are the infinite array element spacings of the magnetic image sources, $\hat{r} = \hat{x}r_x + \hat{y}r_y + \hat{z}r_z$ is the direction of propagation of the bundle of plane waves, $\hat{s} = \hat{x}s_x + \hat{y}s_y + \hat{z}s_z$ is the direction of propagation of the individual plane waves, x' and y' are the displacements of M_m from the reference element of the image array of M_n, and

$$r_x = s_x + n_2\lambda/D_x \quad (16.25)$$

$$r_y = s_y + n_1\lambda/D_y \qquad (16.26)$$

$$r_z = \sqrt{1-(r_x)^2 - (r_y)^2} \qquad (16.27)$$

The reader can use boundary conditions to show that $s_x = \lambda/2D_x$ in (16.25) and $s_y = 0$ in (16.26), as shown in [14].

The function $P(n_1, n_2)$ in (16.24) is the radiation pattern of a magnetic surface source having width w and length $2l$, as shown in Figure 16.6, and is given by

$$P(n_1, n_2) = \frac{4}{\sin\beta l} e^{j\beta(w/2)r_y} \frac{\sin(\beta(w/2)r_y)}{r_y} \frac{[\cos(\beta l r_x) - \cos\beta l]}{1-(r_x)^2} \qquad (16.28)$$

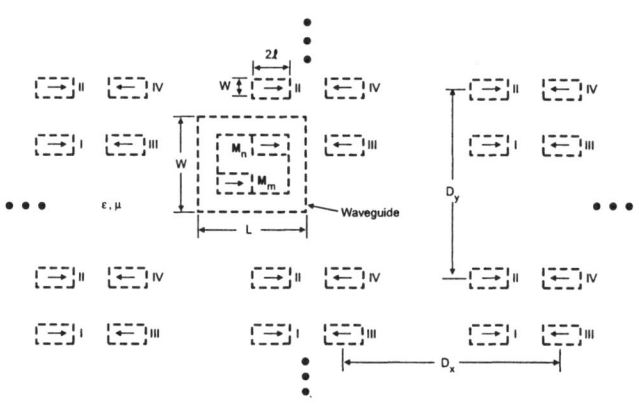

Figure 16.6 Image theory applied to magnetic source M_s in the waveguide results in four interlaced infinite arrays (represented by Roman numerals). © 1982 IEEE [15].

For the present formulation in which the expansion functions are as shown in Figure 16.4, the waveguide mutual admittance of the mth and nth current sources is found by combining the contributions of two image arrays, that is,

$$Y_{mn}^{wg} = Y_{mn(I)} + Y_{mn(II)} \qquad (16.29)$$

Equation (16.29) includes the effects of both propagating and evanescent modes in the waveguide region. It is important to note that although (16.24) is written as a doubly infinite summation, fewer than 100 terms are usually required for convergence. The current excitation matrix elements

$$I_{m,i} = -\frac{<M_m, H_{ti}^{inc}>}{K(m)} \qquad (16.30)$$

are evaluated in a simple manner as will now be discussed. The incident magnetic field for the dominant TE_{10} mode in the ith aperture (in the pth row and qth column of the phased array) is given by [31]

$$H_{x10}^{inc} = -\hat{x}\frac{E_{pq}e^{-j(q\psi_x+p\psi_y)}\sqrt{1-(\lambda/2L)}}{\eta} \cdot \cos(\pi x/L) \qquad (16.31)$$

where $-L/2 < x < L/2$, $0 < y < W$, and where E_{pq} is the excitation amplitude of the pqth waveguide element.

The testing functions are given by

$$M_m = \hat{x}K_m\frac{\sin\beta(L/2-|x|)}{\sin\beta L/2} \qquad (16.32)$$

where $-L/2 < x < L/2$ and $0 < y < w$, and the normalization factor is $K^{(m)} = wK_m$. The above equations are now substituted into (16.30), which yields

$$I_{m,i} = \frac{E_{pq}e^{-j(q\psi_x+p\psi_y)}\sqrt{1-(\lambda/2L)}}{\eta\sin(\beta L/2)}\int_{-L/2}^{L/2}\cos(\pi x/L)\sin\beta(L/2-|x|)dx \qquad (16.33)$$

where the integral over y was equal to w.

The integral over dx in (16.33) can be found from integral tables [32] with the result that

$$I_{m,i} = I_{pq}e^{j(q\psi_x+p\psi_y)} \qquad (16.34)$$

where

$$I_{pq} = \frac{-E_{pq}}{\beta\eta\sin(\beta L/2)\sqrt{1-(\lambda/2L)^2}} \qquad (16.35)$$

is a constant with units of amperes. It is important to note that $I_{m,i}$ is only a function of the aperture position in the array. The excitation currents do not depend on position in the waveguide aperture for this particular formulation. Equation (16.34) along with (16.22) and (16.29) are used in (16.16) to set up simultaneous equations involving the unknown voltage response coefficients V_{jn}. These coefficients are then used in (16.6) to approximate the magnetic current aperture distribution M_s.

The TE_{10} mode reflection coefficient calculation involves finding the magnetic field radiated by M_{sj} in the jth waveguide, denoted by $H_{10}^{wg}(M_{sj})$. The voltage reflection coefficient for the jth aperture (denoted by $\Gamma_{10,j}$) is defined by the negative ratio of the field reflected by the jth aperture to the field incident upon the jth aperture, that is,

$$\Gamma_{10,j} = -\frac{H_{x,10}^{wg}(M_{sj}) + H_{x,10}^{image}}{H_{x,10}^{inc}} \tag{16.36}$$

is the image of the incident field due to the perfect electric conductor covering the jth aperture, and $j = 1, 2, \cdots, N_a$. Note that $H_{x,10}^{wg} = H_{x,10}^{image}$ when the reflection coefficient is evaluated at the waveguide aperture. The magnetic field $H_{x,10}^{wg}(M_{sj})$ can be obtained by imaging M_{sj} in the walls of the jth waveguide and using a plane wave expansion to determine the radiation from the resulting infinite array in free space [14]. A depiction of the infinite array due to M_s is shown in Figure 16.7. The general result in terms of the voltage response coefficients V_m for both propagating and evanescent waves is

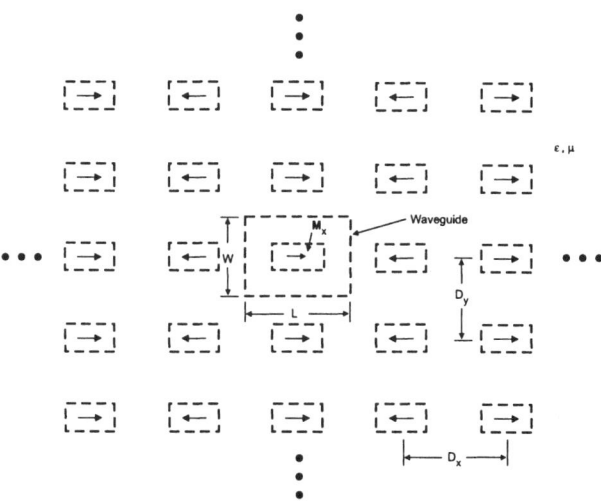

Figure 16.7 Image theory applied to centrally located magnetic surface current M_s in the rectangular waveguide yields one infinite planar array. © 1982 IEEE [15].

$$H_x^{wg}(x, y, z) = -\frac{1}{\eta \beta^2 WL} \sum_{n_2=-\infty}^{\infty} \sum_{n_1=-\infty}^{\infty} (1 - (r_x)^2) P(n_1, n_2)$$

$$\cdot e^{j\beta(x'r_x + y'r_y)} \frac{e^{-j\beta z r_z}}{r_z} \sum_{m=1}^{N_e} V_m e^{j\beta y_m r_y} \tag{16.37}$$

$$r_x = (n_2 + 1/2)\lambda/L \tag{16.38}$$

$$r_y = n_1\lambda/W \tag{16.39}$$

$$r_z = \sqrt{1 - (r_x)^2 - (r_y)^2} \tag{16.40}$$

$$P(n_1, n_2) = \frac{4}{\sin\beta l} e^{j\beta(w/2)r_y} \frac{\sin(\beta(w/2)r_y)}{r_y} \frac{[\cos(\beta(L/2)r_x) - \cos\beta(L/2)]}{1 - (r_x)^2} \tag{16.41}$$

$$w = W/N_e \tag{16.42}$$

A factor of two has been included in (16.37) to account for the image of M_{sj} due to the perfect electric conducting aperture cover. Similar to the mathematical development in Chapter 9, in deriving (16.37), the electric vector potential from an infinite array of magnetic current elements is first expressed as an infinite summation, which is converted to a faster converging series (plane wave expansion) by means of the Poisson Sum Formula. The electric field of the infinite array of magnetic current elements is then computed from the electric vector potential from which the total electric field is obtained by integrating over the magnetic surface source distribtuion. Then the magnetic field is readily computed using a plane wave relationship between the electric and magnetic fields.

The pair of plane waves that produce the TE_{10} mode correspond to $n_1 = 0$ and $n_2 = 0, -1$. This gives $r_x = \pm\lambda/2L$, $r_y = 0$, and $r_z = \sqrt{1 - (\lambda/2L)^2}$ from which

$$P(0,0) = P(0,-1) = -\frac{2\beta w}{(1 - (\lambda/2L)^2)\tan\beta(L/2)} \tag{16.43}$$

The reflection coefficient is calculated at the aperture center where $x = 0$, $y = W/2$, $z = 0$. Equation (16.37) then simplifies to

$$H_{x,10}^{wg}(0, \frac{W}{2}, 0) = \frac{2w}{\eta\beta W L\sqrt{1 - (\lambda/2L)^2}\tan\beta(L/2)} \sum_{m=1}^{N_e} V_m \tag{16.44}$$

from which $\Gamma_{10,j}$ can be computed.

An equation for the far-field pattern of the finite array of waveguides can be expressed as a summation of the product of the pattern factor (denoted by P_b) of the basis function M_n with the voltage responses V_{kn} and the phase factor for M_n at the location (x,y) relative to a reference point $(0,0)$ in the array. That is,

$$P(\theta,\phi) = P_b(\theta,\phi) \sum_{k=1}^{N_a} \sum_{n=1}^{N_e} V_{kn} e^{j\beta_o(xr_x+yr_y)} \tag{16.45}$$

where

$$P_b(\theta,\phi) = \frac{\sin(\beta_o(w/2)r_y)}{r_y} \frac{[\cos(\beta_o(L/2)r_x) - \cos(\beta_o L/2)]}{1-(r_x)^2} \tag{16.46}$$

where $\beta_o = 2\pi/\lambda_o$ is the phase constant of free space, $r_x = \sin\theta\cos\phi$, and $r_y = \sin\theta\sin\phi$.

16.3 RESULTS

The theory of the previous section was implemented in software to analyze various sizes of finite phased arrays: a single element, 3×3, 5×5, 7×7, 9×9, and 11×11 [14]. Aperture reflection coefficients were computed and comparisons were made against published data for finite and infinite arrays. Although only reflection coefficient data is shown in this section, far-field patterns of the array or its elements are readily obtained from (16.45) [33]. The medium inside the waveguides is assumed here to be free space, but the previous equations for the waveguide region are also valid for completely dielectric-filled guides. Several authors have investigated the problem of the reflection coefficient of a single waveguide-fed aperture in an infinite ground plane. Cohen, Crowley, and Levis [34] used a variational formulation and Mautz and Harrington [35] used the method of moments to analyze rectangular and square waveguide-fed apertures. The present moment method formulation is compared against the others in Figure 16.8 for a rectangular waveguide-fed aperture and in Figure 16.9 for a square waveguide-fed aperture. In Figure 16.8 ($L/W = 2.25$) the results obtained with a single piecewise sinusoidal-uniform basis function is in good agreement with published data for $L < 0.6\lambda$. For $L > 0.6\lambda$, seven overlapping piecewise sinusoids were required for good agreement. Similar agreement is obtained in Figure 16.9 for $L/W = 1.0$. In both Figures 16.8 and 16.9, only one uniform (or pulse) basis function was used across the E-plane of the aperture. The reflection coefficient as a function of the number of adjacent pulse expansions for a rectangular waveguide with $L/W = 2.25$ and $L = 0.5714\lambda$ is shown in Figure 16.10. The change in the magnitude of the reflection coefficient is about 20 percent when the number of pulses increases from one to three. Correspondingly, the phase has changed by about $20°$. However, for more than

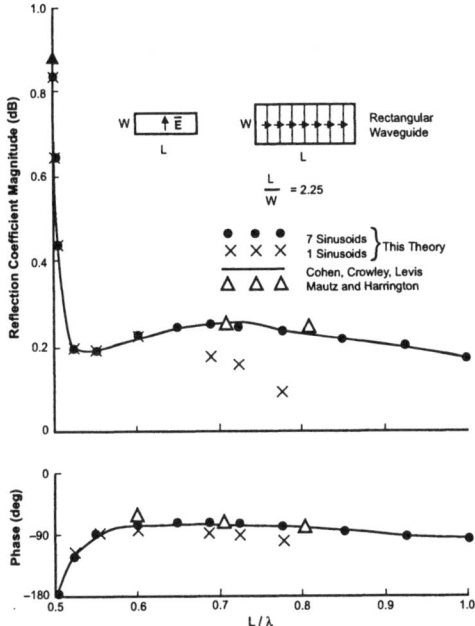

Figure 16.8 Comparison between the reflection coefficient calculated by the present theory and data in the literature for a rectangular waveguide-fed aperture of size $L/W = 2.25$ versus length L in an infinite ground plane. © 1982 IEEE [15].

three pulses the reflection coefficient converges quickly in both magnitude and phase. The E-plane aperture distribution for the case of nine adjacent pulses is shown in Figure 16.11.

The edge singularity is clearly shown and demonstrates that one uniform pulse is inadequate for representing the aperture distribution of an isolated waveguide. In the array calculations that follow, five adjacent pulse functions are used for both rectangular and square waveguide-fed apertures as indicated in Figure 16.4.

Published data were available for comparison with the present technique. Amitay, Galindo, and Wu have analyzed an infinite array of rectangular waveguides where eleven columns of elements (each column is infinite in extent) are uniformly excited to produce scanning in the H-plane [6]. They computed both edge and center element reflection coefficients. The elements were square waveguides with $L = 0.5714\lambda$ and had zero wall thickness, that is, $d_x = d_y = 0.5714\lambda$. Surrounding the eleven infinite columns of excited waveguides are the remaining elements of the infinite array, which are considered to be dummy elements. An 11×11 array was chosen to be a

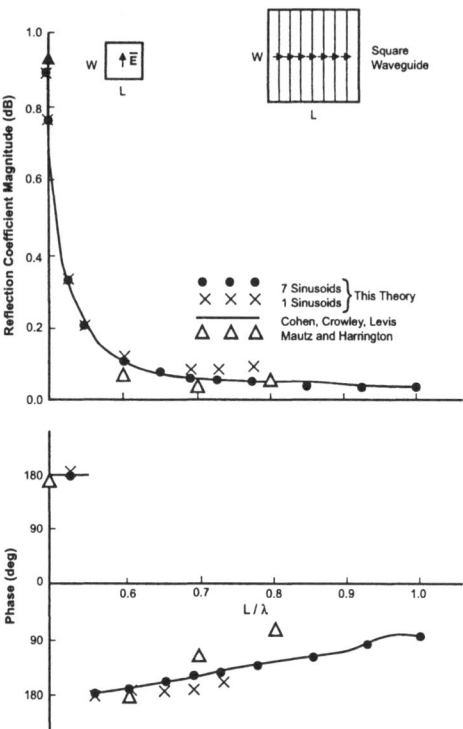

Figure 16.9 Comparison between the reflection coefficient calculated by the present theory and data in the literature for a square waveguide-fed aperture of variable size in an infinite ground plane. © 1982 IEEE [15].

close representation of their semi-infinite array in terms of the center and edge elements in the center row. The center row elements are surrounded by five rows on both sides, which should closely resemble the performance of an infinite array. The other significant difference is that an infinite ground plane surrounds the 11 × 11 array instead of dummy elements as in the semi-infinite excitation case. A comparison of the magnitude of the reflection coefficient for the center element of finite, semi-infinite, and infinite arrays as a function of $\sin\theta$ is shown in Figure 16.12. Agreement is very good out to the occurrence of the grating lobe at $\sin\theta = 0.75$. For $\sin\theta > 0.75$ the difference can probably be attributed to the finiteness of the array. Good agreement was also obtained for the magnitude of the reflection coefficient of the edge element of the center row as a function of H-plane scan angle, as shown in Figure 16.13. For a check on the E-plane scanning performance of the center element of an 11 × 11 array of rectangular waveguides, an infinite array program (called RWED) developed at the Massachusetts Institute of Technology [36] was

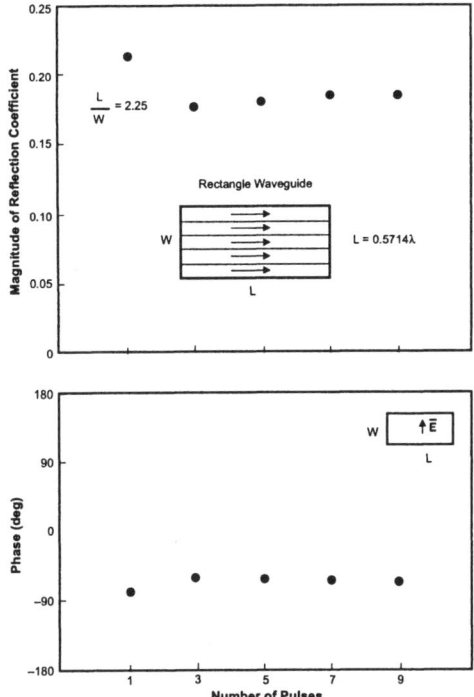

Figure 16.10 Convergence of reflection coefficient as the number of adjacent pulse functions along the waveguide width W is increased for $L/W = 2.25, L = 0.5714\lambda$. © 1982 IEEE [15].

used. The computer program includes the effects of higher order waveguide modes and mutual coupling. Since the center element is surrounded by five rows and five columns of elements, the reflection coefficient should behave much like that of an infinite array. The rectangular waveguides had $L/W = 2.25$, $L = 0.5714\lambda$ with element spacings $d_x = 0.5714\lambda$ and $d_y = 0.254\lambda$. The results for the magnitude and phase of the reflection coefficient for finite and infinite array analyses are given in Figure 16.14, and overall agreement is good. For the $\theta = 30°$ scan case, the array reflection coefficients are given in amplitude and phase in Figure 16.15 for 66 of the 121 elements, the remainder being implied by symmetry. The same array was also scanned in the H-plane and good agreement was also observed, as given in Figure 16.16. The aperture distributions of three elements in the center column (in the E-plane) of the array for $\theta = 0°$ scan are given in Figure 16.17. The edge element clearly shows the edge singularity, while the center element has a relatively uniform amplitude distribution. The edge singularity is not observed at the center of

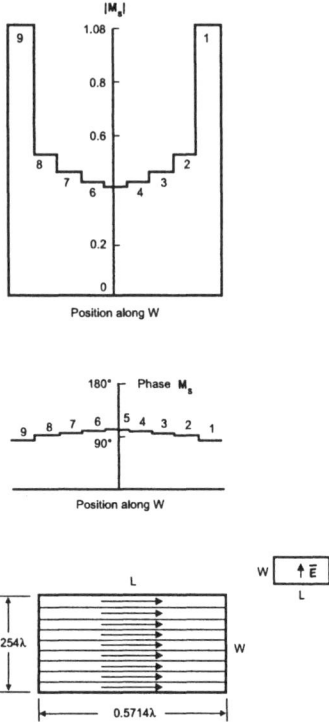

Figure 16.11 Example of the aperture distribution edge singularity that occurs in the E-plane of a rectangular waveguide opening into an infinite ground plane. © 1982 IEEE [15].

the array, apparently because of strong mutual coupling between the closely spaced elements. A case where the waveguides had a finite wall thickness in the H-plane was also considered. The finite array was of the size 11×11 and had the following parameters: $L = 0.5354\lambda$, $W = 0.5714\lambda$, $d_x = 0.5714\lambda$, and $d_y = 0.5714\lambda$. The center element reflection coefficient of the finite array is compared to the results obtained by the RWED computer program as a function of H-plane scan angle in Figure 16.18. The results shown in Figure 16.18 are in good agreement in amplitude and phase out to the grating lobe at $48.5°$. Beyond $\theta = 48.5°$ the phase difference between the infinite array and finite array results is approximately $45°$.

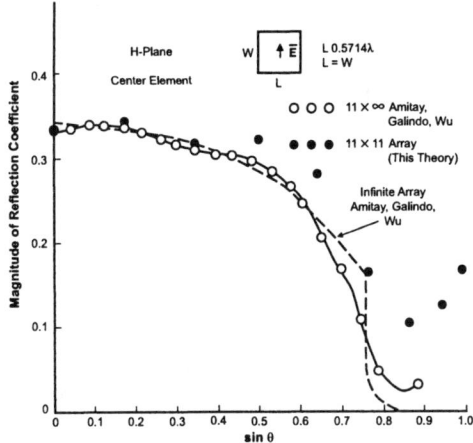

Figure 16.12 Center element reflection coefficient in finite and infinite arrays as a function of H-plane scan angle. Array elements are square waveguide-fed apertures. © 1982 IEEE [15].

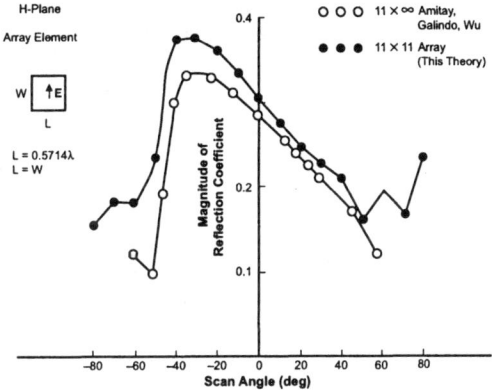

Figure 16.13 Comparison between the magnitude of the edge element reflection coefficient of an 11 × 11 array and an infinite array with eleven infinite columns of elements excited. Array elements are square waveguide-fed apertures. © 1982 IEEE [15].

16.4 SUMMARY

In this chapter, a computationally efficient method of analysis using linearly polarized basis functions has been described for finite planar arrays of rectangular waveguides, provided the apertures have $L < 0.6\lambda$. For larger aperture sizes, two orthogonally polarized basis functions are required to model the higher order waveguide modes. The half-space mutual admittances

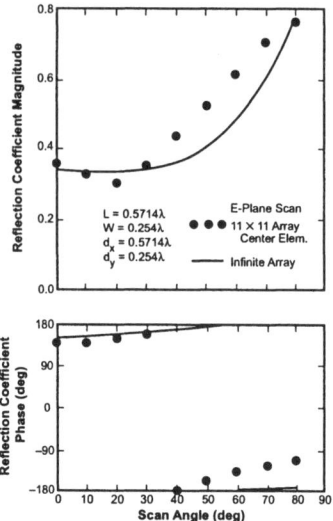

Figure 16.14 Comparison of the reflection coefficient of the center element of an 11 × 11 array and infinite array as a function of E-plane scan angle. Elements are rectangular waveguides. © 1982 IEEE [15].

require only one integration when a single polarization is used, but a double numerical integration must be performed when the cross-polarized basis function is added to the solution. The block-Toeplitz symmetry that exists for rectangular grid arrays has been used to reduce matrix storage requirements and cpu time. For triangular grid arrays, the present formulation applies; however, the admittance matrix is no longer block-Toeplitz. A subroutine such as Crout [22] (for diagonally dominant matrices) would have to be used to solve the system of equations. Modeling of sources in the waveguides with infinite arrays (by image theory) and expressing the radiated field by a plane wave expansion was done for the sake of generality. For example, plane wave reflection and transmission coefficients can be used if a slab of dielectric is inserted in the waveguide. The use of matching sections with variations in the cross-section of the waveguide can be analyzed by solving for the equivalent magnetic currents across each transition aperture. If dielectric sheets cover the face of the array, (16.21) would require modification since the free-space Green's function could not be used. In this case coupling between two elements embedded in a dielectric layer is required. The current excitation vector in (16.17) is arbitrary. For example, if low sidelobes are of interest a variety of aperture amplitude tapers can be used for a given admittance matrix. Furthermore, if a finite ground plane surrounds the array, the geometrical theory of diffraction can be combined with the present moment method

					Center Column
.39 −171°	.44 180°	.45 −173°	.42 180°	.45 −172°	.42 180°
.32 164°	.38 159°	.36 168°	.37 159°	.36 167°	.37 159°
.28 179°	.34 171°	.33 −179°	.33 171°	.33 −180°	.33 172°
.30 174°	.37 165°	.35 175°	.36 166°	.35 174°	.35 167°
.28 169°	.34 161°	.31 173°	.33 163°	.32 170°	.32 164°
.30 169°	.37 162°	.34 171°	.35 164°	.35 169°	.35 164°
.33 168°	.41 161°	.37 167°	.39 162°	.38 166°	.38 163°
.34 175°	.43 167°	.39 173°	.41 169°	.40 172°	.40 169°
.34 −174°	.43 178°	.40 −178°	.41 179°	.41 −179°	.40 179°
.23 −153°	.32 −167°	.30 −162°	.30 −164°	.31 −163°	.30 −164°
.11 −5°	.02 21°	.04 −20°	.04 −2°	.03 −9°	.04 −8°

|Γ| → , ∠Γ →

E-Plane 30° Scan ← L →

Figure 16.15 Reflection coefficients for 66 elements of a uniformly excited 11×11 phased array. Array elements are rectangular waveguide-fed apertures with $L/W = 2.25$, $L = 0.5714\lambda$. © 1982 IEEE [15].

solution [37] to include the coupling due to the edge in Y^{hs} of (16.17).

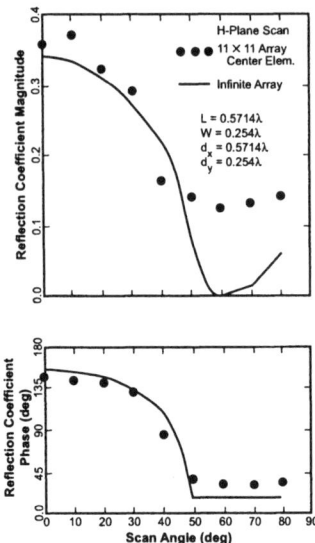

Figure 16.16 Comparison of the reflection coefficient of the center element of an 11 × 11 array and infinite array as a function of H-plane scan angle. Elements are rectangular waveguides. © 1982 IEEE [15].

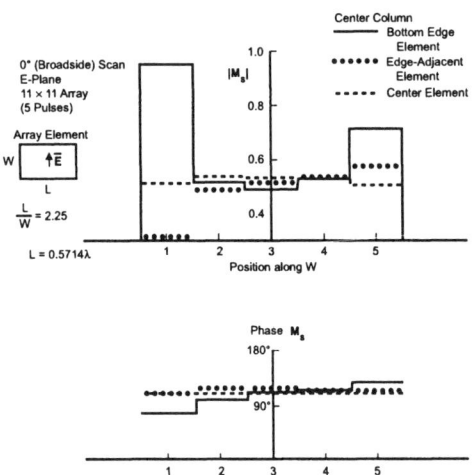

Figure 16.17 Aperture distributions in the vicinity of the array edge along the center column of an 11 × 11 array of uniformly excited rectangular waveguide-fed apertures. Position number 1 refers to the lower edge of each aperture. © 1982 IEEE [15].

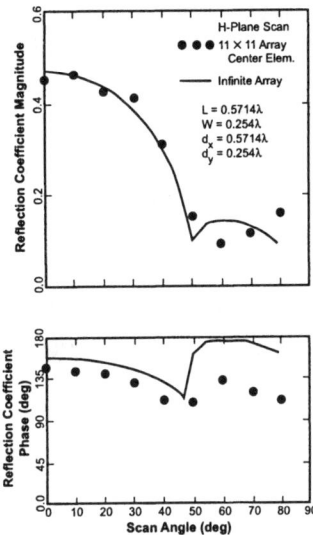

Figure 16.18 Comparison of the reflection coefficient of the center element of an 11 × 11 array and infinite array as a function of H-plane scan angle. Elements are rectangular waveguides with finite wall thickness in the H-plane. © 1982 IEEE [15].

16.5 PROBLEM SET

16.1 Verify that the waveguide admittance function given by (16.13) and the half-space admittance function given by (16.14) have units of Siemens (amperes/volt).

16.2 Verify that the pattern function given by (16.41) reduces to (16.43) for the TE_{10} mode.

16.3 Verify that (16.37) reduces to (16.44) for the TE_{10} mode.

References

[1] Stark, L., "Microwave Theory of Phased-Array Antennas – A Review," *Proc. IEEE*, Vol. 62, No. 12, 1974, pp. 1661-1701.

[2] Borgiotti, G.V., "A Novel Expression for the Mutual Admittance of Planar Radiating Elements," *IEEE Trans. Antennas Propagat.*, Vol. 16, No. 3, 1968, pp. 329-333.

[3] Mailloux, R.J., "First-Order Solutions for Mutual Coupling Between Waveguides Which Propagate Two Orthogonal Modes," *IEEE Trans. Antennas Propagat.*, Vol. 17, No. 6, 1969, pp. 740-746.

[4] Mailloux, R.J., "Radiation and Near-Field Coupling Between Two Collinear Open-Ended Waveguides," *IEEE Trans. Antennas Propagat.*, Vol. 17, No. 1, 1969, pp. 49-55.

[5] Bird, T.S., "Analysis of Mutual Coupling in Finite Arrays of Different-Sized Rectangular Waveguides," *IEEE Trans. Antennas Propagat.*, Vol. 38, No. 2, 1990, pp. 166-172.

[6] Amitay, N., V. Galindo, and C.P. Wu, *Theory and Analysis of Phased Array Antennas*, New York: Wiley, 1972, pp. 149-154.

[7] Wu, C.P., "Analysis of Finite Parallel-Plate Waveguide Arrays," *IEEE Trans. Antennas Propagat.*, Vol. 18, No. 3, 1970, pp. 328-334.

[8] Cha, A.G., and J.K. Hsiao, "A Matrix Formulation for Large Scale Numerical Computation of the Finite Planar Waveguide Array Problem," *IEEE Trans. Antennas Propagat.*, Vol. AP-22, No. 1, January 1974, pp. 106-108.

[9] Hidayet, M.A., "Finite Phased Array Analysis," Ph.D. dissertation, Univ. Michigan, 1974.

[10] Luzwick, J., and R.F. Harrington, "Mutual Coupling Analysis in a Finite Planar Rectangular Waveguide Antenna Array," Dept. Elect. Computer Eng., Syracuse Univ., Tech. Rep., No. 7, June 1978.

[11] Wang, S.S., "Wide-Angle Wide-Band Elements for Phased Arrays," Ph.D. dissertation, Polytechnic Institute of New York, 1975.

[12] Kraus, J.D., *Antennas*, New York: McGraw-Hill, 1950, ch. 4.

[13] Harrington, R.F., and J.R. Mautz, "A Generalized Network Formulation for Aperture Problems," Sci.Rep. no. 8 contract F19628-73-C-0047, A.F. Cambridge Res. Lab., Rep. AFCRL-TR-75-0589, November 1975.

[14] Fenn, A.J., "Moment Method Calculation of Reflection Coefficient for Waveguide Elements in a Finite Planar Phased Antenna Array," Ph.D. dissertation, The Ohio State Univ., 1978.

[15] Fenn, A.J., G.A. Thiele, and B.A. Munk, "Moment Method Analysis of Finite Rectangular Waveguide Phased Arrays," *IEEE Trans. Antennas Propagat.*, Vol. 30, No. 4, 1982, pp. 554-564.

[16] Yavuz, H., and O.M. Buyukdura, "Mutual Coupling Effects of Finite Rectangular Phased Arrays," *IEEE Antennas and Propagat. Soc. Int. Symp. Digest*, 1994, pp. 418-421.

[17] Grassi, P., et al., "Characterization of Finite Waveguide Arrays Using a New Generalized Scattering Matrix Approach," *IEEE Antennas and Propagat. Symposium Digest*, 2004, pp. 125-128.

[18] Diamond, B.L., "Resonance Phenomena in Waveguide Arrays," *IEEE G-AP Int. Symp. Dig.*, 1967, pp. 110-115.

[19] Stark, L., "The Conductance Zero in Element Admittance of Waveguide Phased Arrays," Hughes Aircraft Company, Rep. No. GL75-10-4, May 2, 1975.

[20] Harrington, R.F., *Field Computation by Moment Methods*, New York: Macmillan, 1968.

[21] Sinnott, D.H., "An Improved Algorithm for Matrix Analysis of Linear Antenna Arrays," Australian Defense Sci. Service, Weapons Res. Establishment, Adelaide, South Australia, WRE-TECH, Note-1066(AP), January 1974.

[22] Crout, D., "A Short Method of Evaluating Determinants and Solving Systems of Linear Equations with Real or Complex Coefficients," *AIEE Trans.* (supplement), Vol. 60, 1941, pp. 1235-1241.

[23] Stutzman, W.L., and G.A. Thiele, *Antenna Theory and Design*, New York: Wiley, 1981, pp. 339-343.

[24] Richmond, J.H., "On the Edge Mode in the Theory of TM Scattering by a Strip or Strip Grating," *IEEE Trans. Antennas Propagat.*, Vol. 28, No. 6, 1980, pp. 883-887.

[25] Rahmat-Samii, Y., and R. Mittra, "Electromagnetic Coupling Through Small Apertures in a Conducting Screen," *IEEE Trans. Antennas Propagat.*, Vol. 25, No. 2, 1977, pp. 180-187.

[26] Castello, D., and B.A. Munk, "Table of Mutual Impedance of Identical Dipoles in Echelon," The Ohio State Univ., ElectroScience Laboratory, Dept. Elect. Eng., Rep. 2382-1, prepared under Contract F33615-67-C-1507 for Air Force Avionics Laboratory, Wright-Patterson Air Force Base, OH (AD-822013), October 1967.

[27] Harrington, R.F., *Time-Harmonic Electromagnetic Fields*, New York: McGraw-Hill, 1961, pp. 98-100.

[28] Popovic, Z.D, and B.D. Popovic, "Transformation of Double Integrals Appearing in Variational Formulation of Cylindrical Antenna Problems," publications De la Faculte Dielectrotechnique de L'Universite A. Belgrade, Serie: Electronique, Telecommunications Automatique, No. 64, 1971.

[29] Munk, B.A., G.A. Burrell, and T.W. Kornbau, "A general theory of Periodic Surfaces in Stratified Dielectric Media," The Ohio State University ElectroScience Laboratory, Dept. Elec. Eng., Rep. 784346-1, November 1977, prepared under Contract F33615-76-C-1024 for Aeronautical Systems Division, Wright-Patterson Air Force Base, OH, AFAL-TR-77-219.

[30] Munk, B.A., *Finite Antenna Arrays and FSS*, New York: Wlley, 2003.

[31] Ramo, S., J.R. Whinnery, and T. VanDuzer, *Fields and Waves in Communication Electronics*, New York: Wiley, 1965, p. 425.

[32] Gradshteyn, I.S., and I. W. Ryzhik, *Table of Integrals, Series, and Products,* 4th ed., New York: Academic, 1965, p. 140.

[33] Fenn, A.J. and G.A. Thiele, "A Moment Method Technique for Probe-Fed Cavity-Backed Slot Antennas," The Ohio State University ElectroScience Laboratory, Dept. Elect. Eng., Rep. 4091-3, March 1976, prepared under Contract N00014-75-C-0313 for the Office of Naval Research.

[34] Cohen, M., T. Crowley, and C. Levis, "The Aperture Admittance of a Rectangular Waveguide Radiating into Half-Space," The Ohio State Univ., Antenna Lab. Rep. ac 21114 S.R. No. 22, 1953.

[35] Mautz, J.R., and R.F. Harrington, "Transmission from a Rectangular Waveguide Into Half Space Through a Rectangular Aperture," Syracuse Univ., Dept. Elect. Comput. Eng. Tech. Rep. TR-76-5, 1976.

[36] Tsandoulas, G.N., "Unidimensionally Scanned Phased Arrays," *IEEE Trans. Antennas*

Propagat., Vol. 28, No. 1, 1980, pp. 86-99.

[37] Thiele, G.A., and T.H. Newhouse, "A Hybrid Technique for Combining Moment Methods with the Geometrical Theory of Diffraction," *IEEE Trans. Antennas Propagat.*, Vol. 23, No. 1, 1975, pp. 62-69.

About the Author

Alan J. Fenn is a senior staff member in the Advanced Sensing and Exploitation Group at Lincoln Laboratory, Massachusetts Institute of Technology. He is deputy manager for antenna measurements in the RF Systems Test Facility at Lincoln Laboratory. He has conducted over 30 years of research in the area of phased array antennas. He joined Lincoln Laboratory in 1981 and was a member of the Space Radar Technology Group from 1982 to 1991, where his primary research was in adaptive phased-array antenna design and testing. From 1992 to 1999 he was an assistant group leader in the Radio Frequency Technology Group, where he managed programs involving measurements of atmospheric effects on satellite communications. From 1978 to 1981, he was a senior engineer in the Antenna Systems Design/Analysis Group in the RF Systems Department at Martin Marietta Aerospace, Denver, Colorado. He received a B.S. from the University of Illinois at Chicago in 1974, and an M.S. in 1976 and a Ph.D. in 1978 from The Ohio State University, Columbus, all in electrical engineering.

Dr. Fenn was elected a Fellow of the IEEE in 2000 for his contributions to the theory and practice of adaptive phased-array antennas. He was technical program cochairman of the 2001 IEEE Antennas and Propagation Society Symposium. He has served as an associate editor in the area of adaptive antennas for the *IEEE Transactions on Antennas and Propagation*. In 1990 he was a corecipient of the IEEE Antennas and Propagation Society's H.A. Wheeler Applications Prize Paper Award. He also received the IEEE/URSI-sponsored 1994 International Symposium on Antennas (JINA 94) award. In addition to this book, Dr. Fenn is an author of one book and a coauthor of one book chapter as well as the author of numerous journal articles, patents, and short-course lectures and conference presentations on adaptive phased array antennas.

Index

Adaptive antenna, 1, 112
Adaptive array cancellation, 156
Adaptive array cancellation ratio, 146
Adaptive array output power, 146
Adaptive cancellation of noise source, 173
Adaptive DPCA, 142
Adaptive nulling experiments, 162, 186
Adaptive nulling receiver design, 163
Adaptive signal processor, 4
Adaptive weight vector, 145
Adaptive weights, steady state, 9
Amitay, 360
Amplitude control, 306
Analog-to-digital converter, 165
Antenna pattern match, 54
Antenna quiescent conditions, 140
Antenna test regions, 135
Applebaum-Howells, 8
Array element position errors, 195
Array factor, 194, 195
Array gain loss mechanisms, 198
Array gain pattern, from element patterns, 207
Array gain, versus scan angle, 208
Array mutual coupling, 46, 111, 192, 263, 275, 360
Array mutual coupling, defined, 199
Array mutual impedance, 303
Array mutual-coupling, 302
Array phase calibration, 83
August, 140
Aumann, 140

Balun, 336, 354
Bateman, 234
Bernella, 169
Bird, 359
Blind spot, 202
Blind spot, for dipole arrays, 336
Blind spot, intentional, 204
Block-Toeplitz, 224, 366
Borgiotti, 359
Boroson, 58
Boundary conditions, 338, 361
Brennan, 58
Broadside null radiation pattern, 204

Calibration source, 140
Calibration, of phased array, 204
Cancellation of interference, 126
Cancellation ratio, for interference, 81
Cancellation, of interference, 26
Castello, 221, 367
Cha, 360
Channel mismatch, 162
Channel tracking error, 171
Characteristic impedance, of transmission line, 199
Chebyshev radiation pattern, 151
Circular polarization, 181
Clutter cancellation, 47, 275
Clutter source, simulated, 147
Clutter sources, 149, 153
Cohen, 374
Collinear arrays, 352

Communications users, 177
Compact range, 185
Compact range reflector, 356
Compton, 5
Conical feed horn, 184
Consumption of degrees of freedom, 6
Corporate feed, for array, 203
Corporate-fed phased array, 150
Correlation, between adaptive channels, 80
Covariance matrix, 4, 20, 21, 145
Covariance matrix eigenvalues, 7, 82, 123, 156
Crawford, 284
Cross-correlation, 8
Cross-polarization, 328
Crossed V-dipole antenna, 335
Crowley, 374
Current excitation matrix, for waveguide arrays, 370

Degrees of freedom, 1, 5, 36, 78, 123, 145, 179
Degrees of freedom, consumed, 91
Degrees of freedom, maximally stressed, 25
Diamond, 169, 360
Differential electric field, 236
Digital equalization filter, 166
Digital nulling, 162
Dipole array, 50
Dipole array, linear polarization, 351
Dipole current distribution, 221
Dipole phased array, 150, 202
Dipoles, comparison of straight dipoles and V-dipoles, 341
Directional derivative, 33
Directivity, 196
Directivity pattern, 84
Directivity pattern, far-field, 88
Directivity pattern, near-field, 86
Dispersion, 179
Dispersion model, covariance-matrix based, 78, 112
Dispersion models, basic, 76
Displaced phase center antenna, 43, 141, 275
Doppler shift, 58
DPCA, 149
DPCA (displaced phase center antenna), 43, 275

DPCA beamforming architecture, 280
Dual polarized dipole phased array, 335
Dynamic range, of nulling processor, 24
ECCM (Electronic counter-countermeasures), 5
ECM (Electronic countermeasures), 5
Effective isotropic radiated power, 205
Eigenspace, 11
Eigenvalues, 7, 24, 78
Eigenvalues, equal, 18
Eigenvalues, large, 18, 21
Eigenvectors, orthogonal, 18
EIRP (effective isotropic radiated power), 205
Electric field, 259, 263
Electric field components, 193
Electric field, tangential components, 295
Electronic counter-countermeasures, 5
Electronic countermeasures, 5
Electronic scanning, 191
Element gain pattern, 201, 338
Element gain pattern, measured, 327
Element gain pattern, measured for linear arrays, 330
Element gain pattern, measured for microstrip patch array, 210
Element gain pattern, model for monopole array, 251
Element gain pattern, monopole array, 242
Element transmit power, 206
Embedded element, 200
Equivalent dipole array, 221
Evanescent (reactive) near-field region, 292
Evanescent waves, 229
Exterior element, 238

Far-field adaptive nulling, 78
Far-field interference, 75
Feed horn cluster, for MBA, 181
FF (Far-field), 76
Field point theory, for near-field analysis, 292
Figure of merit, 28, 33
Finite arrays, 46, 359
Finite impulse response equalizer, 166
FIR (finite impulse response), 166
Floquet's theorem, 231
Focal distance, effect on nulling performance, 98

Focused near-field nulling, 111
Focused near-field nulling experiments, 162
Focused near-field pattern, 84
Focused near-field testing, 141
Fractional dissipated power, 200, 209
Free excitation, 223
Fresnel region, 136
Fresnel zone, 88
Friis, 336
Front-to-back ratio, of array, 353
Fully adaptive array, 2, 79

Gabriel, 7
Gain pattern, of an antenna, 226
Gain, maximum of an array, 206
Gain, of antenna, 198
Gain, related to directivity, 87
Galerkin's formulation, 222, 299
Galerkin's method, 365
Galindo, 360
Generator impedance, 224
Geosynchronous altitude, 177
Gradient search, 33, 36
Grating lobe condition, 208, 352
Grating lobe positions, 310
Grating lobe, aliased, 313
Ground test facility, 140
Guard bands, 46, 144
Gupta, 117, 147

Hansen, 351
Harrington, 360
Hermitian, 146
Hermitian matrix, 11
Herper, 217
Hertzian dipoles, 229
Hessel, 217
Hexagonal array, 35, 36
Hidayet, 360
High-resolution nulling, 177
Highly overlapped beams, 177
Hsiao, 360
Hudson, 79

Image theory, for vertical monopoles, 229
Impedance matrix, 222
Impedance mismatch loss, 198
Induced terminal voltage, 239
Infinite array analysis, 229, 369
Input impedance, 225

Input power, 227
INR (interference-to-noise ratio), 5
Interference covariance matrix, 80
Interference power level, 7
Interference sources, 1, 5
Interference sources, equal power, 21
Interference sources, uncorrelated, 20
Interference, narrowband example, 91
Interference, near-field/far-field
 equivalence, 96
Interference, stressing, 18
Interference, wideband example, 91
Interference-to-noise ratio, 5
Interferers, large number, 93
Isotropic point receive antenna elements, 20
Isotropic radiating elements, 194

Jammer, scenario, 5
Jamming, 1
Johnson, J.R., 163
Johnson, R.C., 111
Joy, 285

Ksienski, 147
Kurtze, 218

Least squares, 29
Left-hand circular polarization, 184
Levis, 374
LHCP (left-hand circular polarization), 184
Linear array, example, 196
Loop, 204
Loop-fed slotted cylinder array, 321
Low-sidelobe phased array, 305

Magnetic field, continuity across aperture,
 362
Magnetic vector potential, 231
Mailloux, 359
Mallet, 58
Mautz, 360
Maximum gain, in decibels, 200
Maximum gain, of aperture, 200
Maximum jammer effectiveness, 35
Mayhan, 7, 20, 112
MBA (multiple beam antenna), 177
MBA lens, 179
Method of images, 338
Method of images, magnetic sources, 369
Method of moments, 49, 117, 140, 219,
 221, 261, 298, 338, 361

Method of moments, convergence, 346
Microstrip patch array, example, 208
Miller, 5
Mismatch loss, 198, 227
Mittra, 367
Module illumination errors, 104
Monopole, 204
Monopole array, 50
Monopole array, input impedance of elements, 244
Monopole arrays, effect of array size, 242
Monopole phased array, 120, 215, 260, 293, 299
Monzingo, 5
Multiaperture MBA, 180
Multiple beam antenna, 1, 177
Multiple jammers, 125
Munk, 221, 229
Mutual admittance, between apertures, 359
Mutual admittance, half space region, 369
Mutual admittance, waveguide region, 369
Mutual coupling measurements, 228
Mutual coupling power ratio, 205
Mutual coupling, measured, 327
Mutual coupling, measured and calculated for monopole array, 248
Mutual coupling, measured for microstrip patch array, 209
Mutual impedance, 222
Mutual impedance matrix, 51, 147

Nadir-region clutter, 216
Narrowband interference, 123
Near-field adaptive nulling, 76, 78
Near-field focusing, 260
Near-field interference, 76
Near-field measurements, centerline, 284
Near-field probe pattern compensation, 296
Near-field probe, rectangular waveguide, 307
Near-field region, 136
Near-field region, evanescent, 292
Near-field region, radiating, 292
Near-field source interaction, 156
Near-field testing, 75
Newell, 284
NF (Near-field), 76
Nulling channels, minimum number, 7
Nulling processor, 9

Nulling weight network, 184
Nulls, 5
Omnidirectional pattern coverage, horizontal polarization, 321
Omnidirectional pattern coverage, vertical polarization, 217
Orthogonal interference angles, 22
Orthogonal interference signals, 7

Parabolic cylinder reflector antenna, 352
Parallel plate waveguide array, 360
Parasitic current, 149
Paris, 285
Partially adaptive array, 2
Passively terminated array, 200
Pattern correlation matrix, 53
Pencil beam, 191
Periodic arrays, 6
Periodic surface, 229
Phase center displacement, 45
Phase center separation, 54
Phase focusing, 82, 136, 152
Phase shifter calibration, 171
Phase shifters, 140, 191, 306
Phased array antenna, 191
Phased array design parameters, 193
Piecewise sinusoidal current, 346
Piecewise-sinusoidal current, 50, 261, 299, 364
Planar near-field measurements, 276
Planar near-field scan truncation, 282
Planar near-field scanner, 281
Planar near-field scanning, 291
Plane wave, 76, 111
Plane wave illumination, 148
Plane wave incidence, 112
Plane wave power density, 206
Plane wave spectrum, 291, 296
Poisson sum formula, 232, 234
Power combiners, 354
Power density, of an antenna, 205
Power transmission coefficient, 199
Power transmitted, by an antenna, 205
PRI (pulse repetition interval), 44
Probe compensated near-field pattern, 148
Probe compensated near-field scanning, 293
Projected aperture, 201
Propagating waves, 229

Pulse repetition interval, 44

Quiescent radiation pattern, 186
Quiescent weight vector, 18

Radial component, electric field, 119
Radial polarization, 259
Radiating currents, 195
Radiation intensity, 197, 226
Radiation pattern, quiescent, 8
Rahmat-Samii, 367
Reactive near-field region, 292
Received power, of an antenna, 205
Rectangular waveguide array, 169, 360
Rectangular waveguide, reflection
    coefficient, 374
Reed, 58
Reflection coefficient, 199, 227
Return loss measurements, of elements in
    ultrawideband dipole array, 356
Rexolite, dielectric, 181
RHCP (right-hand circular polarization),
    181
Rhodes, 296
Richmond, 148
Right-hand circular polarization, 181
Ring array, 35, 38

sample matrix inversion, 162
Satellite uplink, 177
SBR (space-based radar), 43, 215
Scan angle, 195
Scan reflection coefficient, 199, 241
Scan reflection coefficient, for waveguide
    arrays, 376
Scan transmission coefficient, 200
Scattering matrix, 228
Scattering parameter, 228
Scharfman, 140
Schelkunoff, 238, 336
Schuman, 217
Sidelobe canceller, 145, 151
Sidelobe canceller array, 84
Sidelobe canceller, example, 104
Signal matrix, extended, 20
Signal matrix, orthogonal, 21
Signal vector, 4
Singular value decomposition, 29
Sinnott, 366
Skewed lattice, 220

SMI (sample matrix inversion), 162
SNR (signal-to-noise ratio), 1, 56
Space-based radar, 43, 133, 215
Spaceborne radar, 133
Spectral theorem, 81
Spherical wave, 76
Spherical wave incidence, 112, 136
Stark, 359, 360
Straight dipole, 336
Subarray, 144, 353
Superposition, of element patterns, 206
Switch tree, for MBA, 179, 183

T/R (transmit/receive module), 44
T/R module error, simulated, 152
Target phase shift, 143
Taylor taper, 264, 310
Thinned arrays, 6
Time delays, 179
Transmission loss, 198
Transmit/receive modules, 44, 169
Transmitting phased array, 192

Ultrawideband dipole array, measured
    radiation patterns, 356
Ultrawideband phased array, 351
Uniform current loop antenna, 322
Uniform pattern coverage, 16
Unit cell area, 206
Unitary matrix, 30
Unitary signal matrix, 19

V-dipole antenna, 303, 336
V-dipole antenna, deriviation of radiation
    pattern, 336
V-dipole array, 343
V-dipole array, measured element gain
    pattern, 346
V-dipole array, measured mutual coupling,
    343
V-dipole probe, 303, 309
Vector wave equation, 293
Voltage excitation matrix, 302
Voltage matrix, 120

Wang, 292
Wavefront dispersion, 78, 138
Wavefront dispersion, far-field, 76
Wavefront dispersion, near-field, 76
Wavefront shape, 145

Waveguide arrays, 360
Weight vector, 4
Weight vector, adaptive, 80
Weight vector, quiescent, 81
Wide angle scanning, 335

Wideband interference, 123
Willwerth, 140
Wu, 360

Yaghjian, 111
Yavuz, 360

Recent Titles in the Artech House Radar Series

David K. Barton, Series Editor

Adaptive Antennas and Phased Arrays for Radar and Communications, Alan J. Fenn

Advanced Techniques for Digital Receivers, Phillip E. Pace

Advances in Direction-of-Arrival Estimation, Sathish Chandran, editor

Airborne Pulsed Doppler Radar, Second Edition, Guy V. Morris and Linda Harkness, editors

Bayesian Multiple Target Tracking, Lawrence D. Stone, Carl A. Barlow, and Thomas L. Corwin

Beyond the Kalman Filter: Particle Filters for Tracking Applications, Branko Ristic, Sanjeev Arulampalam, and Neil Gordon

Computer Simulation of Aerial Target Radar Scattering, Recognition, Detection, and Tracking, Yakov D. Shirman, editor

Design and Analysis of Modern Tracking Systems, Samuel Blackman and Robert Popoli

Detecting and Classifying Low Probability of Intercept Radar, Phillip E. Pace

Digital Techniques for Wideband Receivers, Second Edition, James Tsui

Electronic Intelligence: The Analysis of Radar Signals, Second Edition, Richard G. Wiley

Electronic Warfare in the Information Age, D. Curtis Schleher

ELINT: The Interception and Analysis of Radar Signals, Richard G. Wiley

EW 101: A First Course in Electronic Warfare, David Adamy

EW 102: A Second Course in Electronic Warfare, David L. Adamy

Fourier Transforms in Radar and Signal Processing, David Brandwood

Fundamentals of Electronic Warfare, Sergei A. Vakin, Lev N. Shustov, and Robert H. Dunwell

Fundamentals of Short-Range FM Radar, Igor V. Komarov and Sergey M. Smolskiy

Handbook of Computer Simulation in Radio Engineering, Communications, and Radar, Sergey A. Leonov and Alexander I. Leonov

High-Resolution Radar, Second Edition, Donald R. Wehner

Introduction to Electronic Defense Systems, Second Edition, Filippo Neri

Introduction to Electronic Warfare, D. Curtis Schleher

Introduction to Electronic Warfare Modeling and Simulation, David L. Adamy

Introduction to RF Equipment and System Design, Pekka Eskelinen

Microwave Radar: Imaging and Advanced Concepts, Roger J. Sullivan

Millimeter-Wave Radar Targets and Clutter, Gennadiy P. Kulemin

Modern Radar System Analysis, David K. Barton

Modern Radar System Analysis Software and User's Manual, Version 3.0, David K. Barton

Multitarget-Multisensor Tracking: Applications and Advances Volume III, Yaakov Bar-Shalom and William Dale Blair, editors

Principles of High-Resolution Radar, August W. Rihaczek

Principles of Radar and Sonar Signal Processing, François Le Chevalier

Radar Cross Section, Second Edition, Eugene F. Knott et al.

Radar Evaluation Handbook, David K. Barton et al.

Radar Meteorology, Henri Sauvageot

Radar Reflectivity of Land and Sea, Third Edition, Maurice W. Long

Radar Resolution and Complex-Image Analysis, August W. Rihaczek and Stephen J. Hershkowitz

Radar Signal Processing and Adaptive Systems, Ramon Nitzberg

Radar System Analysis and Modeling, David K. Barton

Radar System Performance Modeling, Second Edition, G. Richard Curry

Radar Technology Encyclopedia, David K. Barton and Sergey A. Leonov, editors

Range-Doppler Radar Imaging and Motion Compensation, Jae Sok Son et al.

Signal Detection and Estimation, Second Edition, Mourad Barkat

Space-Time Adaptive Processing for Radar, J. R. Guerci

Theory and Practice of Radar Target Identification, August W. Rihaczek and Stephen J. Hershkowitz

Time-Frequency Transforms for Radar Imaging and Signal Analysis, Victor C. Chen and Hao Ling

For further information on these and other Artech House titles, including previously considered out-of-print books now available through our In-Print-Forever® (IPF®) program, contact:

Artech House
685 Canton Street
Norwood, MA 02062
Phone: 781-769-9750
Fax: 781-769-6334
e-mail: artech@artechhouse.com

Artech House
46 Gillingham Street
London SW1V 1AH UK
Phone: +44 (0)20 7596-8750
Fax: +44 (0)20 7630-0166
e-mail: artech-uk@artechhouse.com

Find us on the World Wide Web at: www.artechhouse.com